National Center for Construction Education and Research

MW01386073

AWS Entry Level Welder – Phase 1

Upper Saddle River, New Jersey
Columbus, Ohio

National Center for Construction Education and Research

President: Don Whyte
Director of Curriculum Revision and Development: Daniele Dixon
Welding Project Manager: Daniele Dixon
Production Manager: Debie Ness
Editors: Michael Senecal and Rebecca Hassell
Desktop Publisher: Jessica Martin

NCCER would like to acknowledge the contract service provider for this curriculum: Topaz Publications, Liverpool, New York.

Pearson Education, Inc.

Editor in Chief: Stephen Helba
Product Manager: Lori Cowen
Production Editor: Stephen C. Robb
Design Coordinator: Karrie M. Converse-Jones
Text Design Concept: Rebecca Bobb
Cover Designer: Linda Sorrells-Smith
Copy Editor: Sheryl Rose
Scanning Coordinator: Karen L. Bretz
Scanning Technician: Janet Portisch
Production Manager: Pat Tonneman

This book was set in Palatino and Helvetica by Carlisle Communications, Ltd. It was printed and bound by Courier Kendallville, Inc. The cover was printed by Phoenix Color Corp.

This information is general in nature and intended for training purposes only. Actual performance of activities described in this manual requires compliance with all applicable operating, service, maintenance, and safety procedures under the direction of qualified personnel. References in this manual to patented or proprietary devices do not constitute a recommendation of their use.

Pearson Education Ltd.
Pearson Education Singapore Pte. Ltd.
Pearson Education Canada, Ltd.
Pearson Education—Japan

Pearson Education Australia Pty. Limited
Pearson Education North Asia Ltd.
Pearson Educación de Mexico, S.A. de C.V.
Pearson Education Malaysia Pte. Ltd.

10 9 8 7 6 5 4 3 2 1
ISBN 0-13-102577-5

Preface

This volume was developed by the National Center for Construction Education and Research (NCCER) in response to the training needs of the construction, maintenance, and pipeline industries. It is one of many in NCCER's *Contren*™ *Learning Series.* The program, covering training for close to 40 construction and maintenance areas, and including skills assessments, safety training, and management education, was developed over a period of years by industry and education specialists.

NCCER also maintains a National Registry that provides transcripts, certificates, and wallet cards to individuals who have successfully completed modules of NCCER's *Contren*™ *Learning Series,* when the training program is delivered by an NCCER Accredited Training Sponsor.

The NCCER is a not-for-profit 501(c)(3) education foundation established in 1995 by the world's largest and most progressive construction companies and national construction associations. It was founded to address the severe workforce shortage facing the industry and to develop a standardized training process and curricula. Today, NCCER is supported by hundreds of leading construction and maintenance companies, manufacturers, and national associations, including the following partnering organizations:

PARTNERING ASSOCIATIONS

- American Fire Sprinkler Association
- American Petroleum Institute
- American Society for Training & Development
- American Welding Society
- Associated Builders & Contractors, Inc.
- Association for Career and Technical Education
- Associated General Contractors of America

- Carolinas AGC, Inc.
- Carolinas Electrical Contractors Association
- Citizens Democracy Corps
- Construction Industry Institute
- Construction Users Roundtable
- Design-Build Institute of America
- Merit Contractors Association of Canada
- Metal Building Manufacturers Association
- National Association of Minority Contractors
- National Association of State Supervisors for Trade and Industrial Education
- National Association of Women in Construction
- National Insulation Association
- National Ready Mixed Concrete Association
- National Utility Contractors Association
- National Vocational Technical Honor Society
- North American Crane Bureau
- Painting & Decorating Contractors of America
- Plumbing – Heating – Cooling, Contractors Association
- Portland Cement Association
- SkillsUSA
- Steel Erectors Association of America
- Texas Gulf Coast Chapter ABC
- U.S. Army Corps of Engineers
- University of Florida
- Women Construction Owners & Executives, USA

Some features of NCCER's *Contren*™ *Learning Series* are:

- An industry-proven record of success
- Curricula developed by the industry for the industry
- National standardization providing portability of learned job skills and educational credits
- Credentials for individuals through NCCER's National Registry

- Compliance with Apprenticeship, Training, Employer, and Labor Services (ATELS) requirements for related classroom training (CFR 29:29)
- Well-illustrated, up-to-date, and practical information

FEATURES OF THIS BOOK

Capitalizing on a well-received campaign to redesign our textbooks, NCCER and Prentice Hall are continuing to publish select textbooks in color. *AWS Entry Level Welder – Phase 1* incorporates the new design and layout, along with color photos and illustrations, to present the material in an easy-to-use format. Special pedagogical features augment the technical material to maintain the trainees' interest and foster a deeper appreciation of the trade.

Think About It uses "What If?" questions to help trainees apply theory to real-world experiences and put ideas into action.

Hot Tip provides a head start for those entering the trade by highlighting important safety requirements and presenting tricks of the trade from experienced welders.

Case History examines the costly and often devastating consequences of ignoring safe work practices by citing actual examples from the field.

Profile in Success shares the experience of and advice from successful tradespersons in the welding field.

Did You Know? features interesting and sometimes surprising facts about the construction industry.

On-Site offers technical tips from the field.

We're excited to be able to offer you these improvements and hope they lead to a more rewarding learning experience. As always, your feedback is welcome! Please let us know how we are doing by visiting NCCER at www.nccer.org or email us at info@nccer.org.

UNIQUE ASPECTS OF THE NEW AWS ENTRY LEVEL WELDER PROGRAM

NCCER, in cooperation with the American Welding Society (AWS) presents a curriculum more succinctly aligned with the AWS QC10 Specification for the Qualification and Certification for Entry Level Welders for AWS SENSE (Schools Excelling through National Skills Education) programs.

The newly revised *AWS Entry Level Welder – Phase 1* and *AWS Entry Level Welder – Phase 2* (published separately) titles comply with all AWS documents specified in part 3.3 Learning Modules of AWS EG2.0-95 and should aid the participating SENSE school in certifying a trainee as an AWS certified Entry Level Welder. (Note that certification for Entry Level Welders is only available to AWS participating organizations and only through AWS. NCCER cannot issue any certifications on behalf of AWS.)

Performance Acceptance Criteria required by AWS EG2.0-95, "Guide for the Training and Qualification of Welding Personnel – Entry Level Welder" have been included where appropriate in the newly revised *AWS Entry Level Welder* Trainee and Instructor's Guide modules. The new text also includes detailed information on welding symbols and welding assembly drawings approved by AWS and electrical fundamentals per process, as well as AWS wire electrode designations.

This unique version of NCCER/Prentice Hall's welding program allows dual credentialing through both NCCER and AWS, provided all applicable requirements are met by the participating organizations. For more information on AWS's Entry Level Welder program, contact AWS at (305) 443-9353 or visit www.aws.org. For information on NCCER's Accreditation and National Registry, contact NCCER Customer Service at 1-888-622-3720 or visit www.nccer.org.

Acknowledgments

This curriculum was revised as a result of the farsightedness and leadership of the following sponsors:

Applied Welding Technology
Exxon Mobile
Flint Hills Resources
Fluor Daniel Inc.
Kawerak
Lee College
Northern Arizona Vocational
   Institute of Technology /
   Northland Pioneer College

Spec-Weld Technologies, Inc.
Texas A & M University
   System—Texas Engineering
   Extension Service
William H. Turner Adult
   Education Center
Zachry Construction
   Corporation

This curriculum would not exist were it not for the dedication and unselfish energy of those volunteers who served on the Authoring Team. A sincere thanks is extended to:

Tom Atkinson
Rex Ball
Curtis Casey
Bill D. Cherry
John Gault
Terry A. Lowe

David McGrath
John D. Murray
Joe D. Sanders
Jerry Trainor
John R. Yochum

Contents

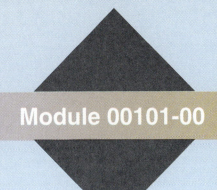

Basic Safety

Course Map

This course map shows all of the modules in the AWS Entry Level Welder—Phase 1 curriculum. The suggested training order begins at the bottom and proceeds up. Skill levels increase as you advance on the course map. The local Training Program Sponsor may adjust the training order.

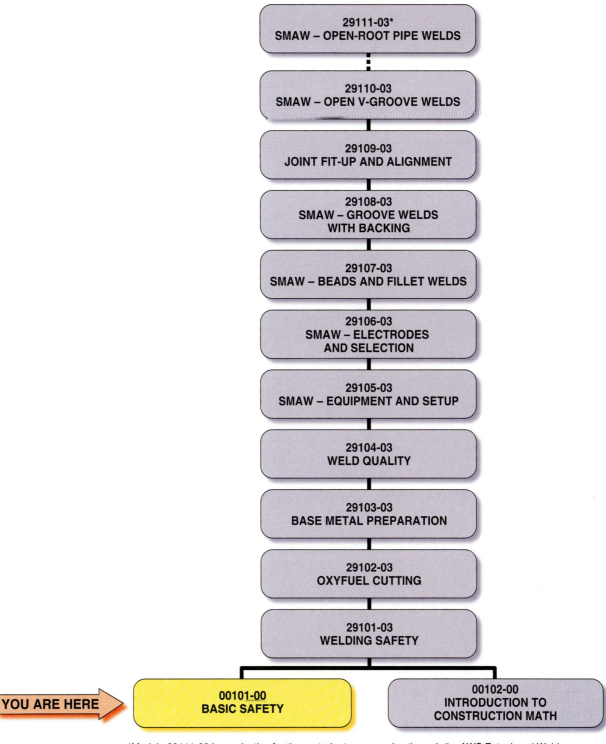

AWS ENTRY LEVEL WELDER—PHASE 1

29111-03*
SMAW – OPEN-ROOT PIPE WELDS

29110-03
SMAW – OPEN V-GROOVE WELDS

29109-03
JOINT FIT-UP AND ALIGNMENT

29108-03
SMAW – GROOVE WELDS
WITH BACKING

29107-03
SMAW – BEADS AND FILLET WELDS

29106-03
SMAW – ELECTRODES
AND SELECTION

29105-03
SMAW – EQUIPMENT AND SETUP

29104-03
WELD QUALITY

29103-03
BASE METAL PREPARATION

29102-03
OXYFUEL CUTTING

29101-03
WELDING SAFETY

YOU ARE HERE

00101-00
BASIC SAFETY

00102-00
INTRODUCTION TO
CONSTRUCTION MATH

*Module 29111-03 is an elective for those students progressing through the AWS Entry Level Welder program.

00101CMAP.EPS

MODULE 00101-00 CONTENTS

Figures

Tables

Basic Safety

Objectives

When you finish this module, you will be able to do the following:

1. Identify the responsibilities and personal characteristics of a professional craftsperson.
2. Explain the role that safety plays in the construction crafts.
3. Describe what job-site safety means.
4. Explain the appropriate safety precautions around common job-site hazards.
5. Demonstrate the use and care of appropriate personal protective equipment.
6. Follow safe procedures for lifting heavy objects.
7. Describe safe behavior on and around ladders and scaffolds.
8. Explain the importance of the HazCom (Hazard Communication Standard) requirement and MSDSs (Material Safety Data Sheets).
9. Describe fire prevention and fire-fighting techniques.
10. Define safe work procedures around electrical hazards.

Materials Required for Trainees

1. This module
2. Pencil and paper

1.0.0 ◆ INTRODUCTION

When you take a job, you have a safety obligation to your employer, co-workers, family, and yourself. In exchange for the benefits of your job and for your own well-being, you agree to work safely. In other words, you are *obligated* to work safely. You are also obligated to make sure anyone you work with is working safely. Your employer is likewise obligated to maintain a safe workplace for all employees. The ultimate responsibility for on-the-job safety, however, rests with you. In this module, you will learn to ensure your safety and the safety of the people you work with by

- Following safe work practices and procedures
- Inspecting safety equipment before use
- Using safety equipment properly

In order for you to take full advantage of the wide variety of training, job, and career opportunities offered by the construction industry, you must first understand the importance of safety. Successful completion of this module will be your first step toward achieving this goal.

1.1.0 Training Opportunities

This program of the National Center for Construction Education and Research (NCCER) was developed *by* the construction industry *for* the construction industry. It is one of the leading nationally accredited, competency-based construction training programs in the United States. A *competency-based* program requires that the trainees show that they can perform specific job-related tasks safely to receive credit. This approach is unlike other apprenticeship programs that merely require a trainee to put in the required number of hours in the classroom and on the job.

The primary goal of the NCCER is to standardize construction craft training throughout the country so that both employers and employees will benefit from the training, no matter where they are located. As a trainee in an NCCER program, you will become part of a national registry. You will receive a certificate for each level of training you complete. If you apply for a job with any participating contractor in the country, a transcript of your training will be available. If your training is incomplete when you make a job transfer, you can pick up where you left off, because every participating contractor is using the same training program. In addition, many technical schools and colleges are using the same program.

1.2.0 Career Opportunities in the Construction Industry

The construction industry employs more people and contributes more to the nation's economy than any other industry. People will always need new homes, roads, airports, hospitals, schools, factories, and office buildings. This means there will always be a source of well-paying jobs and career opportunities for construction trade professionals. As *Figure 1* shows, the opportunities are not limited to work on construction projects.

1.3.0 Characteristics of a Professional Craftsperson

A successful craftsperson is able to use current trade materials, tools, and equipment to produce high-quality products efficiently. A craftsperson must be able to adjust to new situations. The successful craftsperson will continuously train to keep up with the advances in trade materials and equipment. A professional craftsperson never takes chances with regard to personal safety or the safety of others. *Table 1* lists the characteristics of professional behavior.

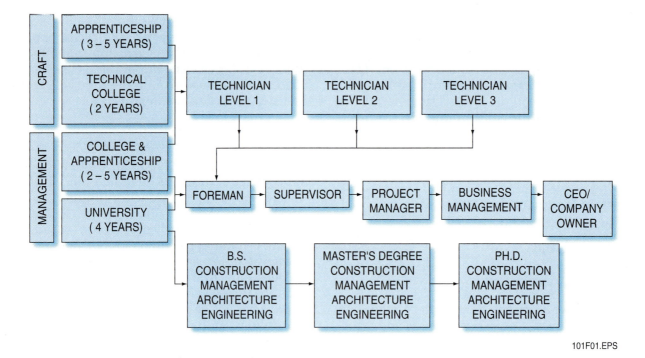

101F01.EPS

Figure 1 ◆ Opportunities in the construction industry.

Table 1 Characteristics of Professional Behavior

Honesty:	Be honest and truthful in all dealings. Conduct business according to the highest professional standards. Faithfully fulfill all contracts and commitments. Do not deliberately mislead or deceive others.
Integrity:	Demonstrate personal integrity and the courage of your convictions by doing what is right even where there is great pressure not to. Do not sacrifice your principles because it seems easier.
Loyalty:	Be worthy of trust. Demonstrate fidelity and loyalty to companies, employers, co-workers, and trade institutions and organizations.
Fairness:	Be fair and just in all dealings. Do not take unfair advantage of another's mistakes or difficulties. Fair people are open-minded and committed to justice, equal treatment of individuals, and acceptance of diversity.
Respect for others:	Be courteous and treat all people with equal respect and dignity, regardless of age, sex, race, or nationality.
Law abiding:	Abide by laws, rules, and regulations relating to all personal and business activities.
Commitment to excellence:	Pursue excellence in performing your duties, be well-informed and prepared, and constantly try to increase your proficiency by gaining new skills and knowledge.
Leadership:	By your own conduct, seek to be a positive role model for others.

2.0.0 ◆ ACCIDENTS: CAUSES AND RESULTS

Your boss might say, "I want my company to have a perfect safety record." What does that mean? A safety record is more than the number of days a company has worked without an accident. Safety is a learned behavior attitude. Safety is a way of working on the job. The time you spend learning and practicing safety procedures can save your life—and the lives of others.

What causes accidents? Either poor behavior or poor work conditions. You can help prevent most accidents by learning safe work habits and understanding what causes accidents. The National Safety Council estimates that the organized safety movement has saved three million lives, or, more accurately, prevented three million tragic deaths. Accidents cost billions of dollars each year and cause much needless suffering. This section of the Basic Safety module examines why accidents happen and how you can help prevent them.

The lessons you will learn in this module will help you work safely. You will be able to spot and avoid hazardous conditions on the job site. By following safety procedures and being aware of the need for safety, you will help keep your workplace free from accidents and protect yourself and your co-workers from injury or even death.

2.1.0 What Causes Accidents?

You may already know some of the main causes of accidents. They include the following:

- Failure to communicate
- Poor work habits
- Alcohol or drug abuse
- Lack of skill
- Intentional acts
- Unsafe acts
- Unsafe conditions
- **Management system** failure

We will discuss each of these causes.

2.1.1 Failure to Communicate

Many accidents happen because of a lack of communication. For example, you may learn how to do things one way on one job, but what happens when you go to a new job site? You need to *communicate* with the people at the new job site to find out if they do things the way you have learned to do them. If you do not communicate, accidents can happen. Remember that different people, companies, and job sites do things different ways.

If you think that people know something without talking with them about it, then you are *assuming* that they know. Assuming that other people know and will do what you think they should know and do can cause accidents.

All work sites have specific markings and signs to identify hazards and provide emergency information (see *Figure 2*). Learn to recognize these types of signs:

- Informational
- Safety
- Caution
- Danger
- Safety tags

Your Attitude Is Showing

Whether you are dealing with basic safety issues or any other area of the job, there is no substitute for a positive, can-do attitude. A positive attitude is essential to your career in construction for many reasons:

- When you are positive, you are usually more energetic, motivated, productive, and alert. Thinking about negative things drains your energy.
- Co-workers and supervisors sense your attitude the first time they meet you. If you send out positive signals, they feel friendly toward you and will like working with you. And research has shown that workers who are friendly with one another have better safety records.
- Your positive attitude contributes to the productivity of others.
- Positive people get more promotions, more recognition, and higher pay.

How can you maintain a positive attitude on the job? Here are a few suggestions:

- Talk about positive things. No one wants to be around a constant complainer.
- Look for the good in the people you work with, especially your supervisors. Nobody is perfect, but almost everyone has a few good qualities. If you focus on the positive, you'll find it will be easier to get along with and work with other people.
- Look for the good things in your company. No job is perfect, but many have excellent things to offer: good pay, good benefits, skilled and likable co-workers, the opportunity to learn a trade and grow as a person. Concentrate on the positive, and success will follow.
- Don't allow co-workers with negative attitudes to trap you into their way of thinking. You may not be able to change these people, but you can protect yourself.
- Think about your attitude and the signals you are sending. Once in a while, ask yourself: Is my attitude positive? How can I improve it?
- Keep thinking positively and always remember that no matter what you do, your attitude is showing.

Source: Adapted from Elwood N. Chapman, *Your Attitude Is Showing: A Primer of Human Relations,* 8th edition (Upper Saddle River, NJ: Prentice Hall, 1996), pp. 21–26.

 WARNING!

Never assume anything! It never hurts to ask, but it can be a disaster if you *don't* ask. For example, do not assume that electrical current is turned off. First, ask if the current is turned off. Then check it yourself to be completely safe.

Informational markings or signs provide general information. These signs are blue. The following are considered informational signs:

- No Admittance
- No Trespassing
- For Employees Only

Safety signs give general instructions and suggestions about safety measures. The background on these signs is white; most have a green panel with white letters. Signs of this kind tell you where to find such important areas as the following:

- First-aid stations
- Emergency eye-wash stations
- Evacuation routes
- **Material Safety Data Sheet (MSDS)** stations
- Exits (usually have white letters on a red field)

Caution markings or signs tell you about potential hazards or warn against unsafe acts. When you see a caution sign, take action to protect yourself against a possible hazard. Caution signs are yellow and have a black panel with yellow letters.

101F02A.EPS
ACCUFORM SIGNS
(A)

101F02B.EPS
ACCUFORM SIGNS
(B)

101F02C.EPS
ACCUFORM SIGNS
(C)

101F02D.EPS
ACCUFORM SIGNS
(D)

Figure 2 ◆ Communication tags/signs. (A) Informational sign. (B) Safety sign. (C) Caution sign. (D) Danger sign.

Signs of this kind may give you the following information:

- Hearing and eye protection are required
- **Respirators** are required
- Smoking is not allowed

Danger markings or signs tell you that an immediate hazard exists and that you must take certain precautions to avoid an accident. Danger signs are red, black, and white. Signs of this kind may indicate the following:

- Defective equipment
- Flammable liquids and compressed gases
- Safety barriers and barricades
- Emergency stop button
- High voltage

Safety tags are temporary warnings of immediate and potential hazards. They are not designed to be used in place of signs or as permanent means of protection. Learn to recognize the standard accident prevention signs and tags (see *Table 2*).

2.1.2 Poor Work Habits

Poor work habits can cause serious accidents. Examples of poor work habits are procrastination, carelessness, and horseplay. Procrastination, or *putting things off*, is a common cause of accidents. For example, putting off the repair, inspection, or cleaning of equipment and tools can cause trouble. If you try to push machines and equipment beyond their operating capacities, you risk injuring yourself and your co-workers.

Machines, power tools, and even a pair of pliers can hurt you if you don't use them safely. It is your responsibility to be careful. Tools and machines

Table 2 Tags and Signs

Basic Stock (background)	Safety Colors (ink)	Message(s)
White	Red panel with white or gray letters	Do Not Operate Do Not Start
White	Black square with a red oval and white letters	Danger Unsafe Do Not Use
Yellow	Black square with yellow letters	Caution
White	Black square with white letters	Out of Order Do Not Use
Yellow	Red/magenta (purple) panel with black letters and includes a radiation symbol	Radiation Hazard
White	Fluorescent orange square with black letters and a biohazard symbol	Biological Hazard

DID YOU KNOW?

True or false? Most accidents are caused by sharp blades.

False. More accidents are caused by dull blades than by sharp ones. If you do not keep your cutting tools sharpened, they won't cut very easily. When you have a hard time cutting, you exert more force on the tool. When that happens, something is bound to slip. And when something slips, you get cut.

don't know the difference between wood or steel and flesh and bone.

Work habits and work attitudes are closely related. If you resist taking orders, you may also resist listening to warnings. If you let yourself be easily distracted, you won't be able to concentrate. If you aren't concentrating, you could cause an accident.

Your safety is affected not only by how you do your work, but also by how you act on the job site. This is why most companies have strict policies for employee behavior. Horseplay and other inappropriate behavior are forbidden. Workers who engage in horseplay and other inappropriate behavior on the job site may be fired.

These strict policies are for *your* protection. There are many hazards on construction sites. Each person's behavior—at work, on a break, or eating lunch—must follow the principles of safety.

The man holding the bag in *Figure 3* may look like he's just having fun by playing a prank on his co-worker. In fact, what he's doing could cause his co-worker serious, even fatal, injury. If you horse around on the job, play pranks, or don't concentrate on what you are doing, you are showing a poor work attitude. That can lead to a serious accident.

101F03.EPS

Figure 3 ◆ Horseplay can be dangerous!

2.1.3 Alcohol and Drug Abuse

Alcohol and drug abuse costs the construction industry millions of dollars a year in accidents, lost time, and lost productivity. The true cost of alcohol and drug abuse is much more than just money, of course. Abuse can cost lives. Just as drunk driving kills thousands of people on our highways every year, alcohol and drug abuse kills on the construction site. How would you like to be the person in *Figure 4*? Would you like to be working near him?

Using alcohol or drugs creates a risk of injury for everyone on a job site. Many states now have laws that prevent workers from collecting insurance benefits if they are injured while under the influence of alcohol or illegal drugs.

Would you trust your life to a crane operator who was stoned? Would you bet your life on the responses of a co-worker on alcohol or drugs? Alcohol or drug abuse has no place in the construction industry. A person on a construction site who is under the influence of alcohol or drugs is an accident waiting to happen—and possibly a fatal accident.

Obviously, people who work while using alcohol or drugs are at risk of accident or injury, and their co-workers are at risk as well. That's why your employer probably has a formal substance abuse policy. You should know that policy and follow it for your own safety.

You don't have to be abusing illegal drugs like marijuana, cocaine, or heroin to create a job hazard. Many prescribed and over-the-counter drugs, taken for legitimate reasons, can affect your ability to work safely. Amphetamines, barbiturates, and antihistamines are only a few of the legal drugs that can affect your ability to work safely or to operate machinery.

CAUTION

If your doctor is prescribing any medication that you think might affect your job performance, *ask* about its effects. Your safety and the safety of your co-workers depend on everyone being alert on the job.

Do yourself and the people you work with a big favor. Be aware of and follow your employer's substance abuse policy. Avoid any substances that can affect your job performance. The life you save could be your own!

2.1.4 Lack of Skill

You should learn and practice new skills under careful supervision. Never perform them alone until you've been checked out by a supervisor.

101F04.EPS

Figure 4 ◆ How many violations can you identify?

Lack of skill can cause accidents quickly. Here's an example: You are told to cut some 2 × 8s with a circular saw, but you aren't skilled with that tool. A basic rule of circular saw operation is never to cut without a properly functioning guard. Because you haven't been trained, you don't know this. You find that the guard on the saw is slowing you down. So you jam the guard open with a small block of wood. The result could be a serious accident.

 WARNING!
Never operate a power tool until you have been trained to use it. You can greatly reduce the chances of accidents by learning safety rules for each task that you perform.

2.1.5 *Intentional Acts*

When someone purposely causes an accident, it is called an intentional act. Sometimes an angry or dissatisfied employee may purposely create a sit-

AWS ENTRY LEVEL WELDER – TRAINEE MODULE 00101-00

uation that leads to property damage or personal injury. If someone you are working with threatens to *get even* or *pay back* someone, let your supervisor know at once.

2.1.6 Unsafe Acts

An unsafe act is a change from an accepted, normal, or correct procedure that usually causes an accident. It can be any conduct that causes unnecessary exposure to a job-site hazard or that makes an activity less safe than usual. Here are examples of unsafe acts:

- Failing to use **personal protective equipment (PPE)**
- Failing to warn co-workers
- Lifting improperly
- Loading or placing equipment or supplies improperly
- Making safety devices (such as saw guards) inoperable
- Operating equipment at improper speeds
- Operating equipment without authority
- Servicing equipment in motion
- Taking an improper working position
- Using defective equipment
- Using equipment improperly

2.1.7 Unsafe Conditions

An unsafe condition is a physical state that is different from the acceptable, normal, or correct condition found on the job site. It usually causes an accident. It can be anything that reduces the degree of safety normally present. Here are examples of unsafe conditions:

- Congested workplace
- Defective tools, equipment, or supplies
- Excessive noise
- Fire and explosive hazards
- Hazardous atmospheric conditions (such as gases, dusts, fumes, vapors)
- Inadequate supports or guards
- Inadequate warning systems
- Poor housekeeping
- Poor lighting
- Poor ventilation
- Radiation exposure

2.1.8 Management System Failure

Sometimes the cause of an accident is failure of the management system that permitted the unsafe act or unsafe condition. The management system should be designed to prevent or correct the acts and conditions that can cause accidents. If the management system did not do these things, that system failure *may* have caused the accident. What traits could mean the difference between a management system that fails and one that succeeds? A few important traits of a *good* management system follow.

- The company puts safety policies and procedures in writing.
- The company distributes written safety policies and procedures to each employee.
- The company reviews safety policies and procedures periodically.
- The company enforces all safety policies and procedures fairly and consistently.
- The company evaluates supplies, equipment, and services to see if they are safe.
- The company provides regular, periodic safety training for employees.

2.2.0 Housekeeping

In construction, housekeeping means keeping your work area clean and free of scraps or spills. It also means being orderly and organized. You must store your materials and supplies safely and label them properly. Arranging your tools and equipment to permit safe, efficient work practices and easy cleaning is also important.

If the work site is indoors, make sure it is well lighted and ventilated. Don't allow aisles and exits to be blocked by materials and equipment. Make sure that **flammable** liquids are stored in safety cans. Oily rags must be placed only in approved self-closing metal containers. Remember that the major goal of good housekeeping is to prevent accidents. Good housekeeping reduces the chances for slips, fires, explosions, and falling objects.

Here are some good housekeeping rules:

- Keep all scrap material and lumber with nails sticking out clear of work areas.
- Clean up spills to prevent falls.
- Remove all **combustible** scrap materials regularly.
- Make sure you have containers for the collection and separation of refuse. Containers for flammable or harmful refuse must be covered.
- Dispose of wastes often.
- Store all tools and equipment when you're finished using them.

Another way of explaining good housekeeping is *pride of workmanship.* If you take pride in what you are doing, you won't let trash build up around you. The old saying, "A place for everything and everything in its place" may sound corny, but it's the right idea on the job site.

2.3.0 Company Safety Policies vs. OSHA Regulations

The mission of the Occupational Safety and Health Administration (OSHA) is to save lives, prevent injuries, and protect the health of America's workers. To accomplish this, federal and state governments must work in partnership with the more than 100 million working men and women and their six and a half million employers who are covered by the Occupational Safety and Health Act of 1970.

Nearly every working man and woman in the nation comes under OSHA's jurisdiction (with some exceptions such as miners, transportation workers, many public employees, and the self-employed).

Section 5(a)1 of the OSHA Act summarizes the intent:

Each Employer—shall furnish to each of his employees employment and a place of employment which are free from recognized hazards that are causing or are likely to cause death or serious physical harm to his employees.

As employers are responsible to OSHA for compliance, so it is for the employee to comply with the company safety policies and rules. Employers are required to identify hazards and potential hazards within the workplace and eliminate them, control them, or provide protection from them. This can only be done through combined efforts from employer and employee. Employers must provide written programs and training on these hazards and the employee must follow the procedures. You, as the employee, must read and understand the OSHA poster located at your site regarding your rights and responsibilities. If you are unsure of the location of the OSHA poster ask your supervisor.

To assist employers in providing a safe workplace OSHA requires companies to provide a competent person to ensure the safety of the employees. By law a competent person is:

"A person who can identify working conditions or surroundings that are unsanitary, hazardous, or dangerous to employees and who has authorization to correct or eliminate these conditions promptly."

In other words a person who is experienced and knowledgeable about the specific operation and has the authority from the employer to correct the problem or shut the operation down until safe is a competent person. The term *competent person* will be an important part of your working career and it is important for you to recognize what and who the competent person is. OSHA requires a competent person for many of the tasks you may be assigned to perform, such as confined space entry, ladder use, trenching, and many more tasks. This may require different individuals to be assigned as a competent person for different tasks according to their expertise. To ensure a long and successful career for you and your co-workers, work closely with your competent person and supervisor.

2.4.0 Reporting Injuries, Accidents, and Incidents

There are three categories of on-the-job events: injuries, accidents, and incidents. An injury is anything that requires treatment, even minor first aid. An accident is anything that causes an injury or property damage. An incident is anything that *could* have caused an injury or damage but, because it was caught in time, did *not* cause an injury or damage.

You must report all on-the-job injuries, accidents, or incidents, no matter how minor, to your supervisor (see *Figure 5*). Some workers think they will get in trouble if they report minor injuries. That's just not true. Often, small injuries, like cuts and scrapes, later become big problems because of infection and other complications.

By analyzing accidents, companies and the government can improve safety policies and procedures. By reporting an accident, you can help keep similar accidents from happening in the future.

Table 3 shows an analysis of the causes of fatal accidents in the construction industry for the years 1997 through 1998.

All other accidents combined accounted for only 10 percent of the total. Because of these findings, OSHA developed its Focused Inspection Program to target these four high-hazard areas. OSHA hopes to reduce accidents, injuries, and fatalities in the construction industry.

Here are explanations of these four leading hazard groups:

- Falls from elevation are accidents involving failure to provide or use appropriate fall protection.
- Struck-by accidents involve unsafe operation of equipment, machinery, and vehicles, as well as

effect when dangerous situations arise, such as fire, chemical spills, and gas leaks. In an emergency situation, you must know the evacuation procedures. You must also know the signal that is used (usually a horn or siren) to tell workers to evacuate.

When you hear the evacuation signal, follow the procedures *exactly.* That usually means taking a certain route to a designated assembly area and telling the person in charge that you are there. If hazardous materials are released into the air, you may have to look at the **wind sock** (see *Figure 6*). The wind sock tells you which way the wind is blowing. Different evacuation routes are planned for different wind directions. Taking the right route will keep you from being exposed to the hazardous material.

101F05.EPS

Figure 5 ◆ All accidents, injuries, or incidents must be reported to your supervisor.

Table 3 Causes of Fatal Accidents

Cause	Percent of Fatalities
Falls from elevation	33%
Struck by . . .	27%
Caught in or caught between . . .	16%
Electrical shock	14%

improper handling of materials, such as through unsafe rigging operations.
- Caught-in or caught-between accidents involve unsafe operation of equipment, machinery, and vehicles, as well as improper safety procedures at **trench** sites and in other confined spaces.
- Electrical shock accidents involve contact with overhead wires, use of defective tools, failure to disconnect power source before repairs, or improper ground fault protection.

2.5.0 Evacuation Procedures

In many work environments, specific evacuation procedures are needed. These procedures go into

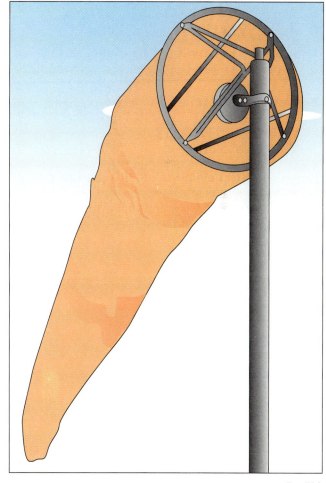

101F06.EPS

Figure 6 ◆ Wind sock.

Review Questions

1. _____ is (are) not a main cause of accidents.
   a. Unsafe acts
   b. Alcohol or drug abuse
   c. Weather
   d. Poor work habits

2. Blue signs or markings that provide general information such as No Trespassing are _____ signs.
   a. caution
   b. information
   c. warning
   d. safety

3. White and green signs or markings that give general instructions or suggestions about first aid stations, exits, and evacuation routes are _____ signs.
   a. safety
   b. danger
   c. information
   d. MSDS

4. If a sign has a white background and a red panel with white or gray letters, you might see _____ on it.
   a. Out of Order
   b. Danger
   c. Biological Hazard
   d. Do Not Start

5. _____ must be reported to the employer.
   a. Only major injuries
   b. Only near misses and major injuries
   c. All injuries and near misses
   d. No injuries

3.0.0 ◆ CONSTRUCTION SITE JOB HAZARDS

It's impossible to list *all* the hazards that can exist on a construction job site. This section describes some of the more common hazards and explains how to deal with them. You may want to make a list of other hazards you think could be present on a job site and discuss them with your instructor or supervisor.

For your safety, you must know the specific hazards where you are working and how to prevent accidents and injuries. Let's look at these now.

3.1.0 Welding

Even if you're not welding, you can be injured when you are around a welding operation. The oxygen and acetylene used in gas welding are very dangerous. The cylinders containing oxygen and acetylene must be transported, stored, and handled very carefully. Always follow these safety guidelines:

- Keep the work area clean and free from potentially dangerous material such as combustible materials and grease or petroleum products.

 WARNING!
Keep oxygen away from sources of flame and combustible materials, especially substances containing oil, grease, or other petroleum products. Oil or grease compressed with oxygen will explode!

- Use great caution when you handle compressed gas cylinders.

 WARNING!
Do not remove the protective cap unless a cylinder is secured. If the cylinder falls over and the nozzle breaks off, the cylinder will shoot off like a rocket, injuring or killing anyone in its path.

- Never look at an **arc welding** operation without wearing the proper eye protection. The glare will burn your eyes.

WARNING!
In an arc welding operation, even a *reflected* **arc** can harm your eyes. It is extremely important that proper safety procedures are maintained at and around all welding operations. Serious eye injury or even blindness can result from unsafe conditions.

- If you are welding, use the proper personal protective equipment (see *Figure 7*), including the following:

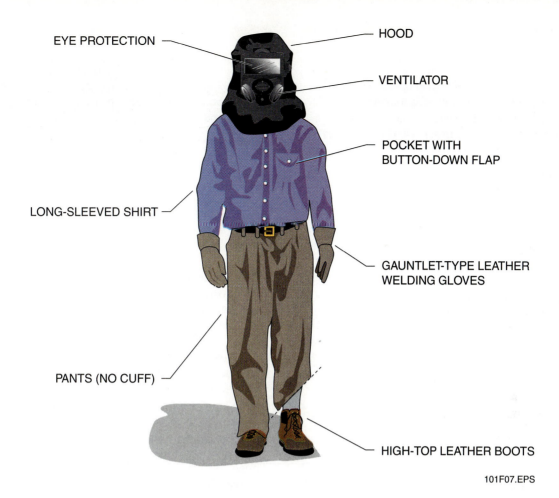

EYE PROTECTION

HOOD

VENTILATOR

POCKET WITH
BUTTON-DOWN FLAP

LONG-SLEEVED SHIRT

GAUNTLET-TYPE LEATHER
WELDING GLOVES

PANTS (NO CUFF)

HIGH-TOP LEATHER BOOTS

101F07.EPS

Figure 7 ◆ Personal protective equipment for welding.

- Snug-fitting cutting goggles that have filter lenses with shade five or six and a clear lens to protect against spatter
- An appropriate long-sleeve shirt with the collar buttoned
- A cap with the bill worn backwards to prevent sparks from falling down the back of your shirt collar
- Earplugs to prevent flying sparks from entering your ears.
- All leather, gauntlet-type welder's gloves
- High-top leather boots to prevent slag from dropping inside your boots
- Cuffless trousers that cover your ankles and boot tops
- A respirator, if necessary

• If you are welding and other workers are in the area around your work, set up welding shields. Make sure everyone wears flash goggles. The flash is the sudden, bright light associated with starting up a welding operation.

 WARNING!
The cutting process results in oxides that mix with molten iron and produce slag. This slag is blown from the cut by the jet of cutting oxygen and can cause severe injury or start fires on contact.

• A welder must be protected when the welding shield worn on the welder's headgear is down, because the shield restricts the welder's field of vision. A helper or monitor must watch the welder and the surrounding area in case of a fire or similar emergency, or rope off the area to prevent collisions and keep other workers away from the area.
• Welded material is hot! Mark it with a sign and stay clear for a while after the welding has been completed.

WARNING!

Post a fire watch when you are welding. One person other than the cutting operator must constantly scan the work area for fires. Fire-watch personnel should have ready access to fire extinguishers and alarms and know how to use them. Cutting operations should never be performed without a fire watch.

WARNING!

Never wear contact lenses while you are welding. The ultraviolet rays may dry out the moisture beneath the contact lens, causing it to stick to your eye.

Pay special attention to the safety guidelines about never looking at the arc without proper eye protection. Even a brief exposure to the ultraviolet light from arc welding can damage your eyes badly, causing a flash burn. You may not notice the symptoms until some time after the exposure. Here are some symptoms of flash burns to the eye:

- Headache
- Feeling of sand in your eyes
- Red or weeping eyes
- Trouble opening your eyes
- Impaired vision
- Swollen eyes

If you think you may have a flash burn to your eyes, seek medical help at once.

Before and during weld operations, you must follow certain safety procedures. As the operator, you must check three things:

- Hoses
- Regulators
- The work area

3.1.1 Hoses

Use the proper hose. The fuel gas hose is usually red (sometimes black) and has a left-hand threaded nut for connecting to the torch. The oxygen hose is green and has a right-hand threaded nut for connecting to the torch.

Hoses showing leaks, burns, worn places, or other defects making them unfit for service must be repaired or replaced. When inspecting hoses, look for charred sections close to the torch. These may have been caused by flashback. Also check that hoses are not taped up to cover leaks (see *Figure 8*).

New hoses contain talc and loose bits of rubber. These materials must be removed from the hoses before the torch is connected. If they are not removed, they will clog the torch needle valves. Common industry practice is to use compressed air to blow these materials out of the hose. Always make sure the regulator valve is turned down to

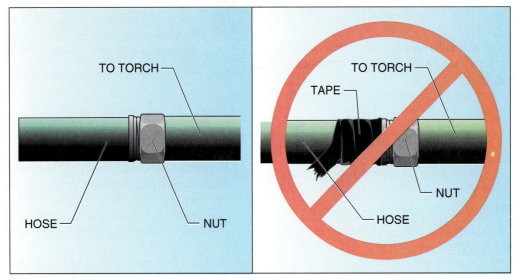

101F08.EPS

Figure 8 ◆ Proper hose connection.

AWS ENTRY LEVEL WELDER – TRAINEE MODULE 00101-00

minimal pressures before using compressed air to clean a hose.

WARNING!

Never point a compressed air hose toward anyone for any reason. Flying debris and particles of dirt may cause serious injury.

3.1.2 Regulators

Regulators are attached to the cylinder valve. They lower the high cylinder pressures to the required working pressures and maintain a steady flow of gas from the cylinder (see *Figure 9*).

To prevent damage to regulators, always follow these guidelines.

- Never jar or shake regulators, because that can damage the equipment beyond repair.
- Always check that the adjusting screw is released before the cylinder valve is turned on.
- Always open cylinder valves slowly.

WARNING!

When opening valves, always stand to the side to avoid injury from flying dirt that may have been stuck in the valve.

- Once cutting or welding has been completed, fully release the adjusting screw to relieve line pressure.
- Never use oil to lubricate a regulator, because that can cause an explosion.

NORTH SAFETY PRODUCTS   101F09.EPS

Figure 9 ◆ Regulator.

- Never use fuel regulators on oxygen cylinders or oxygen regulators on fuel gas cylinders.
- Never use a defective regulator. If a regulator is not working properly, shut off the gas supply and have the regulator repaired by a **qualified person.**
- Never use pliers or channel locks to install or remove regulators.

3.1.3 Work Area

Before beginning a cutting operation, check the area for fire hazards. Cutting sparks can fly 30 feet or more and can fall several floors. Remove or cover any flammable material in the area. Have an approved fire extinguisher available before starting your work.

Always perform cutting operations in a well-ventilated area. Heating and cutting metals with an oxyfuel torch can create toxic fumes. The most common hazardous material is zinc. Zinc is present in galvanized coatings and brass.

WARNING!

Never cut galvanized metal without proper ventilation. The zinc oxide fumes given off as the galvanized material is cut are hazardous. Also, use a respirator when you are cutting galvanized material.

Maintaining a clean and neat work area promotes safety and efficiency. When you are finished welding, be sure to do the following:

- Pick up cutting scraps.
- Sweep up any scraps or debris around the work area.
- Return cylinders and equipment to the proper places.
- Prevent fires by making sure that cut metals and slag are cooled before disposing of them.

3.2.0 Trenches and Excavations

In many construction jobs you will need to work in trenches or **excavations.** Cave-ins and falling objects are hazards in these areas. Obey the following safety rules when working around trenches and excavations:

- Never put tools, materials, or loose dirt or rocks within 2 feet of the edge of a trench. They can easily fall in and injure the people in the trench.

Also, too much weight near the edge of a trench can cause a cave-in.

- Always walk around a trench. Never jump over it or straddle it. You could lose your footing and fall in, or your weight could cause a cave-in.
- Never jump into a trench. Always use a ladder to get in and out.
- Put barricades around all trenches, as shown in *Figure 10*.
- Always follow OSHA regulations and your employer's procedures for **shoring** up a trench to prevent a cave-in. Never work beyond the shoring.

WARNING!

Never work on the face of either sloped excavations or excavations with **benching** at levels above other workers. The workers below you may not be adequately protected from the hazard of falling, rolling, or sliding materials or equipment.

A competent person will inspect excavations daily and decide if there could be any cave-ins or failures of protective systems, and if there are any hazardous atmospheres or other hazardous conditions. The competent person will conduct the inspection before any work begins and as needed throughout the shift. The competent person will also inspect the excavations after every rainstorm or other hazard-increasing incident.

You cannot work in excavations that have standing water or in excavations where water is coming in unless you take precautions to protect yourself. A competent person will know what these precautions are. Always ask your instructor or immediate supervisor if you have any questions about proper safety practices.

3.3.0 Proximity Work

Working near a hazard but not in direct contact with it is called **proximity work.** Proximity work requires extra caution and awareness. The hazard may be hot piping, energized electrical equipment, or running motors or machinery (see *Figure 11*). You must do your work so that you do not come into contact with the nearby hazard.

You may need to put up barricades to prevent accidental contact. Lifting and rigging operations may have to be done in a way that minimizes the risk of dropping things on the hazard. A monitor may watch you and alert you if you are in danger of touching the hazard while you work.

Energized electrical equipment is very hazardous. Regulations and policies will tell you the minimum safe working distance from energized

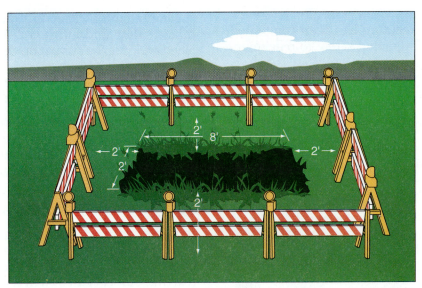

101F10.EPS

Figure 10 ◆ Barricade around a trench.

CEILING   HOT PIPE

WALL

HOT PIPE →

BARRICADE
↓

CONSTRUCTION

101F11.EPS

Figure 11 ◆ Proximity work.

DANGER

CONTENTS
HIGH TEMPERATURE

CONSTR

101F12.EPS

Figure 12 ◆ Avoid touching high-temperature components.

electrical conductors. You'll learn more about working with and around energized electrical equipment later in this module.

3.3.1 *Pressurized or High-Temperature Systems*

In many construction jobs you must work close to tanks, piping systems, and pumps that contain pressurized or high-temperature fluids. Be aware of these two possible hazards:

- Touching a container of high-temperature fluid can cause burns (see *Figure 12*). Many industrial processes involve fluids that are as hot as several thousand degrees.
- If a container holding pressurized fluids is damaged, it may leak and spray dangerous fluids.

Any work around pressurized or high-temperature systems is proximity work. Barricades or a monitor or both may be needed for safety (see *Figure 13*).

3.4.0 Confined Spaces

Construction and maintenance work isn't always done outdoors. A lot of it is performed in **confined spaces.** A confined space is a space that is large enough for a person to work in but has limited means of entry or exit. A confined space is not designed for human occupancy. Examples of confined spaces are tanks, vessels, silos, storage bins, hoppers, vaults, and pits (see *Figure 14*).

A **permit-required confined space** is a type of confined space that has been evaluated and found to have actual or potential hazards. A qualified person classifies such a confined space as a

Figure 13 ◆ Work safely near pressurized or high-temperature systems.

Figure 14 ◆ Examples of confined spaces.

AWS ENTRY LEVEL WELDER – TRAINEE MODULE 00101-00

permit-required confined space. You need written authorization before you can enter a permit-required confined space.

Many confined spaces contain hazardous gases or fluids when equipment is in operation. In addition, the work you are doing may introduce hazardous fumes into the space. Welding is an example of such work. For safety, you must take special precautions before you enter and leave a confined space and while you work there.

Until you have been trained to work in permit-required confined spaces and have taken the needed precautions, you must stay out of them. If you aren't sure whether a confined space requires a permit, ask your supervisor. You must always follow your employer's procedures and your supervisor's instructions. Confined space procedures may include getting clearance from a safety representative before starting the work. You will be told what kinds of hazards are involved and what precautions you need to take. You will also be shown how to use the required protective and emergency equipment. Remember: Better safe than sorry—so ask!

WARNING!

Without proper training, no employee is allowed to enter a permit-required confined space.

3.5.0 *Motorized Vehicles*

Motorized vehicles used on job sites include trucks, forklifts, backhoes, cranes, and **trenchers.** Operators must take care when driving vehicles. Helpers, riggers, and anyone else working nearby must also be careful.

If the vehicle is used indoors, ventilation of the work area is especially important. All internal combustion engines give off carbon monoxide as part of their exhaust. You cannot see, smell, or taste carbon monoxide, but *it can kill you*. Make sure there is good ventilation before you operate any motorized vehicles indoors.

The operator of any vehicle is responsible for the safety of passengers and the protection of the load. Follow these safety guidelines when you operate vehicles on a job site:

- Always wear a seat belt.
- Be sure that each person in the vehicle has a firmly secured seat. (Each person in the front seat of any vehicle *must* have a seat belt.)
- Obey all speed limits. Reduce speed in crowded areas.
- Look to the rear and sound the horn before backing up. If your rear vision is blocked, get a **signaler** to direct you.
- Every vehicle must have a backup alarm. Make sure the backup alarm works.
- Always shut off the engine when you are fueling.
- Turn off the engine and set the brakes before you leave the vehicle.
- Never remain on or in a truck that is being loaded by excavating equipment.
- Keep windshields, rearview mirrors, and lights clean and functional.
- Carry road flares, fire extinguishers, and other standard safety equipment at all times.

WARNING!

Driving a vehicle indoors without good ventilation can sicken or kill you because of the carbon monoxide given off by the exhaust. You cannot see, smell, or taste carbon monoxide.

Cranes and other vehicles used to lift loads require additional precautions. You must have a signaler whenever you cannot see in the direction of the vehicle's travel or the position of the load (see *Figure 15*). You and the signaler must have a clear method of communication. Neither you nor the signaler should let yourselves be distracted.

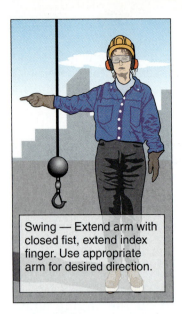

Swing — Extend arm with closed fist, extend index finger. Use appropriate arm for desired direction.

Raise Boom and Lower Load — Extend arm, thumb up, open and close fingers.

Lower Boom and Raise Load — Extend arm, thumb down, open and close fingers.

Travel — Extend arm, palm raised, and motion arm in the direction desired.

Extend Boom — Extend arms in front of body, palms up, fists closed, extend thumbs out to the sides.

Retract Boom — Extend arms in front of body, palms down, fists closed, extend thumbs inward.

Travel Both Tracks — With clenched fists, roll one fist over the other.

Travel One Track — Raise arm, fist clenched, to indicate lock track; roll other fist to travel. Raised hand indicates track to travel.

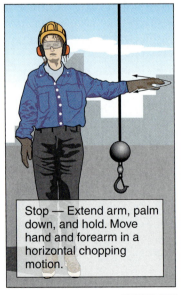

Stop — Extend arm, palm down, and hold. Move hand and forearm in a horizontal chopping motion.

101F15A.EPS

Figure 15 ◆ Communication between operator and signaler.

AWS ENTRY LEVEL WELDER – TRAINEE MODULE 00101-00

Emergency Stop — Same position as for Stop; extend and retract arms rapidly.

Dog Everything — Clasp hands, palm in palm, in front of the body.

Move Slowly — Placing the hand over any signal indicates a slow movement. "Hoist up" is used as an example.

Raise Load or Hoist Up — Fist up with pointer finger pointing straight up. Move hand in small horizontal circles.

Lower Load or Hoist Down — Fist down with pointer finger pointing straight down. Move hand in small horizontal circles.

Use Main Hoist — Rap on hard hat with closed fist.

Use Auxiliary Hoist — Strike open palm with elbow.

Raise Boom — Extend arm with closed fist, thumb extended up.

Lower Boom — Extend arm with closed fist, thumb extended down.

101F15B.EPS

Figure 15 ◆ Communication between operator and signaler.

Review Questions

Section 3.0.0

1. Oil or grease compressed with _____ explodes!
   a. a gas cylinder
   b. acetylene
   c. oxygen
   d. combustible materials

2. Flash burns are caused by exposing your eyes to _____.
   a. ultraviolet light
   b. oxygen and acetylene
   c. petroleum products
   d. zinc oxide fumes

3. A _____ is attached to a cylinder valve to reduce the high cylinder pressure to the required lower working pressure.
   a. safety valve
   b. regulator
   c. torch assembly
   d. compression hose

4. A confined space _____.
   a. has a limited amount of ventilation
   b. has no means of entry
   c. is too small to work in
   d. can be entered by untrained employees

5. All internal combustion engines give off a deadly odorless, tasteless, invisible gas called _____ as part of their exhaust.
   a. carbon trioxide
   b. carbon monoxide
   c. carbon dioxide
   d. carbon cyanide

Effective Teamwork

Construction is a cooperative effort. The construction process is a complex system made up of different crafts working together to build a project. The success of a construction project depends on your dedication to your team's or crew's performance. To be an effective team member, you must do three things:

- Cooperate with your co-workers and other groups.
- Contribute to the completion of assigned tasks.
- Work together to get the job done.

Everyone has different talents and skills. It is the combination of everyone's abilities that makes the team work. Your supervisor has recognized your abilities and grouped you with others who will benefit from working with you. And you will benefit from your team members' experience, talents, and skills.

A good team worker is all of the following things:

- Aware of personal strengths and weaknesses
- Positive
- Motivated
- Cooperative
- Patient
- Tolerant of diversity and varying levels of skill
- Willing to work toward a group goal
- Able to negotiate to solve problems
- Focused
- Forgiving

Successful teams demonstrate the old saying that the whole is greater than the sum of its parts. When team members pull together and pool their talents and skills, the team can do more than anyone can ever do alone.

4.0.0 ◆ WORKING SAFELY WITH JOB HAZARDS

You can safely handle all the job hazards you have learned about if you follow the rules. As long as everyone follows safety procedures, there is little risk of being hurt on the job site. In this section, you will learn about procedures and equipment used on construction sites to ensure worker safety.

4.1.0 Lockout/Tagout

A **lockout/tagout** system safeguards workers from hazardous energy while they work with machines and equipment. A lockout/tagout system protects workers from hazards such as the following:

- Acids
- Air pressure
- Chemicals
- Electricity
- Flammable liquids
- High temperatures
- Hydraulics
- Machinery
- Steam
- Other forms of energy

When people are working on or around any of these hazards, mechanical and other systems are shut down, drained, or de-energized. Tags and locks are placed on each switch, circuit breaker, valve, or other component to make sure that motors aren't started, valves aren't opened or closed, and any other changes that would endanger workers aren't made. Lockouts or tagouts protect workers from all possible sources of energy, including electrical, mechanical, hydraulic, thermal, pneumatic (air), and high temperature.

Generally, each lock has its own key, and the person who puts the lock on keeps the key. That person is the only one who can remove the lock. Tags have the words *DANGER* or *CLEARANCE* (see *Figure 16*).

ACCUFORM SIGNS

101F16.EPS

Figure 16 ◆ Typical safety tags.

Follow these rules for a safe lockout/tagout system:

- *Never* operate any device, valve, switch, or piece of equipment that has a lock or a tag attached to it.
- Use only tags that have been approved for your job site.
- If a device, valve, switch, or piece of equipment is locked out, make sure the proper tag is attached.
- Lock out and tag all electrical systems.
- Lock out and tag pipelines containing acids, explosive fluids, and high-pressure steam.
- Tag motorized vehicles and equipment when they are being repaired and before anyone starts work. Also, disconnect or disable the starting devices.

The exact procedures for lockout/tagout may vary at different companies and job sites. Ask your supervisor to explain the lockout/tagout procedure on your job site. You must know this procedure and follow it. This is for your safety and the safety of your co-workers. If you ever have questions about lockout/tagout procedures, ask your supervisor.

4.2.0 Barriers and Barricades

Any opening in a wall or floor is a safety hazard. There are two types of protection for these openings: they can be **guarded** or they can be covered. Cover any hole in the floor when possible. When it is not practical to cover the hole, use barricades. If the bottom edge of a wall opening is fewer than 3 feet above the floor and would allow someone to fall 4 feet or more, then place guards around the opening. There are several different guard methods:

- *Railings* are used across wall openings or as a barrier around floor openings to prevent falls (see *Figure 17a*).
- *Warning barricades* alert workers to hazards but provide no real protection (see *Figure 17b*). Typical warning barricades are made of plastic tape or rope strung from wire or between posts. The tape or rope is color-coded:
  - Red means danger. No one can enter an area with a red warning barricade. A red barricade is used when there is danger from falling objects or when a load is suspended over an area.
  - Yellow means caution. You can enter an area with a yellow barricade, but be sure you know what the hazard is, and be careful. Yellow barricades are used around wet areas or areas containing loose dust. Yellow with black lettering warns of physical hazards such as bumping into something, stumbling, or falling.

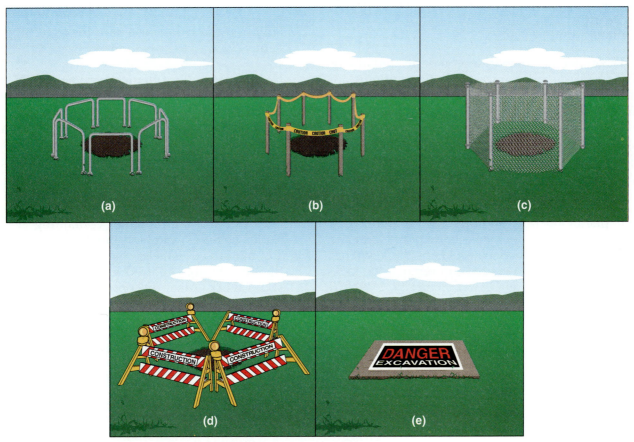

101F17.EPS

Figure 17 ◆ Common types of barriers and barricades.

AWS ENTRY LEVEL WELDER – TRAINEE MODULE 00101-00

- Yellow and purple means radiation warning. No one may pass a yellow and purple barricade. These barricades are often used where piping welds are being X-rayed.
- *Protective barricades* give both a visual warning and protection from injury (see *Figure 17c*). They can be wooden posts and rails, posts and chain, or steel cable. People cannot get past protective barricades.
- *Blinking lights* are placed on barricades so they can be seen at night (see *Figure 17d*).
- *Hole covers* are used to cover open holes in a floor (see *Figure 17e*). They must be secured and labeled. They must be strong enough to support twice the weight of anything that may be placed on top of them.

The types of barriers and barricades you see will vary from job site to job site (see *Figure 17*). There may also be different procedures for when and how barricades are put up. Learn and follow the policies at your particular job site.

 WARNING!
Never remove a barricade unless you have been authorized to do so.

5.0.0 ◆ PERSONAL PROTECTIVE EQUIPMENT

Personal protective equipment is designed to protect you from injury. You must keep it in good condition and use it when you need to. Many workers are injured on the job because they are not using personal protective equipment.

5.1.0 Personal Protective Equipment Needs

You won't see all the potentially dangerous conditions just by looking around a job site. It's important to stop and consider what type of accidents *could* happen on any job that you are about to do. Using common sense and knowing how to use personal protective equipment will greatly reduce your chance of getting hurt.

5.2.0 Personal Protective Equipment Use and Care

The best protective equipment is of no use to you unless you do four things.

- Regularly inspect it.
- Properly care for it.
- Use it properly when it is needed.
- Never alter or modify it in any way.

The sections that follow describe protective equipment commonly used on construction sites and tell how to use and care for each piece of equipment.

5.2.1 Hard Hat

Figure 18 shows a typical hard hat. The outer shell of the hat can protect your head from a hard blow. The webbing inside the hat keeps space between

NORTH SAFETY PRODUCTS   101F18.EPS

Figure 18 ◆ Typical hard hat.

 **ON-SITE**

Hard Hat Care

Inspect your hard hat every time you use it. If there are any cracks or dents in the shell, or if the webbing straps are worn or torn, get a new hard hat. Wash the webbing and headband with soapy water as often as needed to keep them clean.

the shell and your head. Adjust the headband so that the webbing fits your head and there is at least 1 inch of space between your head and the shell. Do not alter your hard hat in any way.

Hard hats used to be made of metal. However, metal conducts electricity, so most hard hats are now made of reinforced plastic or fiberglass.

5.2.2 Safety Glasses, Goggles, and Face Shields

Wear eye protection (see *Figure 19*) wherever there is even the slightest chance of an eye injury. Eye and face protection shall meet the requirements specified in ANSI (American National Standards Institute) 287.1-1968. Areas where there are potential eye hazards from falling or flying objects are usually identified, but you should always be on the lookout for possible hazards.

Regular safety glasses will protect you from falling objects or from objects flying at you from the front. You can add side shields for protection

from the sides. In some cases, you may need a face shield. Safety goggles give your eyes the best protection from all directions.

Welders must use tinted goggles or welding hoods. The tinted lenses protect the eyes from the bright welding arc or flame.

WARNING!
Handle safety glasses and goggles with care. If they get scratched, replace them. The scratches will interfere with your vision. Clean the lenses regularly with lens tissues or a soft cloth.

5.2.3 Safety Harness

Safety harnesses, like those in *Figure 20*, are extra-heavy-duty harnesses that buckle around your body. They have leg, shoulder, chest, and pelvic straps.

Safety harnesses have a D-ring attached to one end of a short section of rope called a **lanyard** (see *Figure 21*). The other end of the lanyard should be attached to a strong anchor point located above the work area. (A qualified person will tell you what a strong anchor point is.) The lanyard should be long enough to let you work but short enough to keep you from falling more than 6 feet.

Use a safety harness and lanyard when you are working in the following situations:

- More than 6 feet above ground or according to company policy
- Near a large opening in a floor
- Near a deep hole
- Near protruding rebar

NORTH SAFETY PRODUCTS   101F19A.EPS
(A)

WARNING!
Never use a safety harness and lanyard for anything except their intended purpose.

Treat a safety harness as if your life depends on it—because it *does!* Carefully inspect the harness each time you use it. Check that the buckles and D-ring are not bent or deeply scratched. Check the harness for any cuts or rough spots. If you find any damage, turn in the harness for testing or replacement.

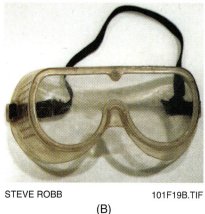
STEVE ROBB   101F19B.TIF
(B)

Figure 19 ◆ Typical (A) safety glasses and (B) goggles.

5.2.4 Gloves

On many construction jobs, you must wear heavy-duty gloves to protect your hands (see *Figure 22*). Construction work gloves are usually made of cloth, canvas, or leather. Never wear cloth gloves around rotating or moving equipment.

Gloves help prevent cuts and scrapes when you handle sharp or rough materials. Heat-resistant gloves are sometimes used for hot materials. Electricians use special rubber-insulated gloves when they work on or around live circuits.

Replace gloves when they become worn, torn, or soaked with oil or chemicals. Electrician's rubber-insulated gloves should be tested regularly to make sure they will protect the wearer.

5.2.5 Safety Shoes

The best shoes to wear on a construction site are steel-toed, steel-soled safety shoes (see *Figure 23*). The steel toe protects your toes from falling

PROTECTA INTERNATIONAL INC.   101F20.EPS

Figure 20 ◆ Typical safety harness.

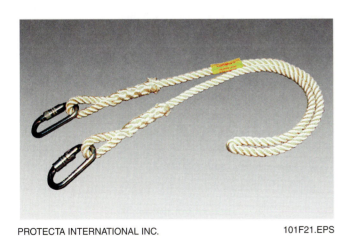

PROTECTA INTERNATIONAL INC.   101F21.EPS

Figure 21 ◆ Lanyard.

101F22.EPS
NORTH SAFETY PRODUCTS

Figure 22 ◆ Work gloves.

101F23.TIF
BARBARA ROBB

Figure 23 ◆ Safety shoes.

objects. The steel sole keeps nails and other sharp objects from puncturing your foot. The next best footwear material is heavy leather. *Never* wear canvas shoes or sandals on a construction site.

Always replace boots or shoes when the sole tread becomes worn or the shoes have holes, even if the holes are on top. Don't wear oil-soaked shoes when you are welding because of the risk of fire.

5.2.6 Hearing Protection

Damage to most parts of the body causes pain. But ear damage does not always cause pain. Exposure to loud noise over a long time can cause hearing loss, even if the noise is not loud enough to cause pain.

Most construction companies follow OSHA rules in deciding when hearing protection must be used. One type of hearing protection is specially designed earplugs that fit into your ears and filter out noise (see *Figure 24*). You need to clean earplugs regularly with soap and water to prevent an ear infection.

Another type of hearing protection is earmuffs, which are large padded covers for the entire ear (see *Figure 25*). You must adjust the headband on earmuffs for a snug fit. If the noise level is very high, you may need to wear both earplugs and earmuffs.

DEWALT INDUSTRIAL TOOL COMPANY   101F25.EPS

Figure 25 ◆ Earmuffs for hearing protection.

NORTH SAFETY PRODUCTS   101F24.EPS

Figure 24 ◆ Earplugs for hearing protection.

Noise-induced hearing loss can be prevented by using noise control measures and personal protective devices. *Table 4* shows the recommended duration of exposure to sound levels rated 90 decibels and higher.

Table 4 Maximum Noise Levels

Sound Level (decibels)	Maximum Hours of Continuous Exposure per Day	Examples
90	8	Power lawn mower
92	6	Belt sander
95	4	Tractor
97	3	Hand drill
100	2	Chain saw
102	1.5	Impact wrench
105	1	Spray painter
110	0.5	Power shovel
115	0.25 or less	Hammer drill

5.2.7 Respiratory Protection

Wherever there is danger of suffocation or other breathing hazards, you must use a respirator. Federal law specifies which type of respirator to use for different types of hazards. There are four general types of respirators (see *Figure 26*):

- Self-contained breathing **apparatus** (SCBA)
- Supplied air mask
- Full facepiece mask with chemical canister (gas mask)
- Half mask or mouthpiece with mechanical filter

The self-contained breathing apparatus, or SCBA, carries its own air supply in a compressed air tank. It is used where there is not enough oxygen or where there are dangerous gases or fumes in the air.

A supplied air mask uses a remote compressor or air tank. A hose supplies air to the mask. Supplied air masks can be used under the same conditions as the SCBA.

Full facepiece masks with chemical canisters are used to protect against brief exposure to dangerous gases or fumes.

A half mask or mouthpiece with a mechanical filter is used in areas where you might inhale dust or other solid particles.

Follow local and OSHA procedures when choosing the type of respirator for a particular job.

It is very important to check a respirator carefully for damage and for a proper fit. A leaking facepiece can be as dangerous as no respirator at all. All respirators must be fitted properly, and their facepiece-to-face seal must be checked each time the respirator is used. When conditions prevent a good seal, the respirator cannot be worn. The following conditions will interfere with the respirator's seal:

- Facial hair (such as sideburns or beards)
- Skullcaps that project under the face piece
- Temple bars on glasses (especially when wearing full-face respirators)
- Absence of upper, lower, or all teeth
- Absence of dentures
- Gum and tobacco chewing

WARNING!
Never wear contact lenses while you are wearing a respirator.

Respirators used by only one person should be cleaned after each day of use, or more often if necessary. Those used by more than one person should be cleaned *and disinfected* (made germ-free) after each use.

WARNING!
When a respirator is required, a personal monitoring device—for example, a carbon monitor—is usually also required.

6.0.0 ◆ LIFTING

You may be surprised to learn that one-fourth of all occupational injuries happen when workers are handling or moving construction materials, especially lifting heavy objects. There is a right way and a wrong way to lift a heavy object. Lifting the wrong way can land you in the hospital. *Figure 27* shows the right way.

Step 1 Move close to the object you are going to lift. Position your feet in a forward/backward stride, with one foot at the side of the object.

NORTH SAFETY PRODUCTS   101F26A.EPS

(A)

NORTH SAFETY PRODUCTS   101F26B.EPS

(B)

NORTH SAFETY PRODUCTS   101F26C.EPS

(C)

BECKI SWINEHART   101F26D.EPS

(D)

Figure 26 ◆ Examples of respirators. (A) Self-contained breathing apparatus. (B) Supplied air mask. (C) Full facepiece mask. (D) Half mask.

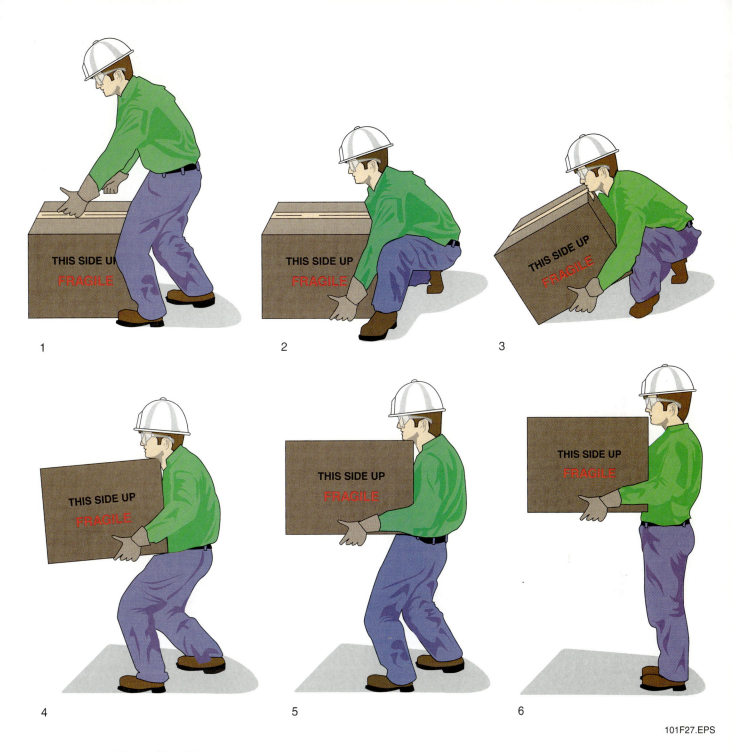

Figure 27 ◆ How to lift safely.

Step 2 Bend your knees and lower your body, keeping your back straight and as nearly upright as possible.

Step 3 Place your hands under the object, wrap your arms around it, or grasp the handles. To get your hands under an object that is flat on the floor, use both hands to lift one corner. Slip one hand under that corner. With one hand under, tilt the object to get the other hand under the opposite side.

Step 4 Draw the object close to your body.

Step 5 Lift by slowly straightening your legs and keeping the object's weight as much as possible over your legs.

Step 6 Pick the object up facing the direction you are going to go, to avoid twisting your knees or back.

These steps let you use your strongest muscles (those in your legs) instead of your weakest ones (those in your back) to lift. Practice with light objects. Then, when you've got it down, move on to heavier ones.

Many employers supply back belts to help reduce back injuries. You should be trained in the right way to use a back belt. Remember, a back belt is no substitute for using the proper lifting techniques.

Review Questions

Sections 4.0.0–6.0.0

1. The _____ keeps the key to a lock used for lockout/tagout.
   a. site supervisor
   b. person who puts the lock on
   c. site safety manager
   d. OSHA inspector

2. A yellow and purple warning barricade means _____.
   a. caution
   b. danger
   c. physical danger
   d. radiation hazard

3. _____ provide the best all-around protection for your eyes.
   a. Welding hoods
   b. Face shields
   c. Safety goggles
   d. Strap-on glasses

4. A _____ has its own clean air supply.
   a. half mask
   b. mouthpiece with mechanical filter
   c. self-contained breathing apparatus
   d. full facepiece mask

5. When lifting heavy objects, keep as much weight as possible over your _____.
   a. hips
   b. shoulders
   c. legs
   d. back

7.0.0 ◆ AERIAL WORK

Working in elevated locations is common in the construction industry. If it is done properly and the proper equipment is used, it is safe. But falls from heights can cause serious injuries or even death. You must always have your supervisor's permission before working in an elevated location. In this section, you will learn about the equipment used for aerial work. You'll learn how to use it, how to inspect it, and how to maintain it.

7.1.0 Ladders and Scaffolds

Ladders and scaffolds are used to perform work in an elevated location. Any time work is performed above ground level, there is a risk of accidents. You can reduce this risk by carefully inspecting ladders and scaffolds before you use them and by using them properly.

Overloading means exceeding the maximum intended load of a ladder. Overloading can cause

Fall Protection

Effective February 1995, fall protection must be provided for working at an elevation above 6 feet or in situations where a fall into a hole or trench is greater than 6 feet. Supervisors must ensure that all walking and working surfaces have the strength and structural integrity to support the workers. Work conducted on an otherwise unprotected side or edge that is 6 feet or more above a lower level must be protected by the use of guardrail systems, safety-net systems, or personal fall-arrest systems.

Effective January 1998, an *acceptable* personal fall-arrest system is a body harness with a lanyard attached to a D-ring in the center of the back. Body belts are *not* acceptable as part of a personal fall-arrest system.

All companies are required by law to have a written Fall Protection Plan that addresses site-specific fall hazards and the preventive steps taken to address each hazard. This information will normally be included in your training or regularly scheduled safety meetings.

ladder failure, which means that the ladder could buckle, break, or topple, among other possibilities. The maximum intended load is the total weight of all employees, equipment, tools, materials, loads that are being carried, and other loads that the ladder can hold at any one time. Check the manufacturer's specifications to determine the maximum intended load. Ladders are usually given a duty rating that indicates the load capacity, as shown in *Table 5*.

(A) 101F28A.EPS   (B) 101F28B.EPS

Figure 28 ◆ Three-point ladder contact.

Table 5 Ladder Duty Ratings and Load Capacities

Duty Ratings	Load Capacities
Type IA:	300 lbs., extra heavy duty/professional use
Type I:	250 lbs., heavy duty/industrial use
Type II:	225 lbs., medium duty/commercial use
Type III:	200 lbs., light duty/household use

WARNING!

Use ladders and scaffolds for their intended uses *only*. Ladders are not interchangeable, and incorrect usage can result in injury or damage.

7.1.1 Portable Straight Ladders

Straight ladders consist of two rails, rungs between the rails, and safety feet on the bottom of the rails (see *Figure 29*). The straight ladders used in construction are made of wood or fiberglass.

Metal ladders conduct electricity and should not be used around electrical equipment. Any portable metal ladder must have "Danger. Do Not Use Around Electrical Installations" stenciled on the rails in 2-inch, red letters. Ladders made of dry wood or fiberglass, which do *not* conduct electricity, should be used instead. Check to make sure that any wooden ladder is, in fact, completely dry before using it around electricity. Even a small amount of water will act as a conductor.

Different types of ladders should be used in different situations (see *Figure 30*). Aluminum ladders are corrosion-resistant and can be used in situations where they might be exposed to the elements. They are also lightweight and can be used in situations where frequently lifting and moving them to another location is required. Wooden ladders, which are heavier and sturdier than fiberglass or aluminum ladders, can be used when heavy loads must be moved up and down. Fiberglass ladders are very durable, so they are useful in situations where some amount of rough treatment is unavoidable. Both fiberglass and aluminum are also easier to clean than wood.

7.1.2 Inspecting Straight Ladders

Always inspect a ladder before you use it. Check the rails and rungs for cracks or other damage. Also, check for loose rungs. If you find any dam-

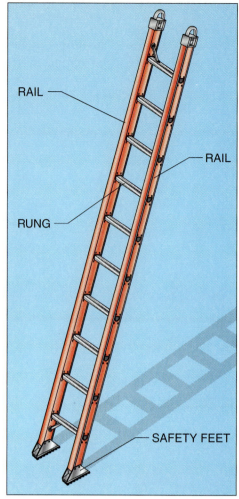

101F29.EPS

Figure 29 ◆ Portable straight ladder.

age, do not use the ladder. Check the entire ladder for loose nails, screws, brackets, or other hardware. If you find any hardware problems, tighten the loose parts or have the ladder repaired before you use it. OSHA requires regular inspections of all ladders and an inspection just before each use.

WARNING!

Wooden ladders should *never* be painted. The paint could hide cracks in the rungs or rails. Clear varnish, shellac, or a preservative oil finish will protect the wood without hiding defects.

Figure 31 shows the safety feet attached to a straight ladder. Make sure the feet are securely attached and that there is no damage or they are

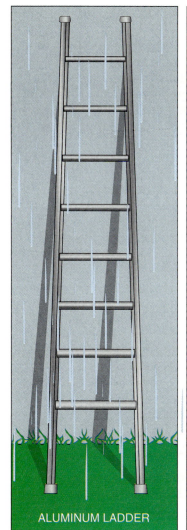

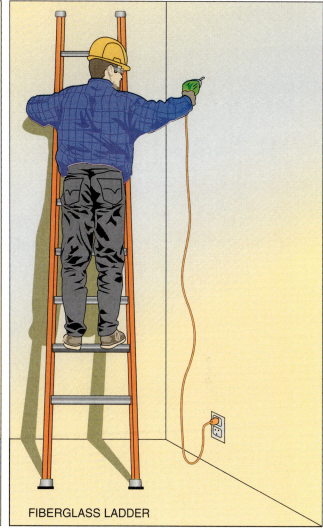

ALUMINUM LADDER | WOODEN LADDER | FIBERGLASS LADDER

101F30.EPS

Figure 30 ◆ Different types of ladders and applications.

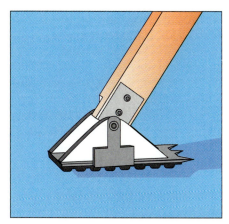

101F31.EPS

Figure 31 ◆ Ladder safety feet.

not worn down. Do not use a ladder if its safety feet are not in good working order.

7.1.3 Using Straight Ladders

It is very important to place a straight ladder at the proper angle before using it. A ladder placed at an improper angle will be unstable and could cause you to fall. *Figure 32* shows a properly positioned straight ladder.

The distance between the foot of a ladder and the base of the structure it is leaning against must be one-fourth of the distance between the ground and the point where the ladder touches the structure. For example, if the height of the wall shown in *Figure 32* is 16 feet, the ladder should be 4 feet

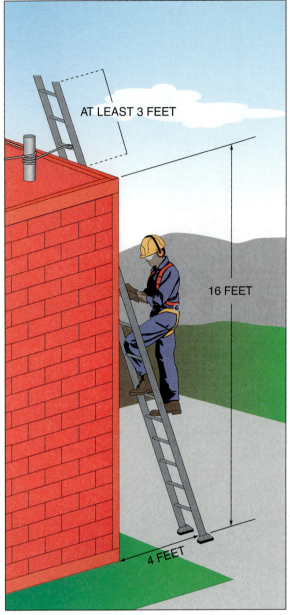

AT LEAST 3 FEET

16 FEET

4 FEET

101F32.EPS

Figure 32 ◆ Proper positioning of a straight ladder.

open. Otherwise, the door could be opened into the ladder.

Ladders are made for vertical use only. *Never* use a ladder as a work platform by placing it horizontally. Make sure the ladder you are about to climb or descend is properly secure before you do so. Check to make sure the ladder's feet are solidly positioned on firm, level ground. Also check to make sure that the top of the ladder is firmly positioned and in no danger of shifting to right or left once you begin your climb. Remember that the addition of your own weight will have an effect on the ladder's steadiness once you mount it. So it is important to test the ladder first by applying some of your weight to it without actually beginning to climb. This way, you will be sure that the ladder remains steady as you ascend.

When climbing a straight ladder, keep both hands on the rails. Always keep your body's weight in the center of the ladder between the rails. Face the ladder at all times. Never go up or down a ladder while facing away from it (see *Figure 34*).

If you are going to use a tool while you are on the ladder, use a **hand line** or tagline. Climb the ladder and then pull up the tool. Don't carry tools in your hands while you are climbing a ladder.

7.1.4 Extension Ladders

An **extension ladder** is actually two straight ladders. They are connected so you can adjust the overlap between them and change the length of the ladder as needed (see *Figure 35*).

7.1.5 Inspecting Extension Ladders

The same rules for inspecting straight ladders apply to extension ladders. In addition, you should inspect the rope that is used to raise and lower the movable section of the ladder. If the rope is frayed or has worn spots, it should be replaced before the ladder is used.

The rung locks (see *Figure 36*) support the entire weight of the movable section and the person climbing the ladder. Inspect them for damage before each use. If they are damaged, they should be repaired or replaced before the ladder is used.

7.1.6 Using Extension Ladders

Extension ladders are positioned and secured following the same rules as for straight ladders. When you adjust the length of an extension ladder, always reposition the movable section from the bottom, not the top, so you can make sure the

from the base of the wall. If you are going to step off a ladder onto a platform or roof, the top of the ladder should extend at least 3 feet above the point where the ladder touches the platform or roof.

Ladders should be used only on stable and level surfaces unless they are secured at both the bottom and the top to prevent any accidental movement (see *Figure 33*). *Never* try to move a ladder while you are on it. If a ladder must be placed in front of a door that opens toward the ladder, the door should be locked or blocked

BOTTOM SECURED

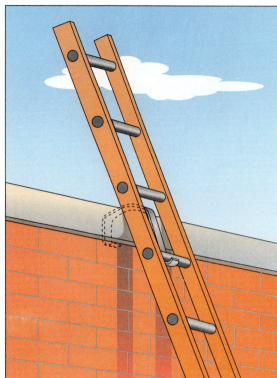
TOP SECURED

101F33.EPS

Figure 33 ◆ Securing a ladder.

- Climb facing the ladder. Center your body between the rails. Maintain a firm grip.

- Always move one step at a time, firmly setting one foot before moving the other.

101F34.EPS

Figure 34 ◆ Moving up or down a ladder.

rung locks are properly engaged after you make the adjustment. Check to make sure the section locking mechanism is fully hooked over the desired rung. Also check to make sure that all ropes used for raising and lowering the extension are clear and untangled.

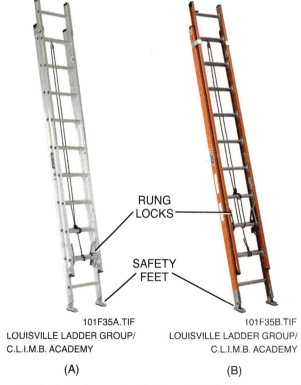

RUNG LOCKS

SAFETY FEET

101F35A.TIF
LOUISVILLE LADDER GROUP/
C.L.I.M.B. ACADEMY

101F35B.TIF
LOUISVILLE LADDER GROUP/
C.L.I.M.B. ACADEMY

(A)   (B)

Figure 35 ◆ Typical extension ladders. (A) Aluminum extension ladder. (B) Fiberglass extension ladder.

RUNG
LOCK

101F36.TIF
LOUISVILLE LADDER GROUP/
C.L.I.M.B. ACADEMY

Figure 36 ◆ Rung locks.

To ensure its strength, an extension ladder needs a certain amount of overlap between the two sections (see *Figure 37*). For ladders up to 36 feet long, the overlap must be at least 3 feet. For ladders 36 to 48 feet long, the overlap must be at least 4 feet. For ladders 48 to 60 feet long, the overlap must be at least 5 feet.

Never stand above the highest safe standing level on a ladder. The highest safe standing level on an extension ladder is the fourth rung from the top. If you stand higher, you may lose your balance and fall. Some ladders have colored rungs to show where you should not stand.

7.1.7 Stepladders

Stepladders are self-supporting ladders made of two sections hinged at the top (see *Figure 38*).

The section of a stepladder used for climbing consists of rails and rungs like those on straight ladders. The other section consists of rails and braces. Spreaders are hinged arms between the sections that keep the ladder stable and keep it from folding while in use. A stepladder may have a pail shelf to hold pails of paint or tools.

7.1.8 Inspecting Stepladders

Inspect stepladders the way you inspect straight and extension ladders. For stepladders, though, pay special attention to the hinges and spreaders to be sure they are in good repair. Also, be sure the rungs are clean. A stepladder's rungs are usually

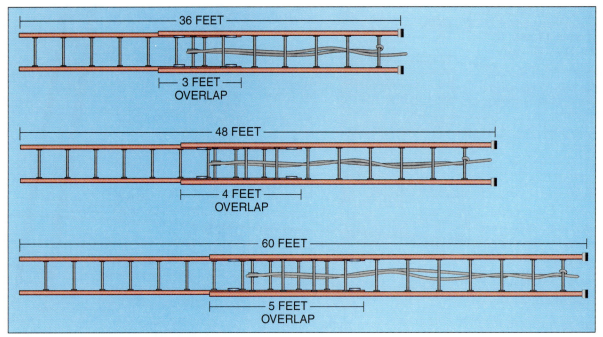

36 FEET

3 FEET
OVERLAP

48 FEET

4 FEET
OVERLAP

60 FEET

5 FEET
OVERLAP

101F37.EPS

Figure 37 ◆ Overlap lengths for extension ladders.

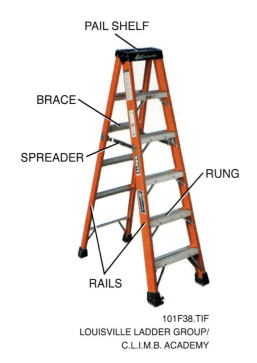

PAIL SHELF

BRACE

SPREADER

RUNG

RAILS

101F38.TIF
LOUISVILLE LADDER GROUP/
C.L.I.M.B. ACADEMY

Figure 38 ◆ Typical fiberglass stepladder.

flat, so oil, grease, or dirt can build up on them and make them slippery.

7.1.9 Using Stepladders

When you position a stepladder, be sure that all four feet are on a hard, even surface. If they're not, the ladder can rock from side to side or corner to corner when you climb it. With the ladder in position, be sure the spreaders are locked in the fully open position.

Never stand on the top step or the top of a stepladder. Putting your weight this high will make the ladder unstable. The top of the ladder is made to support the hinges, *not* to be used as a step. And, although the rear braces may look like rungs, they are not designed to support your weight. *Never* use the braces for climbing. And *never* climb the back of a stepladder. (For certain jobs, however, there are specially designed two-person ladders with steps on both sides.) *Figure 39* shows common *do's* and *don'ts* for using ladders.

7.2.0 Scaffolds

Scaffolds provide safe elevated work platforms for people and materials. Scaffolds are designed and built to comply with high safety standards, but normal wear and tear or accidentally putting too much weight on them can weaken them and make them unsafe. That's why it is important to inspect every part of a scaffold before each use.

Two basic types of scaffolds—manufactured scaffolds and rolling scaffolds—are used in the construction industry, and the rules for safe use apply to both of them.

7.2.1 Manufactured Scaffolds

Manufactured scaffolds are made of painted steel, stainless steel, or aluminum (see *Figure 40*). They are stronger and more fire-resistant than wooden scaffolds. They are supplied in ready-made, individual units, which are assembled on site.

7.2.2 Rolling Scaffolds

A rolling scaffold has wheels on its legs so that it can be easily moved (see *Figure 41*). The scaffold wheels have brakes so the scaffold will not move while workers are standing on it.

7.2.3 Inspecting Scaffolds

 CAUTION

Only a *competent person* has the authority to supervise setting up, moving, and taking down scaffolding. Only a competent person can give the go-ahead for using scaffolding on the job site after he or she completes an inspection of the scaffolding.

Any scaffold that is assembled on the job site should be tagged. These tags indicate whether the scaffold meets OSHA standards and is safe to use. Three colors of tags are used: green, yellow, and red (see *Figure 42*).

A green tag means the scaffold meets all OSHA standards and is safe to use.

A yellow tag means the scaffold does *not* meet all OSHA standards. An example is a scaffold where a railing cannot be installed because of equipment interference. You may use a yellow-tagged scaffold, but you have to wear a safety harness attached to a lanyard. You may have to take other safety measures as well.

Do's

- Be sure your ladder has been properly set up and is used in accordance with safety instructions and warnings.
- Wear shoes with non-slip soles.

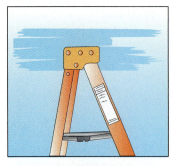

- Keep your body centered on the ladder. Hold the ladder with one hand while working with the other. Never let your belt buckle pass beyond either ladder rail.

- Move materials with extreme caution. Be careful pushing or pulling anything while on a ladder. You may lose your balance or tip the ladder.

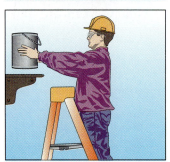

Don'ts

- DON'T stand above the highest **safe standing level.**
- DO NOT stand above the second step from the top of a stepladder and the 4th rung from the top of an extension ladder. You may lose your balance and fall.

- DON'T climb a closed stepladder. It may slip out from under you.
- DON'T climb on the back of a stepladder. It is not designed to hold a person.

- DON'T stand or sit on a stepladder top or pail shelf. They are not designed to carry your weight.
- DON'T climb a ladder if you are not physically and mentally up to the task.

- DON'T exceed the Duty Rating, which is the maximum load capacity of the ladder. Do not permit more than one person on a single-sided stepladder or on any extension ladder.

WERNER LADDER COMPANY

101F39.EPS

Figure 39 ◆ Ladder safety do's and don'ts.

 WARNING!
Never stand or sit on a stepladder's top or pail shelf. They are not designed to carry your weight (see *Figure 39*).

A red tag means a scaffold is being put up or taken down. *Never* use a red-tagged scaffold.

Don't rely on the tags alone. Inspect all scaffolds before you use them. Check for bent, broken, or badly rusted tubes. Check for loose joints where the tubes are connected. Any of these problems must be corrected before the scaffold is used.

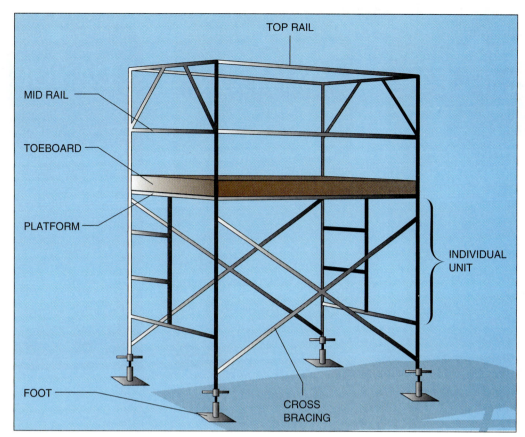

Figure 40 ◆ Typical manufactured scaffold.

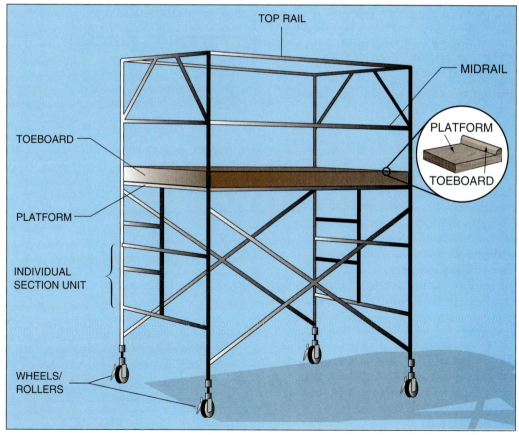

Figure 41 ◆ Typical rolling scaffold.

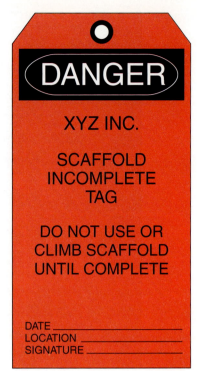

101F42.EPS

Figure 42 ◆ Typical scaffold tags.

Make sure you know the weight limit of any scaffold you will be using. Compare this weight limit with the total weight of the people, tools, equipment, and material you expect to put on the scaffold. Scaffold weight limits must *never* be exceeded.

If a scaffold is more than 10 feet high, check to see that it is equipped with **top rails, mid-rails,** and **toeboards.** All connections must be pinned—that is, they must have a piece of metal inserted through a hole to prevent connections from slipping. **Cross-bracing** must be used. A handrail is *not* the same as cross-bracing. The working area must be completely **planked.**

If it is possible for people to walk under a scaffold, the space between the toeboard and the top rail must be screened. This prevents objects from falling off the work platform.

When you examine a rolling scaffold, check the condition of the wheels and brakes. Be sure that the brakes are working properly and can stop the scaffold from moving while work is in progress. Be sure all brakes are locked before you use the scaffold.

7.2.4   Using Scaffolds

Be sure that a competent person inspects the scaffold before you use it.

There should be firm footing under each leg of a scaffold before you put any weight on it. If you are working on loose or soft soil, you can put matting under the scaffold's legs or wheels.

 WARNING!
Keep scaffolds at least 10 feet away from power lines.

When you move a rolling scaffold, *always* follow these steps:

Step 1 Get off the scaffold.

Step 2 Unlock the wheel brakes.

Step 3 Move the scaffold.

Step 4 Re-lock the wheel brakes.

Step 5 Get back on the scaffold.

 WARNING!
Never unlock the wheel brakes of a rolling scaffold while anyone is on it.

Review Questions

Section 7.0.0

1. Never use a(n) _____ ladder anywhere near electrical current.
   a. fiberglass
   b. aluminum
   c. wooden
   d. straight

2. If you lean a straight ladder against the top of a 16-foot wall, the base of the ladder should be _____ feet from the base of the wall.
   a. 3
   b. 4
   c. 6
   d. 1½

3. With a one-person stepladder, it is safe practice to _____.
   a. stand on the top step
   b. climb the back of it
   c. stand on the rear braces
   d. lock the spreaders in the fully open position

4. Never use a scaffold with a(n) _____ tag.
   a. blue
   b. red
   c. orange
   d. yellow

5. Scaffolds should be at least _____ feet away from power lines.
   a. 5
   b. 2
   c. 10
   d. 12

8.0.0 ◆ HAZARD COMMUNICATION STANDARD

OSHA has a rule that affects every worker in the construction industry. It is called the Hazard Communication Standard (HazCom). You may have heard of it as the "Right to Know" requirement. It requires all contractors to educate their employees about the hazardous chemicals they may be exposed to on the job site. Employees must be taught how to work safely around these materials.

Many people think that there are very few hazardous chemicals on construction job sites. That isn't true. In the OSHA standard, the term *hazardous chemical* applies to paint, concrete, and even wood dust, as well as other substances.

8.1.0 Material Safety Data Sheets

A Material Safety Data Sheet (MSDS) must accompany every shipment of a hazardous substance and must be available to you on the job site. *Figure 43* shows part of a typical MSDS.

The information on an MSDS includes the following:

- The identity of the substance
- Exposure limits
- Physical and chemical characteristics of the substance
- The kind of hazard the substance presents
- Precautions for safe handling and use
- The reactivity of the substance
- Specific control measures
- Emergency first-aid procedures
- Manufacturer contact for more information

8.2.0 Your Responsibilities under HazCom

You have the following responsibilities under HazCom:

- Know where MSDSs are on your job site.
- Report any hazards you spot on the job site to your supervisor.
- Know the physical and health hazards of any hazardous materials on your job site, and know and practice the precautions needed to protect yourself from these hazards.
- Know what to do in an emergency.
- Know the location and content of your employer's written hazard communication program.

The final responsibility for your safety rests with you. Your employer must provide you with the information about the hazard. But you must know this information and follow safety rules.

Review Questions

Section 8.0.0

1. OSHA's Hazard Communication Standard (HazCom) rule requires all contractors to _____ on-site hazardous chemicals.
   a. store
   b. clean up
   c. remove all
   d. educate employees about

Material Safety Data Sheet

May be used to comply with
OSHA's Hazard Communication Standard,
29 CFR 1910. 1200. Standard must be
consulted for specific requirements.

U.S. Department of Labor

Occupational Safety and Health Administration
(Non-Mandatory Form)
Form Approved
OMB No. 1218-0072

IDENTITY *(As Used on Label and List)*	*Note: Blank spaces are not permitted. If any item is not applicable, or no information is available, the space must be marked to indicate that.*

Section I

Manufacturer's Name	Emergency Telephone Number
Address *(Number, Street, City, State, and ZIP Code)*	Telephone Number for Information
	Date Prepared
	Signature of Preparer *(optional)*

Section II — Hazardous Ingredients/Identity Information

Hazardous Components (Specific Chemical Identity; Common Name(s))	OSHA PEL	ACGIH TLV	Other Limits Recommended	% *(optional)*

Section III — Physical/Chemical Characteristics

Boiling Point		Specific Gravity (H_2O = 1)	
Vapor Pressure (mm Hg.)		Melting Point	
Vapor Density (AIR = 1)		Evaporation Rate (Butyl Acetate = 1)	

Solubility in Water

Appearance and Odor

Section IV — Fire and Explosion Hazard Data

Flash Point (Method Used)	Flammable Limits	LEL	UEL

Extinguishing Media

Special Fire Fighting Procedures

Unusual Fire and Explosion Hazards

(Reproduce locally)

OSHA 174, Sept. 1985

101F43.EPS

Figure 43 ◆ Typical MSDS.

2. HazCom classifies all paint, concrete, and wood dust as being _____ materials.
   a. hazardous
   b. common
   c. inexpensive
   d. nonhazardous

3. The information on an MSDS includes _____.
   a. cost and availability
   b. emergency first-aid procedures
   c. point of origin
   d. warranty limitations

4. Under HazCom, if you spot a hazard on your job site you must _____.
   a. report it to your supervisor
   b. leave immediately
   c. notify your co-workers
   d. clean it up

5. Although your employer must provide you with information about hazardous chemicals, the final responsibility for your safety rests with _____.
   a. your immediate supervisor
   b. your site foreman
   c. you
   d. your co-workers

9.0.0 ◆ FIRE SAFETY

Fire is always a hazard on construction job sites. Many of the materials used in construction are flammable, and welding, grinding, and many other construction activities create heat or sparks that can cause a fire. Fire safety involves two elements: fire prevention and fire fighting.

9.1.0 How Fires Start

For a fire to start, three things are needed in the same place at the same time: fuel, heat, and oxygen. If one of these three is missing, a fire will not start.

Fuel is anything that will combine with oxygen and heat to burn. When pure oxygen is present, such as near a leaking oxygen hose or fitting, material that would not normally be considered fuel (including some metals) will burn.

Heat is anything that will raise a fuel's temperature to the **flash point.** The flash point is the temperature at which a fuel gives off enough gases (vapors) to burn. The flash points of many fuels are quite low—room temperature or less. When the burning gases raise the temperature of a fuel to the point at which it ignites, the fuel itself will burn—and keep burning—even if the original source of heat is removed.

Oxygen is always present in the air.

What is needed for a fire to start can be shown as a fire triangle (see *Figure 44*). If any one element of the triangle is missing, a fire cannot start. If a fire has started, removing any one element from the triangle will put it out.

Research has added a fourth side to the fire triangle concept, resulting in the development of a new model called the "Fire Tetrahedron." The fourth element involved in the combustion

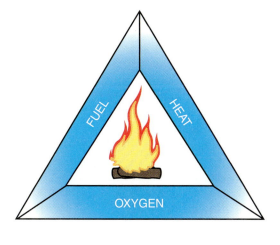

FIRE TRIANGLE

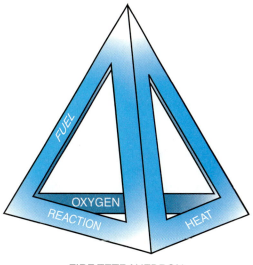

FIRE TETRAHEDRON

101F44.EPS

Figure 44 ◆ Basic fire requirements.

Prevention and Preparation Are the Keys to Safety from Fire

Any fire in the workplace can cause serious injury or property damage. When chemicals are involved, the risks are even greater. *Prevention* is the key to eliminating the hazards of any kind of fire where you work. *Preparation* is the key to controlling any fires that do start. Take the following precautions to make sure you are safe from fire in your workplace.

- Keep work areas clean and clutter free.
- Know how to handle and store chemicals.
- Know what you are expected to do in case of a fire emergency.
- Call professional help immediately. Don't let a fire get out of control.
- Know what chemicals you work with. You might have to tell fire fighters at a chemical fire what kinds of hazardous substances are involved.
- Make sure you are familiar with your company's emergency action plan for fires.

---

process is referred to as the "chemical chain reaction." Specific chemical chain reactions between fuel and oxygen molecules are essential to sustaining a fire once it has begun.

9.2.0 Fire Prevention

Obviously, the best way to provide fire safety is to prevent a fire from starting in the first place. The best way to prevent a fire is to make sure that the three elements needed for fire—fuel, oxygen, and heat—are never present in the same place at the same time. Here are some basic safety guidelines for fire prevention:

- Always work in a well-ventilated area, especially when you are using flammable materials such as shellac, lacquer, paint stripper, or construction adhesives.
- Never smoke or light matches when you are working with flammable materials.
- Keep oily rags in approved, self-closing metal containers.
- Store combustible materials only in approved containers.
- Know where to find fire extinguishers, what kind of extinguisher to use for different kinds of fires, and how to use the extinguishers.
- Make sure all extinguishers are fully charged. Never remove the tag from an extinguisher—it shows the date the extinguisher was last serviced and inspected (see *Figure 45*).

9.2.1 Flammable and Combustible Liquids

Liquids can be flammable or combustible. Flammable liquids have a flash point below 100 degrees

Fahrenheit. Combustible liquids have a flash point at or above 100 degrees Fahrenheit. Fire can be prevented by doing the following things:

- *Removing the fuel*—Liquid does not burn. What burns are the gases (vapors) given off as the liquid evaporates. Keeping the liquid in an approved, sealed container prevents evaporation. If there is no evaporation, there is no fuel to burn.
- *Removing the heat*—If the liquid is stored or used away from a heat source, it will not be able to ignite.
- *Removing the oxygen*—The vapor from a liquid will not burn if oxygen is not present. Keeping safety containers tightly sealed prevents oxygen from coming into contact with the fuel.

9.2.2 Flammable Gases

Flammable gases used on construction sites include acetylene, hydrogen, ethane, and propane (liquid propane gas, or LPG). To save space, these gases are compressed so that a large amount is stored in a small cylinder or bottle. As long as the gas is kept in the cylinder, oxygen cannot get to it and start a fire. The cylinders should be stored away from sources of heat.

Oxygen is classified as a flammable gas, too. If it is allowed to escape and mix with another flammable gas, the result will be a mixture that will explode in the right conditions.

9.2.3 Ordinary Combustibles

The term *ordinary combustibles* means paper, wood, cloth, and similar fuels. The easiest way to

---

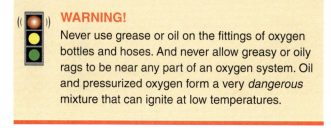

Figure 45 ◆ Fire extinguisher tag.

> **WARNING!**
>
> Never use grease or oil on the fittings of oxygen bottles and hoses. And never allow greasy or oily rags to be near any part of an oxygen system. Oil and pressurized oxygen form a very *dangerous* mixture that can ignite at low temperatures.

prevent fire in ordinary combustibles is to keep a neat, clean work area. If there are no scraps of paper, cloth, or wood lying around, there will be no fuel to start a fire. So, establish and maintain good housekeeping habits. Use approved storage cabinets and containers for all waste and other ordinary combustibles.

9.3.0 Fire Fighting

You are not expected to be an expert fire fighter. But you may have to deal with a fire to protect your safety and the safety of others. You need to know the location of fire-fighting equipment on your job site. You also need to know which equipment to use on different types of fires. However, only qualified personnel are authorized to fight fires.

Most companies tell new employees where fire extinguishers are kept. If you have not been told, be sure to ask. Also ask how to report fires. The telephone number of the nearest fire department should be clearly posted in your work area. If your company has a company fire brigade, learn how to contact them. Learn your company's fire safety procedures.

9.3.1 Classes of Fires

Four classes of fuels can be involved in fires (see *Table 6*). You've already learned about liquids, gases, and ordinary combustibles. Another is metal. (You will learn about preventing electrical fires in another section.) Each class of fuel requires a different method of fire fighting and a different type of extinguisher.

The label on a fire extinguisher clearly shows the class of fire on which it can be used (see *Figure 46*).

Table 6 Classes of Fires

Class	Materials and Proper Fire Extinguisher
Class A fires	Involves ordinary combustibles such as wood or paper. Class A fires are fought by cooling the fuel. Class A fire extinguishers contain water. Using a Class A extinguisher on any other type of fire can be very dangerous.
Class B fires	Involves grease, liquids, and gases. Class B extinguishers contain carbon dioxide (CO_2) or another material that smothers fires by removing oxygen from the fire.
Class C fires	Any fires near or involving energized electrical equipment. Class C extinguishers are designed to protect the fire fighter from electrical shock. Class C extinguishers smother fires.
Class D fires	Involves metals. Class D extinguishers contain a powder that either forms a crust around the burning metal or gives off gases that prevent oxygen from reaching the fire. Some metals will keep burning even though they have been coated with powder from a Class D extinguisher. The best way to fight these fires is to keep using the extinguisher so the fire will not spread to other fuels.

When you check the extinguishers in your work area, you will see that some are rated for more than one class of fire. You can use an extinguisher that has the three codes A, B, and C on it to fight a class A, B, or C fire. But remember, if the extinguisher has only one code letter, *do not* use it on any other class of fire, even in an emergency. You could make the fire worse and put yourself in great danger.

FIRE CLASS RATING

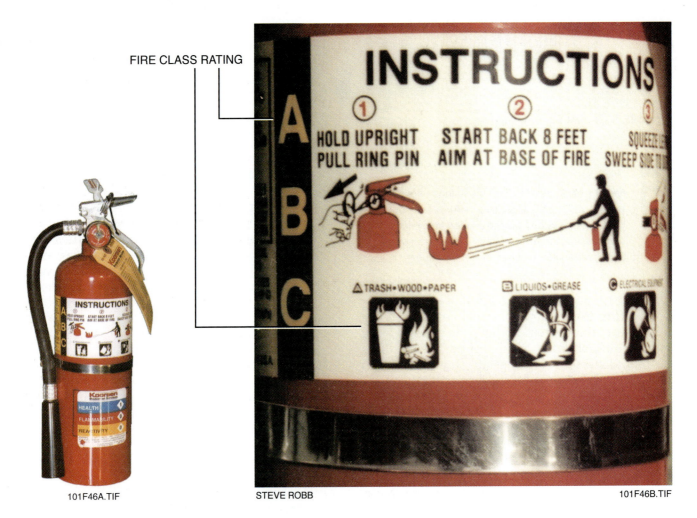

101F46A.TIF

STEVE ROBB

101F46B.TIF

Figure 46 ◆ Typical fire extinguisher labels.

How to Use a Fire Extinguisher

Step 1 Hold the extinguisher upright.

Step 2 Pull the pin, breaking the plastic seal.

Step 3 Stand back at least 10 feet from the fire. Standing any closer may cause burning objects to scatter, spreading the fire.

Step 4 Aim at the base of the fire.

Step 5 Keep the extinguisher upright. Squeeze the handles together to discharge. Sweep from side to side.

Step 6 Move closer as the fire is extinguished (watch for scattering burning material).

Step 7 When the fire is out, watch for reignition.

Review Questions

Section 9.0.0

1. _____ must be present in the same place at the same time for a fire to occur.
   a. Oxygen, carbon dioxide, and heat
   b. Oxygen, heat, and fuel
   c. Hydrogen, oxygen, and wood
   d. Grease, liquid, and heat

2. _____ gas is flammable.
   a. Oxygen
   b. Carbon dioxide
   c. Peroxide
   d. Neon

3. A Class D fire involves _____.
   a. grease
   b. wood
   c. electrical equipment
   d. metal

4. Fire extinguishers that contain water for fighting fires involving ordinary combustibles are _____ extinguishers.
   a. Class A
   b. Class B
   c. Class C
   d. Class D

5. For a grease fire, use a _____ extinguisher.
   a. Class A
   b. Class B
   c. Class C
   d. Class D

10.0.0 ◆ ELECTRICAL SAFETY

Some construction workers think that electrical safety matters only to electricians. But on many of your jobs, no matter what your trade, you will use or work around electrical equipment. Extension cords, power tools, portable lights, and many other day-to-day items use electricity. If you don't use this equipment safely and properly, the result could be death—for you or for someone you work with.

Electricity can be described as the flow of electrons through a conductor. This flow of electrons is called electrical current. Some materials—such as silver, copper, steel, and aluminum—are excellent conductors. This means that electrical current flows easily through them. The human body, especially when it is wet, is also a good conductor.

To create an electrical current, a path must be provided in a circular route, or a circuit. If this circuit is interrupted, the electrical current will try to complete its circular route by flowing along the path of least resistance. This means that it will flow into and through any conductor that is touching it. If it cannot complete its circuit, the electrical current will try to go to **ground.** This means that it will find the path of least resistance that allows it to flow as directly as possible into the earth. All of this takes place in less than the blink of an eye!

If the human body comes in contact with an electrically energized conductor and is also in contact with the ground at the same time, the human body becomes the path of least resistance for the electricity to flow as directly as possible into the ground. This means the electricity flows into and through the human body in less than the blink of an eye. You won't see that it's about to happen, it just happens. That's why safety precautions are so important when working with and around electrical currents. When the human body conducts electrical current and the amount of that current is high enough, the person can be electrocuted, meaning killed, by electric shock. *Table 7* shows the effects of different amounts of electrical current on the human body and lists some common tools that operate using those currents.

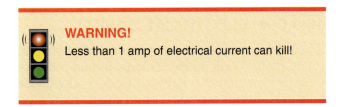

WARNING!
Less than 1 amp of electrical current can kill!

Here's an example. A craftsperson is operating a portable power drill while standing on damp ground. The power cord inside the drill has become frayed, and the electric wire inside the cord comes into contact with the metal drill frame. Three amps of current pass from the wire through the frame, then through the craftsperson's body and into the ground. *Table 7* shows that he will probably die.

Another good method to protect from accidental electrocution is the use of a Ground Fault Circuit Interrupter (GFCI). The GFCI is a fast-acting circuit breaker that senses small (as little as approximately 5 milliamps) imbalances in the circuit caused by current leakage to ground. In as little as ¹⁄₄₀ of a second the GFCI will interrupt the power.

Not all electrical accidents result in death. There are different types of electrical accidents. Any of the following can happen:

- Burns
- Electric shock
- Explosions
- Falls caused by electric shock
- Fires

Electricity can pass for short distances through the air. When it does, the arc and flash generate a great deal of heat. This heat can cause burns, fires, and even explosions.

10.1.0 Basic Electrical Safety Guidelines

OSHA and your company have specific policies and procedures to keep the workplace safe from electrical hazards. There are many things you can do to reduce the chance of an electrical accident. If you ever have any questions about electrical safety on the job site, ask your supervisor.

Table 7 Effects of Electrical Current on the Human Body

Current	Common Tool	Reaction to Current
0.001 amps	Watch battery	Faint tingle
0.005 amps	9-volt battery	Slight shock
0.006–0.025 amps (women) 0.009–0.030 amps (men)	Christmas tree light bulb	Painful shock. Muscular control is lost.
0.050–0.9 amps	Small electric radio	Extreme pain. Breathing stops; severe muscular contractions occur. Death is possible.
1.0–9.9 amps	Jigsaw (4 amps); Sawsall® or Port-a-Band® saw (6 amps); portable drill (3–8 amps)	Ventricular fibrillation and nerve damage occur. Death can occur.
10 amps and above	ShopVac® (15-gallon); circular saw	Heart stops beating; severe burns occur. Death may result.

Here are the basic job site electrical safety guidelines:

- Use three-wire extension cords and protect them from damage. *Never* fasten them with staples, hang them from nails, or suspend them from wires. *Never* use damaged cords.
- Make sure that panels, switches, outlets, and plugs are grounded.
- *Never* use bare electrical wire.
- *Never* use metal ladders near *any* source of electricity.
- *Never* wear a metal hard hat.
- *Always* inspect electrical power tools before you use them.
- *Never* operate any piece of electrical equipment that has a danger tag or lockout device attached to it.
- Use three-wire cords for portable power tools and make sure they are properly connected (see *Figure 47*). The three-wire system is one of the most common safety grounding systems used to protect you from accidental electrical shock. The third wire is connected to a ground. If the insulation in a tool fails, the current will pass to ground through the third wire—and not through your body.
- *Never* use worn or frayed cables.
- Make sure all light bulbs have protective guards to prevent accidental contact (see *Figure 48*).
- Do not hang temporary lights by their power cords unless they are specifically designed for this use.

STEVE ROBB   101F48.TIF

Figure 48 ◆ Work light with protective guard.

- Use only approved **concealed receptacles** for plugs. If different voltages or types of current are used in the same area, the receptacles should be designed so that the plugs are not interchangeable.
- Any repairs or splices in cords must be performed by a qualified person.
- Use a ground fault circuit interrupter (GFCI).

10.2.0 Working Near Energized Electrical Equipment

No matter what your trade, your job may include working near exposed electrical equipment or conductors. This is one example of proximity work. Many times, **electrical distribution panels, switch enclosures,** and other equipment must be left open during construction. This leaves the wires and components in them exposed. Some or all of the wires and components may be energized. Working near exposed electrical equipment can be safe, but only if you keep a safe working distance.

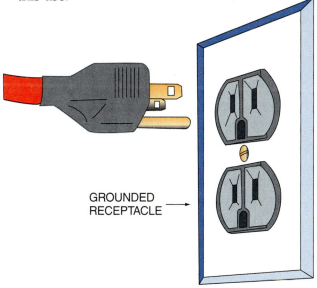

GROUNDED
RECEPTACLE

101F47.EPS

Figure 47 ◆ Three-wire system.

WARNING!

Every electrical cord should have an Underwriter's Laboratory (UL) label attached to it. Check the UL label for specific wattage. Do not plug more than the specified number of watts into an electrical cord.

A cord set not marked for outdoor use is to be used indoors only. Check the UL label on the cord for an outdoor marking.

Do not remove, bend, or modify any metal prongs or pins of an electrical cord.

Extension cords used with portable tools and equipment must be three-wire type and designated for hard or extra-hard use. Check the UL label for the cord's use designation.

Avoid overheating an electrical cord. Make sure the cord is uncoiled, and that it does not run under any covering materials, such as tarps, insulation rolls, or lumber.

Do not run a cord through doorways or through holes in ceilings, walls, and floors, which might pinch the cord. Also, check to see that there are no sharp corners along the cord's path. All of these situations will lead to cord damage.

Regulations and company policies tell you what the minimum safe working distances are from exposed conductors. The safe working distance varies from a few inches to several feet, depending on the voltage. The higher the voltage, the greater the safe working distance.

You must learn the safe working distance for each situation. Make sure you never get any part of your body or any tool you are using closer to exposed conductors than that distance. You can get information on safe working distances from your instructor, your supervisor, company safety policies, and regulatory documents.

10.3.0 If Someone Is Shocked

If you are there when someone gets an electrical shock, you can save a life by taking immediate action. Here's what to do:

Step 1 Immediately disconnect the circuit.

Step 2 If you can't disconnect the circuit, use a dry board, stick, rope, coat, blanket, or any other *nonconducting* material to separate the victim from the circuit. Do not use a metal object or an object that might contain metal. Do not use an object or any material that is wet or contains any liquid (most liquids contain some amount of water).

WARNING!

Do not touch the victim or the electrical source with your hand, foot, or any other part of your body or with any object or material that conducts electrical current. You could become another victim.

Step 3 Once the victim is separated from the circuit, apply first aid and call an ambulance.

Review Questions

Section 10.0.0

1. The _____ system is one of the most common safety grounding systems used with portable power tools.
   a. distribution wire
   b. rubber cord
   c. three-wire cord
   d. insulation plug

2. Observing proper safety precautions when working with and around electrical current is so important because the human body _____.
   a. resists the electricity's path
   b. is a good conductor of electrical current
   c. won't conduct electricity
   d. doesn't offer electricity a circular route

3. The minimum safe working distance from exposed electrical conductors _____.
   a. depends on the voltage
   b. is at least 1 foot
   c. is a few inches
   d. is unlimited

4. If someone is being electrically shocked, the first thing you should try to do is _____.
   a. use a metal pole to separate the victim from the circuit
   b. disconnect the circuit
   c. apply first aid
   d. call your co-workers

5. If someone is being electrically shocked, the second thing you should do is _____.
   a. check your company safety policies
   b. apply first aid
   c. pull the victim away with your hands
   d. use a nonconducting object to separate the victim from the circuit

Summary

Although the typical construction site has many hazards, a construction site does not have to be a dangerous place to work. Your employer has programs in place to deal with potential hazards. There are basic rules and regulations in place that help protect you and your co-workers from unnecessary risks.

This module has presented many of the basic guidelines you must follow to ensure your safety and the safety of the people you work with. These guidelines fall into the following categories:

- Following safe work practices and procedures
- Inspecting safety equipment before use
- Using safety equipment properly

The basic approach to safety is to eliminate hazards in the equipment and the workplace; to learn the rules and procedures for working safely with and around the remaining hazards; and to apply those rules and procedures. The information covered here offers you the groundwork for a safe, productive, and rewarding construction career.

Butch Gibson, Casey Industrial, Inc.

Safety and Training Director
Casey Industrial, Inc., Broomfield, Colorado

Butch Gibson began his construction career with an electrical power company, working with light rigging and cranes. In 1989, he was hired at Casey Industrial, Inc., where, motivated by a desire to learn all he could about his trade and the safety issues surrounding it, he became the company's safety specialist. He received advanced training in rigging and rigging inspection, became an approved OSHA instructor, and is now Casey Industrial's corporate safety and training director.

How did you end up in the construction industry?
In southern Mississippi, where I graduated from high school, there were limited opportunities available to kids who didn't go to college. I decided on a construction career because it offered great wages, benefits, and stability. As a teenager I spent my summers working with local contractors, so I knew the basics of carpentry and equipment operations; but I knew that if I wanted to move ahead, I'd have to learn everything I could, not only about my trade but also about managing a project safely, efficiently, and effectively. So I started reading on my own, and I got lots of help from people who had 20 or 30 years of experience in the construction trades. I wish I'd made the connection between construction and studying basic math and geometry when I was in high school and not waited until I entered the workforce. I use math every day in my work.

How did you decide to become a safety professional?
I really enjoyed working with rigging and crane operations, and I gained an immense appreciation for the level of safety and skill it takes to work in that environment. I learned all I could about the trade and about the safety precautions that are required to be successful. Casey Industrial sent me to advanced training courses, then they started an in-house program where I trained other workers. This led to more training in safety and management, and eventually I became safety and training director—I manage the safety programs for all Casey's projects throughout the United States.

As a safety professional I've trained new employees, conducted safety meetings, conducted job-site inspections and audits, and investigated accidents, and I've helped to develop the hazard-control techniques we use to lessen the likelihood of dangerous situations. Since we started our training program, we've reduced our total incident rate by 60 percent—I'm extremely proud of that.

What do you think it takes to be a success in your trade?
You have to be trainable. This field has specific requirements that translate directly into safety, meaning safe work practices not only for yourself but for all workers and visitors to the project. You have to know why a certain procedure is done in a certain way. If you don't do things the right way, you could cause an accident. You could become a statistic. That's why I believe that training is the key to success.

Once I became proficient at my trade, I found that I liked to train other employees. By helping other people do the best job possible, you raise everyone's level of expertise, which means the company is more productive and profitable. A company with qualified workers who follow safe procedures is a successful company. If you train your people to work safely, you're protecting the company's most valuable asset: the worker.

What are some of the things you do as safety and training director?
I have a dual role at Casey. I'm responsible for the safety of all our employees and for making sure we're in compliance with federal and state laws. I also oversee the craft training program. The NCCER's craft training program is the construction craftsperson's ticket to a better future. These employees are getting both on-the-job experience and classroom training. Four years of craft training puts these tradespeople in an excellent position for success.

In addition to learning trade skills, workers also learn how to do their jobs safely. With all the changes in technology, materials, equipment, and OSHA requirements, I believe that everyone involved in construction, from apprentices to journeymen, needs ongoing training.

What do you like most about your job?
Working industrial projects allows me to travel throughout the United States. Casey Industrial works large, long-term projects and short-term shutdowns. Both kinds of projects have unique challenges that I find very exciting. I enjoy building something permanent, looking at a structure and knowing I helped put it in place. I also like being able to go into communities around the country and create jobs and opportunities for people. It's a great feeling to make a difference.

What would you say to someone entering the trades today?
Well, I have three sons who I anticipate will be involved in the industry. The oldest is an apprentice carpenter and the second is in high school and planning to enter the electrical trade. The third is still too young to know which trade will be right for him. The trades provide a great income and many opportunities for advancement. My advice is to get all the training that's available, and to work hard. Not only will you succeed, you'll make an excellent living.

Trade Terms Introduced in This Module

Apparatus: An assembly of machines used together to do a particular job.

Arc: The flow of electrical current through a gas such as air from one pole to another pole.

Arc welding: The joining of metal parts by fusion, in which the necessary heat is produced by means of an electric arc.

Benching: A safety procedure that involves flattening the area around an excavation or trench, often by laying down concrete slabs, so that soil, debris, and other materials will not slide into the excavation.

Combustible: Capable of easily igniting and rapidly burning; used to describe a fuel with a flash point at or above 100 degrees Fahrenheit.

Competent person: A person who can identify working conditions or surroundings that are unsanitary, hazardous, or dangerous to employees and who has authorization to correct or eliminate these conditions promptly.

Concealed receptacle: The electrical outlet that is placed inside the structural elements of a building, such as inside the walls. The face of the receptacle is flush with the finished wall surface and covered with a plate.

Confined space: An area large enough for a person to work in, but with limited ways of entering and exiting, and not meant for human occupancy. See also *Permit-required confined space.*

Cross-bracing: Braces (metal or wood) placed diagonally from the bottom of one rail to the top of another rail that add support to a structure.

Electrical distribution panel: Part of the electrical distribution system that brings electricity from the street source (power poles and transformers) through the service lines to the electrical meter mounted on the outside of the building to the panel inside the building. The panel houses the circuits that then distribute electricity throughout the structure.

Excavation: Any man-made cut, cavity, trench, or depression in an earth surface, formed by removing earth. Can be made for anything from basements to highways. See also *Trench.*

Extension ladder: A ladder made of two straight ladders that are connected so that the overall length can be adjusted.

Fire watch: A person, other than the one working, who scans the work area for fires and has access to and knowledge of fire extinguishers and alarms.

Flammable: Capable of easily igniting and rapidly burning; used to describe a fuel with a flash point below 100 degrees Fahrenheit.

Flash: A sudden bright light associated with starting up a welding torch.

Flashback: A welding flame that flares up and chars the hose at or near the torch connection. Caused by improperly mixed fuel mixture.

Flash burn: The damage that can be done to eyes after even brief exposure to ultraviolet light from arc welding. A flash burn requires medical attention.

Flash goggles: Eye protective equipment worn during welding operations.

Flash point: The temperature at which fuel gives off enough gases (vapors) to burn.

Ground: The conducting connection between electrical equipment or an electrical circuit and the earth.

Guarded: Enclosed, fenced, covered or otherwise protected by barriers, rails, covers, or platforms to prevent dangerous contact.

Hand line: A line attached to a tool or object so you can pull it up to you after you climb a ladder or scaffold.

Hazard Communication Standard (HazCom): The Occupational Safety and Health Administration standard that requires contractors to educate employees about hazardous chemicals and how to work safely with them.

Lanyard: A short section of rope or strap, one end of which is attached to a worker's safety harness and the other to a strong anchor point above the work area.

Lockout/tagout: A formal procedure for taking equipment out of service and ensuring that it cannot be operated until the lockout or tagout device (such as a lock and/or warning tag) has been removed by a qualified person.

Management system: The organization of a company's management personnel, including reporting procedures, supervisory responsibility, and administration.

Material Safety Data Sheet (MSDS): A document that must accompany any hazardous material. The MSDS identifies the substance and gives the exposure limits, the physical and chemical characteristics, the kind of hazard it presents, precautions for safe handling and use, and specific control measures.

Maximum intended load: The total weight of all employees, equipment, tools, materials, loads that are being carried, and other loads a ladder can hold at any one time.

Mid-rail: Mid-level, horizontal board required on all open sides of scaffolding and platforms that are more than 14 inches from the face of the structure and more than 10 feet above the ground. Placed half-way between the toeboard and the top rail.

OSHA: Occupational Safety and Health Administration, an agency of the U.S. Department of Labor. Also refers to the Occupational Safety and Health Act (OSHA) of 1970, a law that applies to more than five million businesses and more than 65 million workers in the country.

Permit-required confined space: A confined space that has been evaluated and found to have actual or potential hazards, such as a toxic atmosphere or other serious safety or health hazard. You need written authorization before you can enter a permit-required confined space. See also *Confined space.*

Personal protective equipment (PPE): Equipment or clothing designed to prevent or reduce injuries.

Planked: Having pieces of material 2 or more inches thick and 6 or more inches wide used as flooring, decking, or scaffolding.

Proximity work: Work done near a hazard but not actually in contact with the hazard.

Qualified person: A person who, by possession of a recognized degree, certificate, or professional standing, or who by extensive knowledge, training, and experience, has successfully demonstrated the ability to solve or resolve problems relating to a certain subject, work, or project.

Respirator: A device that provides clean, filtered air for breathing, no matter what is in the surrounding air.

Scaffold: An elevated work platform for both workers and materials.

Shoring: Pieces of timber used to support a wall, usually set in a diagonal position, to hold the wall in place temporarily.

Signaler: A person who is responsible for directing a vehicle when the driver's vision is blocked in any way.

Slag: Waste material from welding operations.

Spatter: Flying bits of solder or other molten material produced during welding operations.

Stepladder: A self-supporting ladder consisting of two elements hinged at the top.

Straight ladder: A nonadjustable ladder.

Switch enclosure: Box that houses electrical switches used to regulate and distribute electricity in a building.

Toeboard: Vertical barrier at floor level attached along exposed edges of a platform, runway, or ramp to prevent materials and people from falling.

Top rail: Top-level, horizontal board required on all open sides of scaffolding and platforms that are more than 14 inches from the face of the structure and more than 10 feet above the ground.

Trench: A narrow excavation made below the surface of the ground that is generally deeper than it is wide with a maximum width of 15 feet. See also *Excavation.*

Trencher: An excavating machine used to dig trenches, especially for pipeline and cables.

Welding shield: (1) A protective screen set up around a welding operation designed to safeguard workers not directly involved in that operation. (2) A shield that provides eye and face protection for welders by either connecting to helmet-like headgear or attaching directly to a hardhat; also called a welding helmet.

Wind sock: A cloth cone open at both ends mounted in a high place to show which direction the wind is blowing.

Answers to Review Questions

Section 2.0.0
1. c
2. b
3. a
4. d
5. c

Section 3.0.0
1. c
2. a
3. b
4. a
5. b

Sections 4.0.0–6.0.0
1. b
2. d
3. c
4. c
5. c

Section 7.0.0
1. b
2. b
3. d
4. b
5. c

Section 8.0.0
1. d
2. a
3. b
4. a
5. c

Section 9.0.0
1. b
2. a
3. d
4. a
5. b

Section 10.0.0
1. c
2. b
3. a
4. b
5. d

Additional Resources

This module is intended to present thorough resources for task training. The following reference works are suggested for further study. These are optional materials for continued education rather than for task training.

Construction Site Safety Orientation, Latest Edition. Gainesville, FL: NCCER, University of Florida Press.

Construction Site Safety Technician: Participant Manual, 1998 Edition. Gainesville, FL: NCCER, University of Florida Press.

Construction Safety, 1996 Edition. Jimmie Hinze. Englewood Cliffs, NJ: Prentice Hall.

Construction Safety Manual, 1998 Edition. Dave Heberle. New York: McGraw-Hill.

Construction Safety Council Home Page, http://buildsafe.org/cschome.htm

Occupational Safety and Health Standards for the Construction Industry, Latest Edition. Washington, DC: Occupational Safety and Health Administration, U.S. Department of Labor, U.S. Government Printing Office.

United States Department of Labor, Occupational Safety and Health Administration Home Page, http://www.osha.gov/

The NCCER makes every effort to keep these textbooks up-to-date and free of technical errors. We appreciate your help in this process. If you have an idea for improving this textbook, or if you find an error, a typographical mistake, or an inaccuracy in NCCER's Contren™ textbooks, please write us, using this form or a photocopy. Be sure to include the exact module number, page number, a detailed description, and the correction, if applicable. Your input will be brought to the attention of the Technical Review Committee. Thank you for your assistance.

Instructors – If you found that additional materials were necessary in order to teach this module effectively, please let us know so that we may include them in the Equipment/Materials list in the Instructor's Guide.

Write: Curriculum Revision and Development Department
National Center for Construction Education and Research
P.O. Box 141104, Gainesville, FL 32614-1104

Fax: 352-334-0932

E-mail: curriculum@nccer.org

Craft _____   Module Name _____

Copyright Date _____   Module Number _____   Page Number(s) _____

Description _____

_____

_____

_____

_____

(Optional) Correction _____

_____

_____

(Optional) Your Name and Address _____

_____

_____

_____

Introduction to Construction Math

Course Map

This course map shows all of the modules in the AWS Entry Level Welder—Phase 1 curriculum. The suggested training order begins at the bottom and proceeds up. Skill levels increase as you advance on the course map. The local Training Program Sponsor may adjust the training order.

AWS ENTRY LEVEL WELDER—PHASE 1

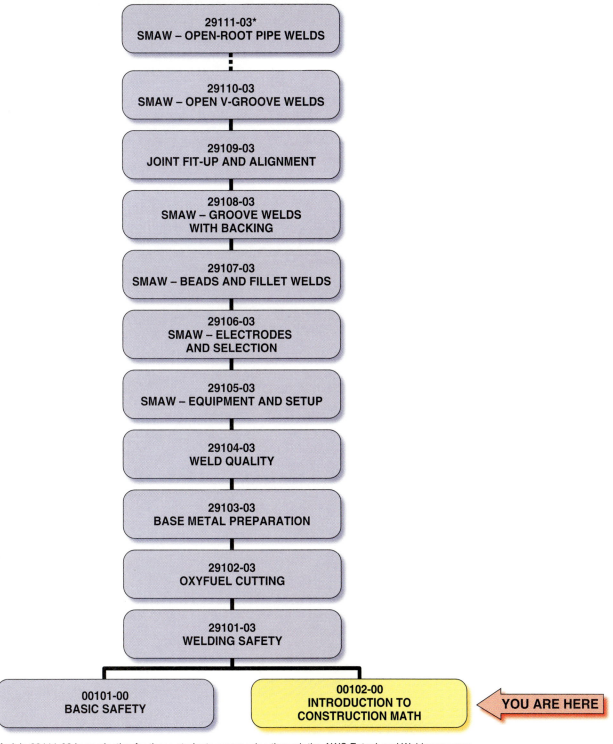

29111-03*
SMAW – OPEN-ROOT PIPE WELDS

29110-03
SMAW – OPEN V-GROOVE WELDS

29109-03
JOINT FIT-UP AND ALIGNMENT

29108-03
SMAW – GROOVE WELDS
WITH BACKING

29107-03
SMAW – BEADS AND FILLET WELDS

29106-03
SMAW – ELECTRODES
AND SELECTION

29105-03
SMAW – EQUIPMENT AND SETUP

29104-03
WELD QUALITY

29103-03
BASE METAL PREPARATION

29102-03
OXYFUEL CUTTING

29101-03
WELDING SAFETY

00101-00
BASIC SAFETY

00102-00
INTRODUCTION TO
CONSTRUCTION MATH

YOU ARE HERE

*Module 29111-03 is an elective for those students progressing through the AWS Entry Level Welder program.

00102CMAP.EPS

MODULE 00102-00 CONTENTS

Figures

Tables

Introduction to Construction Math

Objectives

When you finish this module, you will be able to:

1. Add, subtract, multiply, and divide whole numbers, with and without a calculator.
2. Use a standard ruler and a metric ruler to measure.
3. Add, subtract, multiply, and divide fractions.
4. Add, subtract, multiply, and divide decimals, with and without a calculator.
5. Convert decimals to percents and percents to decimals.
6. Convert fractions to decimals and decimals to fractions.
7. Explain what the metric system is and how it is important in the construction trade.
8. Recognize and use metric units of length, weight, volume, and temperature.
9. Recognize some of the basic shapes used in the construction industry and apply basic geometry to measure them.

Prerequisites

Before you begin this module, it is recommended that you successfully complete Module 00101-00.

Materials Required for Trainees

1. This module
2. Standard ruler (with $\frac{1}{16}$-inch markings)
3. Metric ruler (with centimeters [cm] and millimeters [mm])
4. Machinist's rule
5. Calculator
6. Pencil and paper

1.0.0 ◆ INTRODUCTION

Remember back in school when the teacher said, "You may need to use this mathematical operation some day"? Today is the day! In the construction trade, workers use math all the time. When you measure a length of material, fill a container with a specified amount of liquid, or calculate the dimensions of a room, you will be using mathematical operations. This module will provide practice in some basic mathematical procedures and give you an idea of how they might apply in the construction industry. Along the way, it will also introduce you to some construction careers and their use of mathematics.

2.0.0 ◆ WHOLE NUMBERS

In this section, you will learn how to work with **whole numbers.** Whole numbers are complete units *without* **fractions** or **decimals.**

The following are whole numbers:

| 1 | 5 | 67 | 335 | 2,654 |

The following are *not* whole numbers:

| ½ | ¾ | 7⅛ | 0.45 | 4.25 |

In this section, you will work only with whole numbers. Later, you will work with other kinds of numbers.

2.1.0 Parts of a Whole Number

Let's look at the parts of a whole number.

The whole number shown in *Figure 1* has seven **digits.** A digit is any of the numerical symbols from 0 to 9. The seven digits in *Figure 1* are the

Everyday Whole Numbers

You use whole numbers daily, probably without stopping to think about it. Math is such a common part of everyday life that we take it for granted. Knowing math makes it possible for you to fill the following order from your supervisor.

Tuesday: Need for the job (get 5-gallon buckets when possible)

10 gallons red paint

5 gallons blue paint

20 gallons gray paint

6 gallons yellow paint

How many 5-gallon buckets should you get?

2 buckets of red

1 bucket of blue

4 buckets of gray

1 bucket of yellow (plus a 1-gallon can)

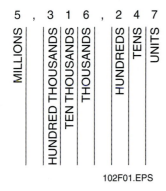

102F01.EPS

Figure 1 ◆ Place values.

numbers shown across the top of that figure. If you read this seven-digit whole number out loud, you would say "five million, three hundred sixteen thousand, two hundred forty-seven."

Each of this whole number's seven digits represents a **place value.** Each digit has a value that depends on its place, or location, in the whole number. In this whole number, for example, the place value of the 5 is five million, and the place value of the 2 is two hundred.

Numbers larger than zero are called **positive (+) numbers** (such as 1, 2, 3). Numbers less than zero are **negative (−) numbers** (such as −1, −2, −3). Zero is neither positive nor negative. All other numbers without a minus sign in front of them are positive.

Don't forget that some whole numbers may contain the digit zero—for instance, the whole number 7,093 has a zero in the hundreds place. When you read that number out loud, you would say "seven thousand ninety-three."

Review Questions

Section 2.1.0

1. Look at the description of the number below.
    Digit in the units place: 4
    Digit in the tens place: 6
    Digit in the hundreds place: 9
    Digit in the thousands place: 3

    This number would be written as _____.
    a. 4,693
    b. 3,964
    c. 30,964
    d. 39,064

2. In the number 25,718, the numeral 5 is in the _____ place.
    a. tens place
    b. thousands place
    c. units place
    d. hundreds place

3. The number for the words eighty-five is _____.
    a. 58
    b. 508
    c. 805
    d. 85

4. The number for the words one hundred twenty-two is _____.
   a. 212
   b. 221
   c. 122
   d. 1,022

5. The number for the words two thousand four hundred ninety-seven is _____.
   a. 2,479
   b. 4,297
   c. 2,079
   d. 2,497

2.2.0 Adding Whole Numbers

To *add* means to combine the values of two or more numbers together. The total when you add two or more numbers together is called the **sum.** The sign for addition is the plus sign (+). Addition problems can be written vertically (up and down) or horizontally (left to right). Here are two examples:

$$\begin{array}{r} 6 \\ + 3 \\ \hline 9 \end{array} \quad 6 + 3 = 9 \qquad \begin{array}{r} 5 \\ + 2 \\ \hline 7 \end{array} \quad 5 + 2 = 7$$

In these examples, the sum of 6 + 3 is 9. The sum of 5 + 2 is 7.

2.2.1 Carrying in Addition

Adding single numbers probably seems pretty easy at this point. But what happens when you want to add larger numbers, such as 58 + 34? You still add up single numbers, but you do it in a certain order, and you may need to *carry* (see *Figure 2*) the tens part of a sum from the units column to the tens column and add it there.

Look at the following example:

$$\begin{array}{l} 58 \text{ pounds of nails} \\ + 34 \text{ pounds of nails} \\ \hline \end{array}$$

To add these numbers, start with the two digits on the right-hand side (8 + 4), which add up to 12, right? But you don't write down 12 in the sum line at the bottom—you write down just the 2, which is the units amount. (Remember that the units place is the farthest right-hand column.) Then you *carry* the 1 (the tens amount from the 12) to the top of the tens column, which is the 5 + 3 column. That column then becomes 1 + 5 + 3. Those numbers add up to 9, and you put the 9 below that column, where the sum goes. Using 58 + 34,

Think		Write	
5 tens	+ 8 units	58	*(1 ten carried to*
+ 3 tens	+ 4 units	+ 34	*tens column,*
(8 tens)	+ (1 ten + 2 units)	92	*2 units*
			remaining in
			units column)

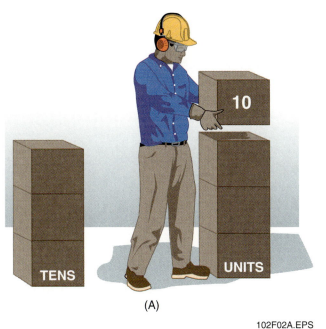

(A)

102F02A.EPS

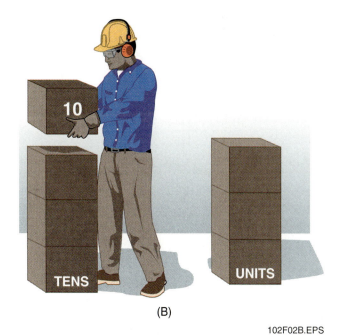

(B)

102F02B.EPS

Figure 2 ◆ Carrying in addition.

Adding It Up!

It is your first day on the job. Your supervisor asks you to distribute the materials to 3 job sites. She asks you to take the following to each site:

Job 1

4 boxes of 10d nails

6 boxes of drywall screws

18 rolls of R-19 insulation

12 bags of mortar mix

Job 2

10 boxes of 10d nails

9 boxes of drywall screws

23 rolls of R-19 insulation

4 bags of mortar mix

Job 3

13 boxes of 10d nails

11 boxes of drywall screws

5 rolls of R-19 insulation

14 bags of mortar mix

How much of each material do you load on the truck?

Boxes of 10d nails (27)

Boxes of drywall screws (26)

Rolls of insulation (46)

Bags of mortar mix (30)

So 58 + 34 = 92. You can use the same techniques for carrying when you are adding even larger numbers. If a column adds to 10 or more, put only the right-hand digit below that column and carry the other digit to the next column to the left. If you have a total of 13 in the hundreds column, for example, put the 3 at the bottom of that column and carry the 1 to the thousands column.

Review Questions

Section 2.2.1

1. The answer to the following addition problem is _____.

   32 pounds of nails
   +75 pounds of nails

   a. 108
   b. 107
   c. 97
   d. 100

2. The answer to the following addition problem is _____.

   73 in. of molding
   +45 in. of molding

   a. 128
   b. 118
   c. 78
   d. 108

3. The answer to the following addition problem is _____.

   83 ft. of cable
   +53 ft. of cable

   a. 139
   b. 86
   c. 123
   d. 136

4. The answer to the following addition problem is _____.

   452 ft of baseboard
   + 74 ft of baseboard

   a. 4,126
   b. 526
   c. 1192
   d. 528

5. The answer to the following addition problem is _____.

   323 yds of binding
   +758 yds of binding

   a. 981
   b. 1,071
   c. 1,081
   d. 1,094

2.2.2 Problem-Solving Steps

On the job, you will be using addition to solve construction-related problems. For example, suppose you need to cut the following lengths from a bar of hot-rolled steel: 19 inches, 39 inches, and 8 inches. How many total inches of bar will you cut

AWS ENTRY LEVEL WELDER – TRAINEE MODULE 00102-00

off? (Note that the symbol " is often used instead of the word *inches* when you are writing out mathematical problems. So 19 inches would appear as 19" and 39 inches as 39".)

It may help to break down a problem like this into steps, as shown in *Figure 3*.

Step 1 Read the entire problem, and create a picture in your mind (or on paper) of the problem. [For the sample problem above, picture cutting first a 19-inch piece, then a 39-inch piece, then an 8-inch piece from a bar of steel. Then picture laying all three pieces end to end so you can see, and measure, the total amount cut.]

Step 2 Decide what the problem is. [This problem is to find out the total inches of steel that have been cut off.]

Step 3 Decide what information the problem gives to help you solve it, and what unit of measure (for example, inches, feet, or dollars) it asks for. [For this problem, you know the numbers of inches of each of the three separate pieces you need to cut off, but not the total number of inches to be cut off. The answer will be in inches.]

Step 4 Decide which math operation or combination of operations are needed for the solution. [Because you are looking for the total number of inches, you would add.]

Step 5 Estimate a reasonable answer. [Make a guess as to the approximate number of inches. With these numbers, a good guess might be 70 inches.]

Step 6 Solve the problem.

$$\begin{array}{r} 19" \\ 39" \\ + \ 8" \\ \hline 66" \end{array}$$

Step 7 Check the answer against your estimate. Does it make sense? [Our estimate was 70 inches. The actual answer is 66 inches. The answer makes sense!]

Step 8 Check the answer against the problem. Does it answer the question? Does it make sense? [66 total inches of bar were cut off. If you think again of the pieces cut (19 inches, 39 inches, and 8 inches), this answer makes sense.]

Section 2.2.2

1. A project has 8 workers on one job, 35 on another, and 18 on a third. The total number of people working on all three jobs is _____.
    a. 53
    b. 61
    c. 133
    d. 71

2. A bricklayer lays 649 bricks the first day, 632 the second day, and 478 the third day. The bricklayer lays _____ bricks in three days.
    a. 1,760
    b. 1,659
    c. 1,769
    d. 1,759

3. Four walls of a bathroom require 31, 46, 49, and 16 tiles. You will need _____ tiles to tile all four walls.
    a. 132
    b. 142
    c. 140
    d. 152

4. For eight different jobs, you need to fill a total of eight boxes with different numbers of a specified size of brightwood screw. The numbers of screws you need to put in the boxes are 142, 57, 35, 79, 32, 79, 53, and 95. You will need a total of _____ screws to fill all eight boxes.
    a. 472
    b. 572
    c. 471
    d. 571

5. You have already used 36,000 bricks, and your supervisor orders 1,500 more to complete the job. The job requires a total of _____ bricks.
    a. 37,000
    b. 35,500
    c. 51,000
    d. 37,500

2.3.0 Subtracting Whole Numbers

Subtraction means finding the **difference** between two numbers, or taking away one number from another. The subtraction sign (−) is also called the minus sign. The result (answer) of a subtraction problem is called the difference.

For example, you have a total of nine sockets to install today. You have installed five so far. How many more do you have to install today?

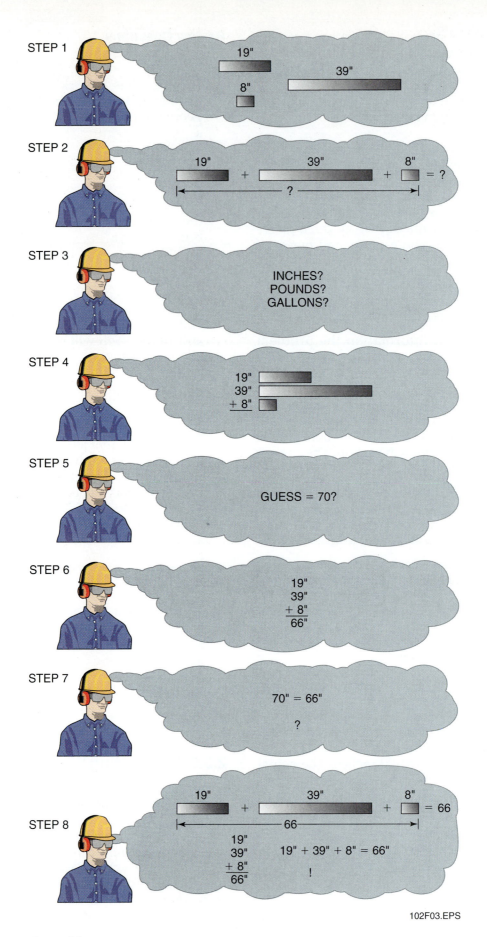

Figure 3 ◆ How to solve problems.

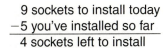

9 sockets to install today
−5 you've installed so far
4 sockets left to install

In some subtraction problems, you may have to subtract a larger digit from a smaller digit, such as in the following problem:

76
−48

As with adding, when you are subtracting, you start with the units column, which is the right-hand side. In the units column in this problem, you have to subtract a larger number, 8, from a smaller number, 6. How can you do this? By **borrowing** from the tens place.

Problem	Think	Write
76 −48	7 tens + 6 units −4 tens + 8 units	76 −48

Step 1 Notice that there are not enough units to subtract from (you cannot subtract 8 from 6).

Step 2 Borrow 1 ten from the tens column.

(6 tens + 1 ten) + (6 units)   76
−4 tens   + 8 units   −48

Step 3 Add the borrowed 10 to the 6 in the units column (10 + 6 = 16), so that you have the larger number (16) you need there. Visualize a two-digit number in the units column. In the tens column, you are now left with 6 tens.

6 tens + 16 units   76
−4 tens   8 units   −48

Step 4 You now have enough in both columns to subtract from.

6 tens + 16 units   76
−4 tens +   8 units   −48
2 tens   8 units   28

Review Questions

Section 2.3.0

1. The answer to the following subtraction problem is _____.

87 sockets
−38 sockets

a. 59
b. 49
c. 51
d. 40

2. The answer to the following subtraction problem is _____.

26 connectors
−17 connectors

a. 19
b. 9
c. 8
d. 10

3. The answer to the following subtraction problem is _____.

92 hours
−34 hours

a. 62
b. 57
c. 58
d. 68

4. The answer to the following subtraction problem is _____.

246 ft of cable
− 18 ft of cable

a. 238
b. 232
c. 228
d. 227

5. The answer to the following subtraction problem is _____.

826 bricks
−717 bricks

a. 111
b. 109
c. 119
d. 19

2.4.0 Multiplying Simple Whole Numbers

You have eight construction sites. You need to send four wrenches to each site. How many wrenches will you need in total? You *could* count out four wrenches eight times, put them in eight piles, and then count them all. *Figure 4* shows how you might do this.

Obviously, this is the hard way to add—you can see it's worth learning your *times-table* through the 12s.

Multiplication is the quick way to add the same number together many times. It would be much easier, and more efficient, to multiply 4 times 8 than to add 4 eight times. In this example, you use

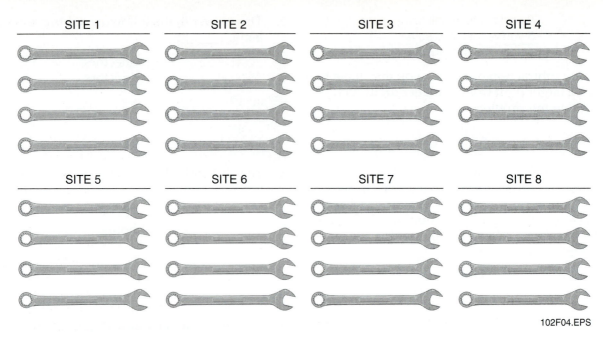

Figure 4 ◆ Counting out the wrenches.

multiplication to figure out 8 (construction sites) multiplied by 4 (wrenches), or 8 times 4.

The symbol for multiplication is the × sign. In this example, the operation would be written in either of the following two ways:

$$8 \text{ construction sites}$$
$$\times\ 4 \text{ wrenches}$$
$$\overline{32 \text{ wrenches needed}}$$

$$4 \times 8 = 32$$

DID YOU KNOW?

Numeral Systems

People in ancient Egypt, Babylon, Greece, and Rome developed different numeral systems or ways of writing numbers. Some of these early systems were very complex and difficult to use. A new numeral system came into use around 750 c.e. Originally developed by the Hindus in India, the system was spread by Arab traders. The Hindu-Arabic system uses only ten symbols and is still in use today: 1, 2, 3, 4, 5, 6, 7, 8, 9, 0. These ten symbols or numerals can be combined to write any number imaginable.

The answer to the problem is 32. To work successfully in the construction industry, it is very helpful to know the answer to the most basic multiplication equations without having to write them out or think too hard about them. It is a good idea to spend some time memorizing the most basic multiplication table, which means knowing automatically that, for example, $2 \times 4 = 8$, and $5 \times 9 = 45$, and $12 \times 12 = 144$. Appendix A is a basic multiplication table that will help you.

Review Questions

Section 2.4.0

1. Adding $4 + 4 + 4 + 4 + 4$ can be simplified to multiplying _____.
   a. 4×5
   b. 4×4
   c. 12×8
   d. 20×4

2. The answer to the following multiplication problem is _____.

   $$9 \text{ doors installed by}$$
   $$\times 8 \text{ workers}$$

   a. 78
   b. 72
   c. 17
   d. 81

3. The answer to the following multiplication problem is ____.

$$\begin{array}{r} 9 \text{ bolts in} \\ \times 6 \text{ frames} \end{array}$$

    a. 45
    b. 63
    c. 15
    d. 54

4. The answer to the following multiplication problem is ____.

$$\begin{array}{r} 7 \text{ fixtures} \\ \times 2 \text{ plumbers} \end{array}$$

    a. 9
    b. 14
    c. 21
    d. 15

5. The answer to the following multiplication problem is ____.

$$\begin{array}{r} 8 \text{ beams hung by} \\ \times 4 \text{ carpenters} \end{array}$$

    a. 48
    b. 12
    c. 32
    d. 23

2.4.1 Multiplying Larger Whole Numbers

These multiplication problems may have been easy, but what happens when you have a more difficult problem involving larger numbers, such as the following example? On a job, a pipefitter installs 75 feet of pipe in one hour. How many feet of pipe can this pipefitter install in 16 hours?

To solve that problem, you need to multiply 75 by 16. You do this by going through these steps:

Step 1 Write the numbers on two lines. As you did with addition and subtraction, place units under units, and tens under tens.

$$\begin{array}{r} 75 \\ \times 16 \end{array}$$

Step 2 Start with the digit in the units place of the bottom number (the 6 in the 16).

$$\begin{array}{r} 75 \\ \times 16 \end{array}$$

Step 3 Multiply every number in the top number (the 75) by the number in the units place of the bottom number (the 6), starting on the right with the units place of the top number (the 5).

carry the
$$\begin{array}{r} 3 \\ 75 \\ \times\ 16 \\ \hline 450 \end{array}$$

6 times 5 is 30, or 3 tens and 0 units. Place the 0 in the units place. Carry the 3 tens to the tens place.
6 times 7 is 42, or 4 hundreds and 2 tens. Add the 3 tens that you carried, which gives you 45.

Step 4 Now multiply every individual number in the top number (the 75) by the number in the tens place of the bottom number (the 1). Start with the number in the units place (on the right) of the top number (the 5).

$$\begin{array}{r} 75 \\ \times\ 16 \\ \hline 450 \\ 75 \end{array}$$

One ten times 5 units = 5 tens. Place a 5 in the tens place.
One ten times 7 tens = 7 hundreds. Place a 7 in the hundreds place.

Step 5 Now add the numbers in each column, beginning with the units place.

$$\begin{array}{r} 75 \\ \times\ 16 \\ \hline 450 \\ +\ 75 \\ \hline 1,200 \end{array}$$

The answer to this multiplication problem is 1,200. In solving this problem, you have found that the pipefitter can install 1,200 feet of pipe in 16 hours.

Review Questions

Section 2.4.1

1. The answer to the following multiplication problem is ____.

$$\begin{array}{r} 12 \text{ wheelbarrows at} \\ \times 21 \text{ sites} \end{array}$$

    a. 252
    b. 240
    c. 36
    d. 360

2. The answer to the following multiplication problem is ____.

$$\begin{array}{r} 11 \text{ ladders at} \\ \times 15 \text{ sites} \\ \hline \end{array}$$

  a. 115
  b. 150
  c. 165
  d. 105

3. The answer to the following multiplication problem is ____.

$$\begin{array}{r} 30 \text{ barrels at} \\ \times 25 \text{ sites} \\ \hline \end{array}$$

  a. 210
  b. 600
  c. 300
  d. 750

4. The answer to the following multiplication problem is ____.

$$\begin{array}{r} 452 \text{ square feet used at} \\ \times \quad 4 \text{ sites} \\ \hline \end{array}$$

  a. 1,008
  b. 1,808
  c. 8,108
  d. 456

5. The answer to the following multiplication problem is ____.

$$\begin{array}{r} 162 \text{ bricks at} \\ \times \quad 52 \text{ sites} \\ \hline \end{array}$$

  a. 2,140
  b. 1,134
  c. 11,024
  d. 8,424

2.5.0 Dividing Whole Numbers

Division is the opposite of multiplication. Instead of adding a number several times ($5 + 5 + 5 = 5 \times 3 = 15$), when you are dividing you subtract a quantity several times. But just as with using multiplication instead of addition, you can solve a problem much faster by using division instead of subtracting the same number over and over.

For example, you have 60 wrenches to distribute equally among 10 construction sites, and you need to find out how many wrenches should go to each site. You *could* make 10 piles (one for each site) and then count the number of wrenches in each pile. But it would be much easier and more efficient to divide 10 into 60. (Another way to state this is that you divide 60 by 10. Or you can say you are going to see how many times 10 goes into 60.) The answer, 6, would be the number of wrenches to send to each site.

DID YOU KNOW?

In a division problem, the number you are dividing by is called the **divisor,** and the **dividend** is the number being divided. In the example used in this section, the number 10 is the divisor and the number 60 is the dividend.

In this easy example, the numbers came out as whole numbers—60 divided by 10 equals 6. Some problems don't come out evenly, such as this one: You have 10 feet of cable and are asked to cut it into 4-foot pieces. How many 4-foot pieces will you have? How much of the 10-foot cable will be left over? (Note that the symbol ' is often used instead of the word *foot,* or *feet,* when you are writing out mathematical operations. This means that 4 feet will appear as 4' and that 10 feet will appear as 10'.)

Step 1 A division problem is written like this:

$$4\,)\overline{10}$$

Four goes into 10 two times. Place a 2 on the line above the 10.

$$4\,)\overline{10}^{\,2}$$

Step 2 Multiply 2 times 4. Place the answer, 8, under the 10.

$$\begin{array}{r} 2 \\ 4\,)\overline{10} \\ 8 \end{array}$$

Step 3 Subtract 8 from 10. The answer is 2. This is called the **remainder,** meaning it is the amount left over. The answer to this problem would be written 2 r2.

By dividing, you have found that you will have two 4-foot pieces of cable, with 2 feet of cable left over.

Review Questions

Section 2.5.0

1. The answer and the remainder (if there is one) for the following division problem is _____.

    $$15 \div 3$$

    a. 45
    b. 5
    c. 3
    d. 5 r3

2. The answer and the remainder (if there is one) for the following division problem is _____.

    $$36 \div 4$$

    a. 9
    b. 12
    c. 8
    d. 8 r2

3. The answer and the remainder (if there is one) for the following division problem is _____.

    $$54 \div 5$$

    a. 11
    b. 10
    c. 1 r4
    d. 10 r4

4. The answer and the remainder (if there is one) for the following division problem is _____.

    $$6\overline{)17}$$

    a. 11
    b. 9
    c. 3 r1
    d. 2 r5

5. The answer and the remainder (if there is one) for the following division problem is _____.

    $$7\overline{)39}$$

    a. 7
    b. 5 r4
    c. 6 r3
    d. 4 r11

2.5.1 Dividing More Complex Whole Numbers

Now let's take a look at solving a more complicated division problem: Twenty-four people were a part of a lottery pool that just won $2,638 (after taxes). They decided to split it up evenly among themselves and then give whatever was left over (the remainder) to charity. How much did each person get? How much went to charity?

To solve this problem, you would use what is called **long division.**

The problem is $2,638 divided by 24 people.

Step 1 Estimate the answer. 24 is close to 25. 2,638 is close to 2500. 25 goes into 2,500 one hundred times. Estimate: approximately $100 for each person.

Step 2 Now do the actual division, using the following steps:

24 goes into 26 one time, with a remainder of 2.

$$
\begin{array}{r}
1?? \\
24\overline{)2638} \\
-\ 24 \\
\hline
02
\end{array}
$$

Bring down the next number, 3. Can 24 go into 23? It cannot, so put a 0 in the answer line at the top.

$$
\begin{array}{r}
10? \\
24\overline{)2638} \\
-\ 24 \\
\hline
023 \\
000
\end{array}
$$

Subtract 0 from 23. The answer is 23. Bring down the next number, the 8.

$$
\begin{array}{r}
10? \\
24\overline{)2638} \\
-\ 24 \\
\hline
023 \\
000 \\
\hline
0238
\end{array}
$$

How many times can 24 go into 238? To figure this, think: 24 is close to 25. 238 is close to 250. How many times can 25 go into 250? About 9 or 10. Let's try 9. Nine times 24 is 216. Subtract 216 from 238. The remainder is 22.

$$
\begin{array}{r}
109 \\
24\overline{)2638} \\
-\ 24 \\
\hline
023 \\
-\ 000 \\
\hline
0238 \\
-\ 0216 \\
\hline
0022
\end{array}
$$

So how much money did each of the 24 people get? $109 each. How much went to charity? $22. (Note that the answer, $109, is close to your estimate of $100.)

Setting Up the Site

Electricians are working on four service box installations. Each electrician needs the following materials:

 1 service box

 20 circuit breakers

 150 feet of wire

Wire is sold in 200-foot rolls. How many rolls do you need? Is there any left over?

$$150 \times 4 = 600 \text{ feet}$$
$$600 \div 200 = 3 \text{ rolls, nothing left over}$$

Review Questions

Section 2.5.1

1. The answer and the remainder (if there is one) for the following division problem is _____.

$$12 \overline{)263}$$

 a. 21 r11
 b. 22
 c. 201
 d. 22 r11

2. The answer and the remainder (if there is one) for the following division problem is _____.

$$16 \overline{)4218}$$

 a. 207 r6
 b. 263 r10
 c. 363 r10
 d. 26 r58

3. The answer and the remainder (if there is one) for the following division problem is _____.

$$15 \overline{)4532}$$

 a. 32 r2
 b. 32
 c. 302 r2
 d. 302

4. Your supervisor sends you to the truck for 150 feet of electrical wire. When you get there, you find that all the coils of wire come in 15-foot lengths. You will need to bring back _____ coils of wire.
 a. 100
 b. 10
 c. 3
 d. 5

5. A plumbing job requires 100 feet of plastic pipe available in 20-foot sections. You will need _____ sections.
 a. 12
 b. 5
 c. 10
 d. 6

2.6.0 Using the Calculator to Add, Subtract, Multiply, and Divide Whole Numbers

It is important to be able to perform calculations (such as addition and subtraction) in your head even if you have a calculator. It allows you to estimate the answer before you use the calculator. Why is this important? So you can double-check the answer against your estimate. If you press a wrong key on the calculator, you might be out hundreds of dollars or off by several inches.

The calculator is a marvelous tool for saving time. Let's look at the most frequently used operations of the calculator: adding, subtracting, multiplying, and dividing whole numbers.

2.6.1 Parts of the Calculator

Figure 5 shows the parts of a common calculator.

2.6.2 Using the Calculator to Add Whole Numbers

Adding numbers is easy with a calculator. Just follow the steps below, using 5 + 4 to practice.

Step 1 Turn the calculator *on*. A zero (0) appears in the display.

Step 2 Press 5. A 5 appears in the display.

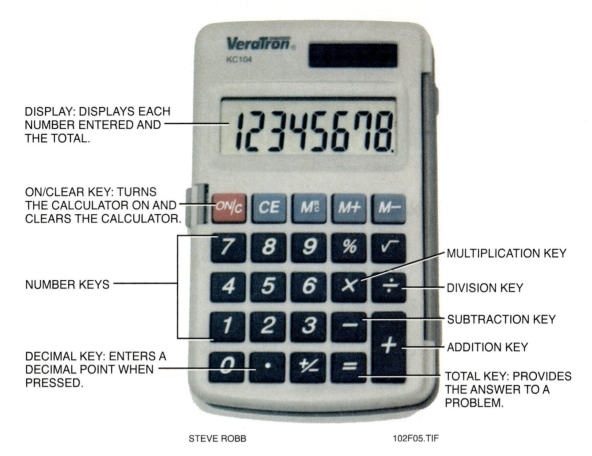

DISPLAY: DISPLAYS EACH NUMBER ENTERED AND THE TOTAL.

ON/CLEAR KEY: TURNS THE CALCULATOR ON AND CLEARS THE CALCULATOR.

NUMBER KEYS

DECIMAL KEY: ENTERS A DECIMAL POINT WHEN PRESSED.

MULTIPLICATION KEY

DIVISION KEY

SUBTRACTION KEY

ADDITION KEY

TOTAL KEY: PROVIDES THE ANSWER TO A PROBLEM.

STEVE ROBB

102F05.TIF

Figure 5 ◆ Parts of a common calculator.

Step 3 Press the + key. The 5 is still displayed.

Step 4 Press the 4 key. A 4 is displayed.

Step 5 Press the = key. The sum, 9, appears in the display.

Step 6 Press the *ON/C* key to clear the calculator.

Review Questions

Section 2.6.2

Practice using your calculator by finding the answers to these addition problems:

1.   12 (press +)
     24 (press +)
    +33 (press =)

The answer is _____.

a. 69
b. 21
c. 45
d. 39

2.   67
     46
    +96

The answer is _____.

a. 209
b. 117
c. 17
d. 182

3.   83
     35
    +50

The answer is _____.

a. 2
b. 98
c. 168
d. 186

4.
```
      34
     938
      24
   +  63
```

The answer is _____.

a. 996
b. 1,059
c. 214
d. 885

5.
```
      67
     774
     983
    +532
```

The answer is _____.

a. 1,824
b. 1,292
c. 2,359
d. 2,356

2.6.3 Using the Calculator to Subtract Whole Numbers

Subtracting with a calculator is as easy as adding with one. Here are the steps, using the problem $25 - 5$ to practice.

Step 1 Turn the calculator *on*. A zero (0) appears in the display.

Step 2 Press the 2 key and then the 5 key. A 25 appears in the display.

Step 3 Press the − key. The 25 is still displayed.

Step 4 Press the 5 key. A 5 is displayed.

Step 5 Press the = key. The difference, 20, appears in the display.

Step 6 Press the *ON/C* key to clear the calculator.

Review Questions

Section 2.6.3

Practice using your calculator by finding the answers to these subtraction problems.

1.
```
      97
   −   5
```

The answer is _____.

a. 102
b. 20
c. 2
d. 92

2.
```
     452
    −414
```

The answer is _____.

a. 38
b. 866
c. 138
d. 46

3.
```
   1,254
   −  557
```

The answer is _____.

a. 684
b. 1,811
c. 697
d. 1,197

4.
```
   4,593
   −4,247
```

The answer is _____.

a. 8840
b. 356
c. 346
d. 354

5. At 2:00 P.M., the thermometer showed 99 degrees. At 7:00 P.M., it showed 75 degrees. The temperature fell by _____ degrees between 2 P.M. and 7 P.M.

a. 42
b. 24
c. 25
d. 204

2.6.4 Using the Calculator to Multiply Whole Numbers

Multiplying with a calculator is as easy as adding and subtracting with one. Here are the steps, using the problem 6×5 to practice.

Step 1 Turn the calculator *on*. A zero (0) appears in the display.

Step 2 Press 6. A 6 appears in the display.

Step 3 Press the × key. The 6 is still displayed.

Step 4 Press the 5 key. A 5 is displayed.

Step 5 Press the = key. The answer, 30, appears in the display.

Step 6 Press the *ON/C* key to clear the calculator.

Review Questions

Section 2.6.4

Practice using your calculator by finding the answers to these multiplication problems:

1. 1,254
    × 57

    The answer is ____.

    a. 71,478
    b. 1,197
    c. 22
    d. 8,778

2. 943
    × 93

    The answer is ____.

    a. 2,829
    b. 87,699
    c. 36,777
    d. 1,036

3. 4,593
    × 47

    The answer is ____.

    a. 215,871
    b. 4,640
    c. 97
    d. 4,546

4. A machine produces 465 screws in one hour. It will produce ____ screws in 16 hours.
    a. 37
    b. 7,440
    c. 46,500
    d. 481

5. There are 12 inches to a foot. There are ____ inches in 18 feet.
    a. 122
    b. 216
    c. 200
    d. 180

2.6.5 Using the Calculator to Divide Whole Numbers

Dividing with a calculator is as easy as the other operations. Here are the steps, using 12 ÷ 4 to practice:

Step 1 Turn the calculator *on*. A zero (0) appears in the display.

Step 2 Press the 1 key and then the 2 key. A 12 appears in the display.

Step 3 Press the ÷ key. The 12 is still displayed.

Step 4 Press the 4 key. A 4 is displayed.

Step 5 Press the = key. The answer, 3, appears in the display.

Step 6 Press the ON/C key to clear the calculator.

2.6.6 Expressing a Remainder as a Whole Number

When one number does not go into another number evenly, you are left with a remainder. For example, use your calculator to figure this problem: There is a piece of wood 6 feet long. How many 4-foot pieces can a worker cut from it? How many feet will be left over?

Step 1 Turn the calculator *on*. A zero (0) appears in the display.

Step 2 Press 6. A 6 appears in the display.

Step 3 Press the ÷ key. The 6 is still displayed.

Step 4 Press the 4 key. A 4 is displayed.

Step 5 Press the = key.

Step 6 The total, 1.5, appears in the display (.5 is a decimal, a part of a number, represented by digits to the right of a point called the decimal point).

Step 7 Press the ON/C key to clear the calculator.

Step 8 To express the .5 as a whole number rather than a decimal, multiply it by the number you divided by (4). The remainder expressed as a whole number is 2 feet (see *Figure 6*).

The answer to the problem, "How many 4-foot pieces can a worker cut from a 6-foot piece of wood and how many feet will be left over?" is one piece with 2 feet left over.

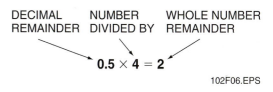

DECIMAL REMAINDER   NUMBER DIVIDED BY   WHOLE NUMBER REMAINDER

0.5 × 4 = 2

102F06.EPS

Figure 6 ◆ From decimal remainder to whole number remainder.

Review Questions

Section 2.6.6

Practice using your calculator by finding the answers to these division problems.

1. 20 ÷ 5 = ____
   a. 4
   b. 10
   c. 3
   d. 4 r3

2. 54 ÷ 9 = ____
   a. 486
   b. 16
   c. 6
   d. 46

3. 16 ÷ 3 = ____
   a. 2 r8
   b. 5 r1
   c. 3 r5
   d. 11

4. You have 100 gallons of liquid. You need to put the liquid into 20-gallon containers. You can fill ____ 20-gallon containers.
   a. 5
   b. 6
   c. 20
   d. 10

5. You need to cut two 3-foot lengths out of a board 8 feet long. You will have a piece ____ feet long left over when you are done.
   a. 2
   b. 5
   c. 0.66
   d. 3

3.0.0 ◆ WORKING WITH MEASUREMENTS

In the construction trade, you will need to use a ruler to measure various objects. There are two types of rulers you may see on the job: the **standard, or English, ruler** (see *Figure 7*) and the **metric ruler** (see *Figure 8*). The markings on the standard ruler measure in inches and feet, while the markings on the metric ruler measure in millimeters and centimeters. A *yardstick* is a standard ruler that is one yard, or three feet, long.

In this section, you are working only with the standard ruler. Later you will work with the metric ruler. The standard ruler is divided into whole inches and then halves, fourths, eighths, and sixteenths. Some rulers are divided into thirty-seconds, and some into sixty-fourths. These represent fractions of an inch. In this unit, you will be working with a standard ruler and fractions to solve problems.

STEVE ROBB   102F07.TIF

Figure 7 ◆ The standard, or English, ruler.

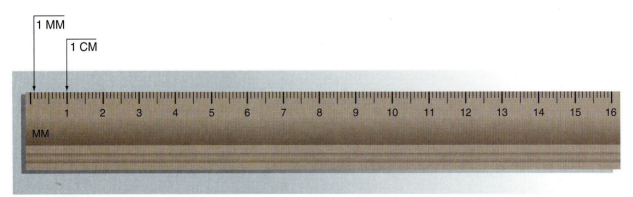

102F08.EPS

Figure 8 ◆ The metric ruler.

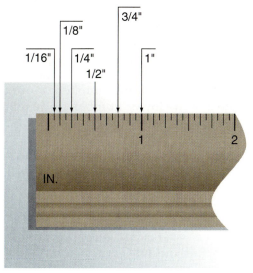

Figure 9 ◆ Distances on the standard ruler.

3.1.0 Using the Standard Ruler

You must learn to recognize and identify the distances shown (see *Figure 9*) on the standard ruler.

Use the ruler shown in *Figure 10* to find the answers to the following questions.

1. A is at the ___-inch mark.
   a. ½
   b. 1½
   c. 1
   d. 10

2. B is at the ___-inch mark.
   a. 2½
   b. 2¾
   c. 2¼
   d. 2

3. C is at the ___-inch mark.
   a. 1½
   b. 1¾
   c. 1⅝
   d. 1⅞

4. D is at the ___-inch mark.
   a. ¾
   b. ½
   c. ¼
   d. ⅞

5. E is at the ___-inch mark.
   a. ⅛
   b. ¼
   c. 1⁵⁄₁₆
   d. ⁵⁄₁₆

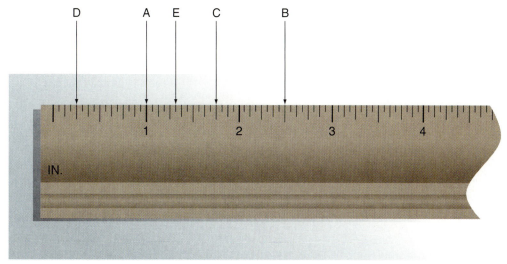

Figure 10 ◆ Review questions ruler.

4.0.0 ◆ WHAT ARE FRACTIONS?

A fraction divides whole units into parts. Common fractions are written as two numbers, separated by a slash or by a horizontal line, like this:

$$1/2 \quad \text{or} \quad \frac{1}{2}$$

The slash or horizontal line means the same thing as the ÷ sign. So think of a fraction as a division problem. The fraction ½ means 1 divided by 2, or one divided into two equal parts. Read this fraction as "one-half."

DID YOU KNOW?

Pipefitters

Pipefitters install and repair high- and low-pressure pipe systems used in manufacturing, in generation of electricity, and in heating and cooling buildings. Specializations include steamfitters, who install pipe systems to move liquids or gases under high pressure, and sprinkler fitters, who install automatic fire sprinkler systems in buildings.

Pipefitters work from blueprints or drawings. They lay out the job and perform such tasks as measuring, marking, cutting, and threading pipe. Pipefitters must have good mathematical skills so that they can measure piping and fittings accurately to connect pipe systems.

The lower number (**denominator**) of the fraction tells you the number of parts by which the upper number (**numerator**) is being divided. The upper number is a whole number that tells you how many parts are going to be divided. In the fraction ½, the 1 is the upper number, or numerator, and the 2 is the lower number, or denominator. These numbers are also referred to as the *terms* of the fraction.

What measurement is the arrow in *Figure 11* pointing to?

Circle the correct answer:

A. ⁸⁄₁₆ B. ¾ C. ⅛ D. ½ E. *All* are correct!

The correct answer is E . . . all of the answers are correct! Let's find out why.

4.1.0 Finding Equivalent Fractions

You can see that ½ inch = ¾ inch = ⅛ inch = ⁸⁄₁₆ inch. These fractions are called **equivalent fractions.** Equivalent means that they have the same value or that they are equal. If you cut off a piece of wood ⁸⁄₁₆-inch long, and I cut off a piece ½-inch long, we would each have the same length of wood.

When you measure objects, you often need to record all measurements as common fractions, such as sixteenths of an inch. Doing it this way allows you to easily compare, add, and subtract fractional measurements. This is why you need to know how to find equivalent fractions.

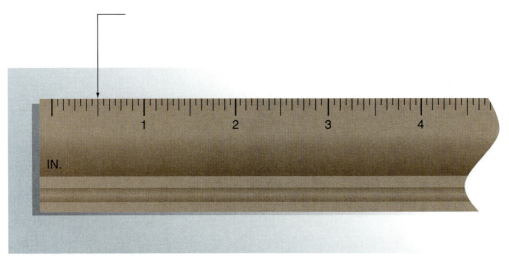

102F11.EPS

Figure 11 ◆ What's the measurement?

Fractions on the Job

The first day you report to work on a construction site, you will realize how important it is to understand fractions. In the real world, most measurements are not whole numbers. Typically you will be measuring and cutting pipe or lumber to fractional lengths such as ⅜, ⁵⁄₁₆, and ¾. Being comfortable working with fractions is an important job skill.

To find out how many sixteenths of an inch are equal to ½ inch, for example, you need to multiply both the numerator and the denominator by the same number. (Remember that a fraction in which both the numerator and the denominator are the same number is equal to 1.) Ask yourself what number you would multiply by two to get 16. The answer is eight, so you multiply both numbers by eight. (Remember that the fraction ⅛ is equal to 1.)

$$\frac{1 \times 8 = 8}{2 \times 8 = 16}$$

The answer is that ½ inch is equivalent to ⁸⁄₁₆ inch.

Review Questions

Section 4.1.0

Find the equivalents of the following measurements:

1. ¼ inch = ___/16 inch
    a. 2
    b. 4
    c. 6
    d. 8

2. ²⁄₁₆ inch = ___/32 inch
    a. 1
    b. 2
    c. 4
    d. 8

3. ¾ inch = ___/8 inch
    a. 2
    b. 4
    c. 8
    d. 6

4. ¾ inch = ___/64 inch
    a. 48
    b. 32
    c. 16
    d. 12

5. ³⁄₁₆ inch = ___/32 inch
    a. 2
    b. 4
    c. 6
    d. 8

4.2.0 Reducing Fractions to Their Lowest Terms

If you find that the measurement of something is ⁴⁄₁₆, you may want to reduce the measurement to its lowest terms so the number is easier to work with. To find the lowest terms of ⁴⁄₁₆, you will use division:

Step 1 To reduce a fraction, ask yourself, what is the largest number that I can divide evenly into both the numerator and the denominator? If there is no number (other than 1) that will divide evenly into both numbers, the fraction is already in its lowest form.

Step 2 Divide the numerator and the denominator by the same number.

In this example, you could divide both the numerator and the denominator by 4.

$$\frac{4 \div 4 = 1}{16 \div 4 = 4}$$

DID YOU KNOW?
When you cannot think of a larger number to divide by, as long as the numerator and the denominator are both even, you can always divide each by 2.

Review Questions

Section 4.2.0

Find the lowest form of each of the fractions below:

1. $\frac{2}{16} =$ _____
   a. 1
   b. $\frac{1}{4}$
   c. $\frac{1}{8}$
   d. $\frac{1}{2}$

2. $\frac{2}{8} =$ _____
   a. 4
   b. $\frac{1}{2}$
   c. $\frac{1}{16}$
   d. $\frac{1}{4}$

3. $\frac{12}{32} =$ _____
   a. $\frac{1}{2}$
   b. $\frac{6}{16}$
   c. $\frac{2}{6}$
   d. $\frac{3}{8}$

4. $\frac{4}{8} =$ _____
   a. $\frac{1}{2}$
   b. $\frac{1}{4}$
   c. $\frac{2}{4}$
   d. $\frac{1}{8}$

5. $\frac{4}{64} =$ _____
   a. $\frac{1}{2}$
   b. $\frac{1}{16}$
   c. $\frac{1}{8}$
   d. $\frac{4}{8}$

4.3.0 Comparing Fractions and Finding the Lowest Common Denominator

Which measurement is larger, $\frac{3}{4}$ or $\frac{5}{8}$?

To find the answer, think about this question: Would you have more pizza if you had three pieces from a pie that was cut up in four equal slices (see *Figure 12*) or if you had five pieces of the same size pie that was cut up in eighths (see *Figure 13*)?

As you can see, it's hard to compare fractions that do not have common denominators, just as it's hard to compare pizzas that are cut up differently. Using our pizza pie, remember that we are trying to determine which amount of pie is larger:

$$\frac{3}{4} \quad \text{or} \quad \frac{5}{8}$$

To compare, you need to find a common denominator for the pizza slices. The common denominator is a number that both denominators can go into evenly.

Step 1 Multiply the two denominators together ($4 \times 8 = 32$). This is a common denominator between the two fractions.

You found a common denominator so that you can compare the pieces more easily. Now **convert** the two fractions so that they will have the same denominator.

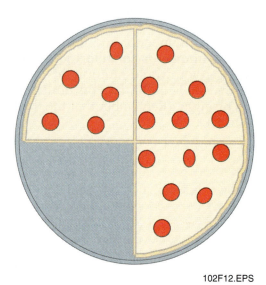

102F12.EPS

Figure 12 ◆ $\frac{3}{4}$ of a pizza.

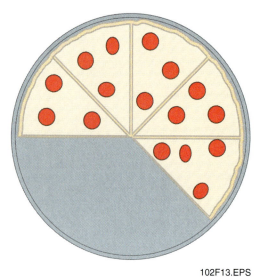

102F13.EPS

Figure 13 ◆ $\frac{5}{8}$ of a pizza.

Step 2 Convert each of the two fractions to fractions having the common denominator of 32.

$$\frac{3 \times 8 = 24}{4 \times 8 = 32}$$

$$\frac{5 \times 4 = 20}{8 \times 4 = 32}$$

Now it's easy to compare the two fractions to see which is larger. You'd have more pizza if you chose ¾ (you'd have ²⁴⁄₃₂ as opposed to ²⁰⁄₃₂ of the pizza).

You have found the common denominator for the pizza problem. However, working with fractions like ²⁴⁄₃₂ or ²⁰⁄₃₂ is difficult. To make this problem easier, you can find the lowest common denominator, which means reducing the fractions to their lowest terms.

To find the lowest common denominator, follow these steps:

Step 1 Reduce each fraction to its lowest terms.

Step 2 Find the lowest common multiple of the denominators. Sometimes this is as simple as one denominator already being a *multiple* of the other, meaning you can multiply by a whole number to get the larger number. If this is the case, all you have to do is find the equivalent fraction for the term with the smaller denominator.

Step 3 If neither of the denominators is a multiple of the other, you can then multiply the denominators together to get a common denominator.

Let's look at the pizzas again. In this example, ¾ and ⅝ are already in their lowest terms. When you look at the denominators, you see that 8 is a multiple of 4. So you should find the equivalent fraction for ¾ that has a denominator of 8.

$$\frac{3 \times 2 = 6}{4 \times 2 = 8}$$

You can now compare ⁶⁄₈ to ⅝, and see that ⁶⁄₈ (which is the same as ¾) is the larger fraction.

Whether you find the lowest common denominator, or just multiply the denominators to find a common denominator will depend on the situation. In some applications, you may want all fractions involved to have a particular denominator. But by finding the lowest common denominator this way, you can decrease the amount of multiplying you need to do, and reduce the chances of making a mathematical error.

Review Questions

Section 4.3.0

Find a common denominator for the following pairs of fractions:

1. $\dfrac{2}{6}, \dfrac{3}{4}$

   The answer is _____.

   a. 6
   b. 10
   c. 12
   d. 16

2. $\dfrac{1}{4}, \dfrac{3}{8}$

   The answer is _____.

   a. 4
   b. 8
   c. 2
   d. 18

3. $\dfrac{1}{8}, \dfrac{3}{6}$

   The answer is _____.

   a. 3
   b. 9
   c. 12
   d. 8

4. $\dfrac{1}{4}, \dfrac{3}{8}$

   The answer is _____.

   a. 18
   b. 8
   c. 14
   d. 20

5. $\dfrac{4}{32}, \dfrac{5}{8}$

   The answer is _____.

   a. 45
   b. 6
   c. 10
   d. 32

4.4.0 Adding Fractions

How many *total inches* will you have if you add $\frac{3}{4}$ of an inch plus $\frac{5}{8}$ of an inch? To answer this question, you will have to add two fractions using the following steps.

Step 1 Find the common denominator of the fractions you wish to add. A common denominator for $\frac{3}{4}$ and $\frac{5}{8}$ is 32.

Step 2 Convert the fractions to equivalent fractions with the same denominator. This is how to convert the fractions to equivalent fractions:

$$\frac{3 \times 8}{4 \times 8} = \frac{24}{32}$$

$$\frac{5 \times 4}{8 \times 4} = \frac{20}{32}$$

Step 3 Add the numerators of the fractions. Place this sum over the denominator.

$$\frac{24}{32} + \frac{20}{32} = \frac{44}{32}$$

Step 4 Reduce the fraction to its lowest terms. When you have done this, there will be no number other than 1 that can go evenly into both the numerator and the denominator. Also, the fraction will not be an **improper fraction** (meaning the numerator will not be larger than the denominator).

When you reduce $\frac{44}{32}$ to $\frac{11}{8}$, it becomes a fraction in which no number other than 1 will go evenly into both the numerator and the denominator. But it is an improper fraction, because the numerator is larger than the denominator.

In this case, you need to reduce the improper fraction $\frac{11}{8}$ to its lowest terms. We will soon learn how to convert improper fractions to **mixed numbers.** For now, try solving the problems with the fractions in the review questions for this section.

Don't forget! The fractions will need common denominators before they can be added.

Review Questions

Section 4.4.0

Find the answers to the following addition problems. Remember to reduce the sum to the lowest terms.

1. $\frac{1}{8} + \frac{4}{16} =$ _____
  a. $\frac{5}{16}$
  b. $\frac{3}{8}$
  c. $\frac{5}{8}$
  d. $\frac{1}{2}$

2. $\frac{4}{8} + \frac{6}{16} =$ _____
  a. $\frac{10}{16}$
  b. $\frac{7}{8}$
  c. $\frac{10}{24}$
  d. $\frac{16}{24}$

3. $\frac{2}{4} + \frac{1}{4} =$ _____
  a. $\frac{3}{4}$
  b. $\frac{3}{8}$
  c. $\frac{1}{4}$
  d. $\frac{3}{16}$

4. $\frac{3}{4} + \frac{2}{8} =$ _____
  a. $\frac{5}{32}$
  b. $\frac{5}{8}$
  c. $\frac{4}{12}$
  d. $\frac{4}{4}$

5. $\frac{14}{16} + \frac{3}{8} =$ _____
  a. $\frac{17}{16}$
  b. $\frac{17}{4}$
  c. $\frac{5}{4}$
  d. $\frac{17}{8}$

4.5.0 Subtracting Fractions

Subtracting fractions is very much like adding fractions. You must find a common denominator before you subtract. Say you have a piece of wood $\frac{7}{8}$ of a foot long. You use $\frac{1}{4}$ of a foot. How much do you have left?

Step 1 Find the common denominator. In this case it is 8.

$$\frac{7}{8} \quad \frac{1}{4}$$

Step 2 Multiply each term of the fraction ($\frac{1}{4}$) by 2 to get a fraction with the common denominator of 8.

$$\frac{1 \times 2}{4 \times 2} = \frac{2}{8}$$

Step 3 Subtract the numerators. $\frac{5}{8}$ of a foot is left.

$$\frac{7}{8} - \frac{2}{8} = \frac{5}{8}$$

Review Questions

Section 4.5.0

Find the answers to the following subtraction problems and reduce the differences to lowest terms:

1. $\frac{3}{8} - \frac{5}{16}$ = _____

a. $\frac{8}{16}$
b. $\frac{1}{16}$
c. $\frac{1}{8}$
d. $\frac{1}{2}$

2. $\frac{11}{16} - \frac{5}{8}$ = _____

a. $\frac{6}{16}$
b. $\frac{3}{4}$
c. $\frac{1}{8}$
d. $\frac{1}{16}$

3. $\frac{3}{4} - \frac{2}{6}$ = _____

a. $\frac{5}{12}$
b. $\frac{1}{6}$
c. $\frac{1}{2}$
d. $\frac{20}{24}$

4. $\frac{11}{12} - \frac{4}{8}$ = _____

a. $\frac{5}{12}$
b. $\frac{7}{8}$
c. $\frac{8}{24}$
d. $\frac{7}{12}$

5. $\frac{11}{16} - \frac{1}{2}$ = _____

a. $\frac{10}{16}$
b. $\frac{3}{16}$
c. $\frac{1}{4}$
d. $\frac{5}{16}$

4.5.1 Subtracting a Fraction from a Whole Number

Sometimes you must subtract a fraction from a whole number. For example, you need to take $\frac{1}{4}$ of a day off from a five-day workweek. How many days will you be working that week?

Here is how to set up this type of problem:

Step 1 To subtract a fraction from a whole number, borrow 1 from the whole number to make it into a fraction.

$$\begin{array}{rcl} 5 & = & 4 + 1 \\ -\frac{1}{4} & & -\frac{1}{4} \\ \hline \end{array}$$

Step 2 Convert the 1 to a fraction having the same denominator as the number you are subtracting.

$$\begin{array}{rcl} 5 & = & 4 + \frac{4}{4} \\ -\frac{1}{4} & & -\frac{1}{4} \\ \hline \end{array}$$

Step 3 Subtract and reduce to the lowest terms.

$$\begin{array}{rcl} 5 & = & 4 + \frac{4}{4} \\ -\frac{1}{4} & & -\frac{1}{4} \\ \hline & & 4 + \frac{3}{4} \text{ working days} \end{array}$$

Review Questions

Section 4.5.1

Find the answers to the following subtraction problems and reduce the fractions to the lowest terms.

1. $\begin{array}{r} 8 \\ -\frac{3}{4} \\ \hline \end{array}$

The answer is _____.

a. $7\frac{1}{4}$
b. $\frac{5}{4}$
c. $7\frac{3}{4}$
d. $5\frac{1}{4}$

2. $\begin{array}{r} 12 \\ -\frac{5}{8} \\ \hline \end{array}$

The answer is _____.

a. $12\frac{5}{8}$
b. $\frac{3}{8}$
c. $11\frac{5}{8}$
d. $11\frac{3}{8}$

3. If two punches, one $4\frac{1}{64}$ inches long and the other $4\frac{3}{32}$ inches long, are made from a bar of stock $9\frac{7}{16}$ inches long, _____ inches of stock are not used.

a. $1\frac{1}{16}$
b. $1\frac{10}{32}$
c. $4\frac{21}{64}$
d. $1\frac{21}{64}$

4. If you saw $12\frac{1}{16}$ inches off a board $20\frac{3}{4}$ inches long, you'll have _____ inches left over.

a. $8\frac{1}{4}$
b. $8\frac{11}{16}$
c. $7\frac{1}{4}$
d. $16\frac{11}{8}$

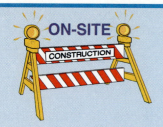

ON-SITE

Fractions at Work

Fractions are parts of a whole, and you will have to use them often on the job. For example, when you mix solutions the measurements of the different parts are fractions. To create a solution for cleaning the surface of a concrete slab, you will mix parts together to make a whole cleaning solution. When preparing a sealer for treating an exterior surface, you will be mixing parts to create a whole.

5. A rough opening for a window measures $36\frac{3}{8}$ inches. The window to be placed in the rough opening measures $35\frac{15}{16}$ inches. The total clearing that will exist between the window and the rough opening will be _______ inch(es).
   a. 1
   b. $1\frac{7}{16}$
   c. $\frac{7}{16}$
   d. $1\frac{12}{16}$

4.6.0 Multiplying Fractions

Multiplying and dividing fractions is very different from adding and subtracting fractions. You do not have to find a common denominator when you multiply or divide fractions.

In a word problem, you usually know to multiply by the word *of*. If a problem asks "What is $\frac{2}{8}$ of 9?" then think of the problem this way: $\frac{2}{8} \times \frac{9}{1}$. Note that any number (except 0) over 1 equals itself.

Using $\frac{4}{8} \times \frac{5}{6}$ as an example, follow these steps:

Step 1 Multiply the numerators together to get a new numerator. Multiply the denominators together to get a new denominator.

$$\frac{4 \times 5}{8 \times 6} = \frac{20}{48}$$

Step 2 Reduce if possible ($\frac{20}{48}$ reduces to $\frac{5}{12}$).

Although you can multiply fractions without first reducing them to their lowest terms, keep in mind that you *can* reduce them before you multiply. This will sometimes make the multiplication easier, since you will be working with smaller numbers. It will also make it easier to reduce the product to the lowest terms. What may seem like an extra step can save you time in the long run.

Review Questions

Section 4.6.0

Find the answers to the following multiplication problems and reduce them to their lowest terms.

1. $\frac{4}{16} \times \frac{5}{8} =$ _____
   a. $\frac{20}{128}$
   b. $\frac{5}{32}$
   c. $\frac{5}{2}$
   d. $\frac{5}{64}$

2. $\frac{3}{4} \times \frac{7}{8} =$ _____
   a. $\frac{21}{32}$
   b. $\frac{5}{6}$
   c. $\frac{21}{12}$
   d. $\frac{6}{7}$

3. $\frac{2}{8}$ of 15 is _____
   a. $\frac{30}{8}$
   b. 60
   c. $\frac{1}{4}$
   d. $\frac{15}{4}$

4. $\frac{5}{16}$ of 12 is _____.
   a. $\frac{17}{16}$
   b. $\frac{15}{4}$
   c. $\frac{60}{16}$
   d. $\frac{60}{4}$

5. $\frac{3}{4}$ of 24 is ____.
   a. 72
   b. 12
   c. 18
   d. 6

4.7.0 Dividing Fractions

Dividing fractions is very much like multiplying fractions, with one difference. You must **invert,** or flip, the fraction you are dividing by. Using $\frac{1}{2} \div \frac{3}{4}$ as an example, follow these steps:

Step 1 Invert the fraction you are dividing by ($\frac{3}{4}$).

$$\frac{3}{4} \text{ becomes } \frac{4}{3}$$

Step 2 Change the division sign ($\div$) to a multiplication sign ($\times$).

$$\frac{1 \div 3}{2 \div 4} = \frac{1 \times 4}{2 \times 3}$$

Step 3 Multiply the fraction as instructed earlier.

$$\frac{1 \times 4}{2 \times 3} = \frac{4}{6}$$

Step 4 Reduce if possible.

$$\frac{4}{6} \text{ reduces to } \frac{2}{3}$$

If you are working with a mixed number (for example, $2\frac{1}{3}$), you must convert it to a fraction before you invert it. Do this by multiplying the denominator by the whole number (3×2), adding the numerator $[(3 \times 2) + 1]$, and placing the result over the denominator. It looks like the following:

$$2\frac{1}{3} = \frac{(3 \times 2) + 1}{3} = \frac{7}{3}$$

When dividing by a whole number, place the whole number over 1 and then invert it. Remember that $\frac{4}{1}$ is the same as 4.

For example: $\frac{1}{2} \div 4 =$

$\frac{1}{2} \div \frac{4}{1} =$

$\frac{1}{2} \times \frac{1}{4} = \frac{1}{8}$

Review Questions

Section 4.7.0

Find the answers to the following division problems and reduce them to lowest terms (no improper fractions allowed).

1. $\frac{3}{8} \div 3 =$ _____
   a. $\frac{3}{24}$
   b. $\frac{3}{4}$
   c. $\frac{1}{8}$
   d. $\frac{9}{8}$

2. $\frac{5}{8} \div \frac{1}{2} =$ _____
   a. $\frac{5}{16}$
   b. $\frac{6}{10}$
   c. $1\frac{2}{8}$
   d. $1\frac{1}{4}$

3. $\frac{3}{4} \div \frac{3}{8} =$ _____
   a. $\frac{24}{12}$
   b. $\frac{9}{32}$
   c. 2
   d. $\frac{8}{24}$

4. On a scale drawing, if $\frac{1}{4}$ of an inch represents a distance of 1 foot, then a line on the drawing measuring $8\frac{1}{2}$ inches long represents _____ feet.
   a. 34
   b. $2\frac{1}{8}$
   c. 17
   d. $8\frac{1}{2}$

5. You can cut _____ $\frac{7}{8}$-inch lengths from a 7-inch strip.
   a. $6\frac{1}{8}$
   b. 7
   c. 14
   d. 8

5.0.0 ◆ DECIMALS

Decimals represent values less than one whole unit. You are already familiar with decimals in the form of money.

25¢ = 0.25 or 25/100
10¢ = 0.10 or 10/100
50¢ = 0.50 or 50/100

On the job, you may need to use decimals to read instruments or calculate flow rates. Look at the scale on a typical **machinist's rule,** as shown in *Figure 14*. Each number shows the distance, in inches, from the squared end of the rule. The marks between the numbers divide each inch into ten equal parts. These ten parts are referred to as tenths.

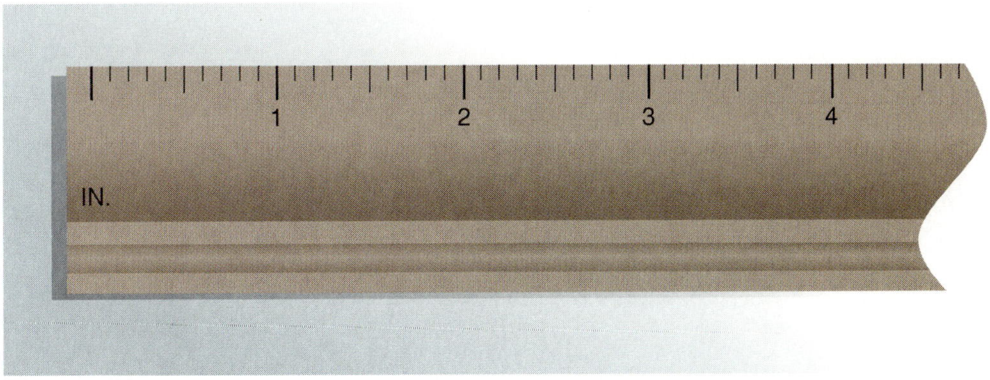

102F14.EPS

Figure 14 ◆ The machinist's rule (divided into tenths).

In *Figure 15,* the nail spans one whole inch plus three-tenths of a second inch. It is one and three-tenths of an inch long. This is written as 1.3 inches.

Review Questions

Section 5.0.0

Use the machinist's rule in *Figure 16* to find the answers to the following questions:

1. P is at the _____ tenths of an inch mark.
   a. 9
   b. 99
   c. 8
   d. 1.9

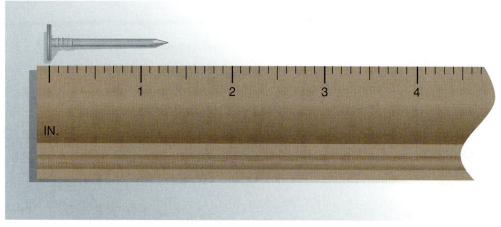

102F15.EPS

Figure 15 ◆ Showing 1.3 inches on a machinist's rule.

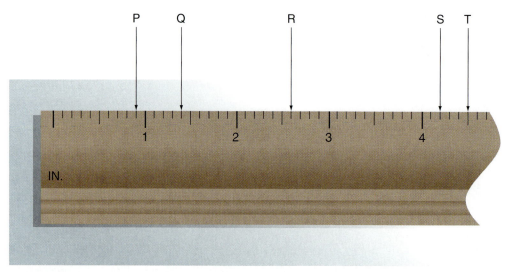

102F16.EPS

Figure 16 ◆ Review questions machinist's rule.

2. Q is at the _____-inch mark.
   a. .4
   b. 1.5
   c. 1.4
   d. 4

3. R is at the _____ tenths of an inch mark.
   a. 260
   b. 26
   c. 6
   d. 16

4. S is at the _____-inch mark.
   a. 42
   b. 4.2
   c. 24
   d. 2.4

5. T is at the _____-inch mark.
   a. 5
   b. ⅘
   c. ⁴⁵⁄₁₀
   d. ⁴⁰⁄₁₀

5.1.0 Comparing Whole Numbers with Decimals

The following chart compares whole number place values with decimal place values:

Whole Numbers		Decimals	
1	ones		
10	tens	.1	tenths
100	hundreds	.01	hundredths
1000	thousands	.001	thousandths

To read a decimal, say the number as it is written and then the name of its place value. For example, read 0.56 as "fifty-six hundredths."

Mixed numbers also appear in decimals. You read 15.7 as "fifteen and seven-tenths." Notice the use of the word "and" to separate the whole number from the decimal.

Review Questions

Section 5.1.0

For the following problems, find the words that mean the same as the decimal or the decimal equivalent of the words.

1. 0.4 = _____
   a. four
   b. four-tenths
   c. four-hundredths
   d. four-thousandths

2. 0.05 = _____
   a. five
   b. five-tenths
   c. five-hundredths
   d. five-thousandths

3. 2.5 = _____
   a. two and five-tenths
   b. two and five-hundredths
   c. two and five-thousandths
   d. twenty-five-hundredths

4. eighteen-hundredths = _____
   a. 1.8
   b. .18
   c. .018
   d. .0018

5. five and eight-tenths = _____
   a. 58.1
   b. 5.8
   c. 5.08
   d. 5.008

ON-SITE

Measuring the Thickness of a Coating

Coating thickness is important because either too little or too much can cause problems. A coating such as paint needs a minimum thickness to prevent corrosion, withstand abrasion, and look good. A coating that is too thick may crack, flake, blister, or not cure properly.

Many jobs have requirements that specify the thickness of the coating applied to an object or surface. To ensure that the specifications are met, periodic checks of the wet-film thickness can be made using a wet-film thickness gauge. Typically these gauges use measurements in mils or microns. For example, the required wet-film thickness for a coat of paint may be 10.25 mils.

5.2.0 Comparing Decimals with Decimals

Which decimal is the larger of the two?

0.4 or 0.42?

Here's how to compare decimals:

Step 1 Line up the decimal points of all the numbers.

0.4

0.42

Step 2 Place zeros to the right of each number until all numbers end with the same place value.

0.40

0.42

Step 3 Compare the numbers.

0.42 (42 hundredths) is larger than 0.40 (40 hundredths).

Review Questions

Section 5.2.0

For the following problems, put the decimals in order from *smallest* to *largest*.

1. 0.400, 0.004, 0.044, 0.404

   The answer is _____.

   a. 0.400, 0.004, 0.044, 0.404
   b. 0.004, 0.044, 0.404, 0.400
   c. 0.004, 0.044, 0.400, 0.404
   d. 0.404, 0.044, 0.400, 0.004

2. 0.567, 0.059, 0.56, 0.508

   The answer is _____.

   a. 0.508, 0.56, 0.567, 0.059
   b. 0.059, 0.56, 0.508, 0.567
   c. 0.567, 0.059, 0.56, 0.508
   d. 0.059, 0.508, 0.56, 0.567

3. 0.320, 0.032, 0.302, 0.003

   The answer is _____.

   a. 0.003, 0.032, 0.302, 0.320
   b. 0.320, 0.302, 0.032, 0.003
   c. 0.302, 0.320, 0.003, 0.032
   d. 0.003, 0.032, 0.320, 0.302

4. 0.867, 0.086, 0.008, 0.870

   The answer is _____.

   a. 0.870, 0.867, 0.086, 0.008
   b. 0.008, 0.086, 0.867, 0.870
   c. 0.086, 0.008, 0.867, 0.870
   d. 0.008, 0.870, 0.867, 0.086

5. 0.626, 0.630, 0.616, 0.641

   The answer is _____.

   a. 0.616, 0.641, 0.630, 0.626
   b. 0.616, 0.626, 0.630, 0.641
   c. 0.061, 0.616, 0.626, 0.630
   d. 0.630, 0.616, 0.626, 0.641

5.3.0 Adding and Subtracting Decimals

There is only one major rule to remember when adding and subtracting decimals:

Keep your decimal points lined up!

Suppose you want to add 4.76 and 0.834. Line up the problem like this:

```
  4.760      You can add a 0 to help keep
+0.834       the numbers lined up.
  5.594
```

The same thing is true for subtraction of decimals. Line up the decimal points.

```
  5.6       5.600     Notice that two zeros were
- 2.724    -2.724     added to the end of the first
            2.876     number to make it easier to
                      see where you need to borrow.
```

DID YOU KNOW?

Place-Value Systems

As mathematics advanced, counting systems became more efficient for performing calculations and solving problems. These systems made it easier to represent large numbers and to simplify the process of computing. These were called place-value systems. In a place-value system, the value of a particular symbol depends not only on the symbol but also on its position or place in the number.

In our decimal system, each place value is ten times greater than the place to the right. Place values in the decimal system are:

1000	100	10	1	
5	3	4	9	= 5,349

Review Questions

Section 5.3.0

Find the answers to these addition and subtraction problems, and don't forget to line up the decimal points.

1. 2.50
    4.20
    +5.00

    The answer is _____.

   a. 11.7
   b. 10.7
   c. 117
   d. 170

2. $1.82 + 3.41 + 5.25 =$ _____
   a. 9.48
   b. 10.48
   c. 9.148
   d. .948

3. $6.43 + 86.4 =$ _____
   a. 92.83
   b. 250.7
   c. 25.07
   d. 92.8

4. The combined thickness of a piece of sheet metal 0.078-inch thick and a piece of band iron 0.25-inch thick is _____.
   a. 1.03
   b. .328
   c. 3.28
   d. .103

5. Yesterday, a lumberyard contained 6.7 tons of wood. Since then, 2.3 tons were removed. The lumberyard now contains ____ tons of wood.
   a. 9
   b. 4.4
   c. .9
   d. 44

5.4.0 Multiplying Decimals

While unloading wood panels, you measure one panel as 4.5 feet wide. You have seven panels the same width. What is the total width if you put the panels side-by-side?

Step 1 Set up the problem just like the multiplication of whole numbers.

$$\begin{array}{r} 4.5 \\ \times\ 7 \\ \hline \end{array}$$

Step 2 Proceed to multiply.

$$\begin{array}{r} 4.5 \\ \times\ 7 \\ \hline 315 \end{array}$$

Step 3 Once you have the answer, count the number of digits to the right of the decimal point in both numbers being multiplied. (In this example, there is only one decimal point (4.5) and only one number to the right of it.)

Decimals at Work

When are you going to use decimals on the job? Here are two examples of using decimals to get your work done:

- You are installing a boiler to specifications on its concrete pad. When measuring with your level, you find that one corner of the boiler is level within .003 inch. Another corner is .005 too high, and a third corner is .001 too high. The concrete pad is not as even as it should be. To adjust for this, you must place **shims** between the boiler base and the concrete pad. From your box of shim stock, how many shims will it take to make the boiler level? [Shims are in .001- and .002-inch widths.]
- You are working on a conveyor system. You have checked the lubricant by taking an oil sample and measuring the metal particles with a **micrometer.** You found metal particles that are .0001 in size. The system currently has a 20-micron filter installed to filter pieces bigger than .0002 inch. What size filter do you need to filter the .0001 metal particles out of the oil?

Step 4 In the answer, count over the same number of digits (from right to left) and place the decimal point there.

$$
\begin{array}{r}
4.5 \\
\times \quad 7 \\
\hline
31.5
\end{array}
$$

(count one total digit to the right of the decimal point in the two numbers)

(count in one digit from right to left in the answer, and place the decimal point there)

Note

You may have to add a zero if there are more digits to the right of the decimal points than there are in the answer, as shown in the following example.

$$
\begin{array}{r}
0.507 \\
\times 0.022 \\
\hline
1014 \\
1014 \\
000 \\
\hline
11154 = .011154
\end{array}
$$

(Add the total digits to the right of the decimal point in the two numbers: There are six.)

(Count six digits from right to left in the product. In this case, you'll need to add a zero.)

Review Questions

Section 5.4.0

Use the following to answer questions 1 and 2. You are machining a part. The starting thickness of the part is 6.18 inches. You take three cuts. Each cut is three-tenths of an inch.

1. You have removed _____ inches of material.
   a. 6.09
   b. .09
   c. .9
   d. .18

2. The remaining thickness of the part is _____ inches.
   a. 6.10
   b. 6.08
   c. 6.15
   d. 5.28

Use the following to answer questions 3 and 4. An electrician wants to know if a light circuit is overloaded. The circuit supplies two different machines.

3. The first machine has 11 bulbs lit. Each bulb uses 4.68 watts. The lights on the first machine need _____ watts.
   a. 5.148
   b. 51.48
   c. 514.8
   d. 468

4. The second machine has 7 bulbs lit. Each of these bulbs uses 5.14 watts. The lights on both machines need a total of _____ watts.
   a. 87.46
   b. 874.6
   c. 35.98
   d. 176.76

5. Ceramic tile weighs 4.75 pounds per square foot. Therefore, 128 square feet of ceramic tile weighs _____ pounds.
   a. 6,080
   b. 608
   c. 90.8
   d. 908

5.5.0 Dividing with Decimals

When would you divide with decimals? An example is if you need to cut a 44.5-inch pipe into as many 22-inch pieces as possible. How many 22-inch pieces will you be able to cut? How much will be left over?

There are three types of division problems involving decimals:

- Those that have a decimal point in the number being divided (the dividend)

$$22 \overline{)44.5}$$

- Those that have a decimal point in the number you are dividing by (the divisor)

$$0.22 \overline{)4,450}$$

- Those that have decimal points in both numbers (the dividend and the divisor)

$$0.22 \overline{)44.5}$$

5.5.1 Dividing with a Decimal in the Number Being Divided

For the first type of problem, let's use 44.5 ÷ 22 as our example.

Step 1 Place a decimal point directly above the decimal point in the dividend.

$$22 \overline{)44.\overset{.}{5}}$$

Step 2 Divide as usual.

$$\begin{array}{r} 2.0 \\ 22 \overline{)44.5} \\ -44 \\ \hline 00.5r \end{array}$$

How many 22-inch pieces of pipe will you have? The answer: two, with a little (.5 inch) left over.

Review Questions

Section 5.5.1

Find the answers to the following division problems, and don't go any further than the hundredths (.01) place, unless otherwise noted.

1. 45.36 ÷ 18 = _____
   a. .025
   b. .25
   c. 2.52
   d. 25.20

2. 4.536 ÷ 18 = _____
   a. .025
   b. .25
   c. 2.52
   d. 25.20

3. 0.4536 ÷ 18 = _____ [To nearest thousandths (.001) place]
   a. .025
   b. .252
   c. 2.520
   d. 25.205

4. $25 \overline{)10.20}$
   a. .48
   b. .40
   c. 4.08
   d. 40.08

5. $6 \overline{)31.2}$
   a. 5.02
   b. .52
   c. 5.2
   d. 5.22

Equivalents

Forty-four and one-half is the same as forty-four point five. If you measure a piece of pipe and the tape measure says it is 44½ inches, another way to say this is "forty-four point five," which is the decimal equivalent. The chart below shows some other decimal equivalents.

Chart of Equivalents

½ = .5	¼ = .25	⅛ = .125	¹⁄₁₆ = .0625

5.5.2 Dividing with a Decimal in the Number You Are Dividing By

For the second type of problem, let's use 4,450 ÷ .22 as our example.

Step 1 Move the decimal point in the divisor to the right until you have a whole number.

$$.22\,\overline{)\,4450.}$$ (The decimal point in the divisor will be moved two places to the right.)

Step 2 Move the decimal point in the dividend the same number of places to the right. You may have to add zeros first. Then divide as usual.

```
            20227.2
22 ) 4450.00.0      (After adding zeros, move
    −44             decimal in dividend two
      0050          places to right so number
     −0044          becomes 445,000.)
       00060
      −00044
        000160
       −000154
         0000060
        −0000044
          0000016r
```

Review Questions

Section 5.5.2

Perform the following division problems. Don't go any further than the hundredths (.01) place in your answer:

1. 282 ÷ 14.1 = _____
   a. .2
   b. 2
   c. 20
   d. 200

2. 694 ÷ 3.2 = _____
   a. 216.87
   b. 21.68
   c. 5.2
   d. 25.28

3. 99 ÷ .45 = _____
   a. 44.55
   b. 2.2
   c. 220
   d. 22

4. $2.5\,\overline{)\,102}$
   a. 408
   b. 40.8
   c. 4.8
   d. 4.08

5. $0.6\,\overline{)\,312}$
   a. 1,872
   b. 187.2
   c. 52
   d. 520

5.5.3 Dividing with Decimals in Both Numbers

For the third type of problem, let's use 44.5 ÷ .22 as our example.

Step 1 Move the decimal point in the divisor to the right until you have a whole number.

$$.22\,\overline{)\,44.5}$$ (Moving the decimal point in the divisor to the right, .22 becomes 22.)

Step 2 Move the decimal point in the dividend the same number of places to the right. Then divide as usual.

$$\begin{array}{r} 202. \\ 22\,\overline{)\,4450.} \end{array}$$ (Moving the decimal point in the dividend two places to the right, 44.5 becomes 4,450.)

Now you can see that 44.5 ÷ .22 is 202.

Review Questions

Section 5.5.3

Find the answers to the following division problems, and don't go any further than the hundredths (.01) place.

1. 20.82 ÷ 4.24 = _____
   a. 4.91
   b. .49
   c. 49.1
   d. 491.03

2. 38.9 ÷ 3.7 = _____
   a. 14.39
   b. 143.93
   c. 1.05
   d. 10.51

3. 9.9 ÷ .45 = _____
   a. 2.2
   b. 9.45
   c. 22
   d. 4.45

4. $.25\,\overline{)10.20}$
   a. .02
   b. 40.8
   c. 4.08
   d. .09

5. $0.6\,\overline{)31.2}$
   a. 52
   b. 520
   c. 5.2
   d. .52

5.6.0 Rounding Decimals

Sometimes the answer is a bit more precise than you require. For example, you need to cut a 107.5-inch pipe into as many 4.25-inch pieces as possible. How many 4.25-inch pieces will you be able to cut?

The precise answer is 25.29411764. But you probably only need to measure it to the nearest tenth. What would you do? For this exercise, you will round 25.29411764 to the nearest tenth (0.1):

Step 1 Underline the place to which you are rounding.

25.29411764

Step 2 Look at the digit one place to its right.

25.29411764

Step 3 If the digit to the right is 5 or more, you will round up by adding 1 to the underlined digit. If the digit is less than 5, leave the underlined digit the same. In this example, the digit to the right is 9, which is more than 5, so you round up by adding 1 to the underlined digit.

25.39411764

Step 4 Drop all other digits to the right.

25.3

Review Questions

Section 5.6.0

Solve these problems to practice rounding decimals.

DID YOU KNOW?

When calculating a division problem by hand, it is usually only necessary to divide an answer out to one more digit than you want in your final answer. Then you can round the final number out to the required number of places.

1. You need to cut a 90.5-inch pipe into as many 3.75-inch pieces as possible. You will be able to cut _____ 3.75-inch pieces. (Round your answer to the nearest tenth.)
   a. 4
   b. 24.1
   c. 240
   d. 20.4

2. If you drove your car 622 miles on 40.1 gallons of gas, you got ___ miles per gallon. (Round your answer to the nearest tenth.)
   a. 15.5
   b. 15.511
   c. 1.5
   d. 155.1

3. If wire costs $4.30 per pound and you pay a total of $120.95, then you have purchased _____ pounds of wire. (Round your answer to the nearest tenth.)
   a. 28.11
   b. 2.8
   c. 281.1
   d. 28.1

4. Vent pipe is on sale at XYZ Supply Company this week for $.37 per linear foot. If you spend $115.38 you will purchase _____ linear feet of pipe. (Round your answer to the nearest tenth.)
   a. 38.11
   b. 31.2
   c. 311.8
   d. 318

5. Vent pipe at XYZ Supply normally costs $.48 per linear foot. If you spend the same amount of money ($115.38) when vent pipe is not on sale, you will purchase _____ linear feet of pipe. (Round your answer to the nearest tenth.)
   a. 240
   b. 240.4
   c. 24.4
   d. 240.3

5.7.0 Using the Calculator to Add, Subtract, Multiply, and Divide Decimals

Performing operations on the calculator using decimals is very much like performing the operations on whole numbers. Follow these steps using the problem 45.6 + 5.7 as an example.

Step 1 Turn the calculator *on*. A zero (0) appears in the display.

Step 2 Press 45.6. A 45.6 appears in the display.

Step 3 Press the + key. The 45.6 is still displayed.

Note

For this step, press whichever operation key the problem calls for: + to add, − to subtract, × to multiply, ÷ to divide.

Step 4 Press 5.7. A 5.7 is displayed.

Step 5 Press the = key. After you press the = key, whether you are adding, subtracting, multiplying, or dividing, the answer will appear on your display.

$$45.6 + 5.7 = 51.3$$
$$45.6 - 5.7 = 39.9$$
$$45.6 \times 5.7 = 259.92$$
$$45.6 \div 5.7 = 8$$

Step 6 Press the *ON/C* key to clear the calculator.

Review Questions

Section 5.7.0

Use your calculator to find the answers to the following problems, and round your answers to the nearest hundredth (.01).

1. 45.89
    + 7.85

    The answer is _____.

    a. 38.04
    b. 360.24
    c. 5.85
    d. 53.74

2. 7.6
    × .12

    The answer is _____.

    a. 7.72
    b. .91
    c. 7.48
    d. 63.33

3. 685.79
    − 56.266

    The answer is _____.

    a. 742.06
    b. 12.19
    c. 629.524
    d. 629.52

4. 6.45 ÷ 3.25 =

    The answer is _____.

    a. 1.98
    b. 2
    c. 3.2
    d. 20.96

5. 34.76
    + 3.64

    The answer is _____.

    a. 31.12
    b. 9.55
    c. 38.4
    d. 3.84

6.0.0 ◆ CONVERSION PROCESSES

Sometimes you will be faced with a situation in which you need to convert some of the numbers you want to work with so that all your numbers appear in the same form. For example, you may have some numbers that appear as decimals, some that appear as **percents,** and some that appear as fractions. Decimals, percents, and fractions are all just different ways of expressing the same thing. The decimal .25, the percent 25%, and the fraction ¼ all mean the same thing. In order to work successfully with the different forms of numbers like these, you will need to know how to convert them from one form into another.

6.1.0 Converting Decimals to Percents and Percents to Decimals

What are percents? Think of a whole number divided into 100 parts. You can express any part of the whole as a percent. Let's look at an example: The tank shown in *Figure 17* has a capacity of 100 gallons. It is now filled with 50 gallons. What percent of the tank is filled?

If you answered 50 percent (50%), you are correct. Percent means out of 100. How many gallons out of 100 does the tank contain? It contains 50 out of 100, or 50 percent. Percents are an easy way to express parts of a whole. Decimals and fractions also express parts of a whole. Let's look at the relationship among percents, decimals, and fractions.

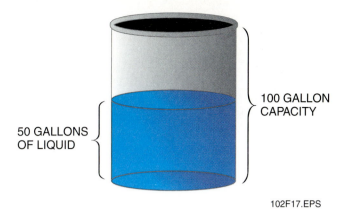

Figure 17 ◆ 100-gallon-capacity tank.

The tank in *Figure 17* is 50 percent full. If you expressed this as a fraction you would say it was ½ full. You could also express this as a decimal and say it's 0.50 full.

Sometimes you may need to express decimals as percents or percents as decimals. Suppose you are preparing a gallon of cleaning solution. The mixture should contain 10 to 15 percent of cleaning agent. The rest should be water. You have 0.12 gallon of cleaning agent. Will you have enough to prepare a gallon of the solution? To answer the question, you must convert a decimal (0.12) to a percent. You will change 0.12 to a percent for this exercise:

Step 1 Multiply the decimal by 100. (Move the decimal point two places to the right.)

$$0.12 \times 100 = 12$$

Step 2 Add a % sign.

12%

You have enough cleaning agent to make the solution. Recall that the mixture should be from 10 to 15 percent cleaning agent. You have 12 percent.

You may also need to convert percents to decimals. Let's say that another mixture should contain 22 percent of a certain chemical by weight. You're making 1 pound of the mixture. You weigh the ingredients on a digital scale. How much of the chemical should you add? To answer this, you must convert a percent (22%) to a decimal. You will change 22 percent to a decimal in the following exercise:

Step 1 Drop the % sign.

22

Step 2 Divide the number by 100. (Move the decimal point two places to the left.)

$$22 \div 100 = 0.22$$

The answer to the problem is that you would add 0.22 pounds of the chemical to .88 pounds of the other ingredient to make a 22 percent mixture.

Review Questions

Section 6.1.0

Find the answers to the following conversion problems:

1. 0.62 = _____
   a. 6.2 percent
   b. 62 percent
   c. 620 percent
   d. 0.62 percent

2. 0.475 = _____
   a. 4.75 percent
   b. 40.75 percent
   c. 475 percent
   d. 47.5 percent

3. 0.7 = _____
   a. .7 percent
   b. 7 percent
   c. 70 percent
   d. 700 percent

4. 72% = _____
   a. .072
   b. .72
   c. 7.2
   d. 72

5. 12.5% = _____
   a. .125
   b. .0125
   c. 1.25
   d. 12.5

6.2.0 Converting Fractions to Decimals

You will often need to change a fraction to a decimal. For example, you need ¾ of a dollar. How do you convert ¾ to its decimal equivalent?

Step 1 Divide the numerator of the fraction by the denominator.

$$4\overline{)3.0}$$

In this example, you need to put the decimal point and the zero after the number 3, because you need a number large enough to divide by 4.

Step 2 Put the decimal point directly above its location within the division symbol.

$$\overset{.?}{4\overline{)3.0}}$$

Step 3 Once the decimal point is in its proper place above the line, you can divide as you normally would. The decimal point *holds* everything in place.

$$
\begin{array}{r}
.75 \\
4\overline{)3.00} \\
-2.8 \\
\hline
0.20 \\
-0.20 \\
\hline
0.00
\end{array}
$$

Step 4 Read the answer. The fraction ¾ converted to a decimal is 0.75. In relation to the earlier problem: ¾ of a dollar is the same as $0.75.

Review Questions

Section 6.2.0

Find the answers to the following conversion problems:

1. ¼ = _____
   a. .25
   b. .75
   c. 2.5
   d. 1.4

2. ¾ = _____
   a. 3.25
   b. .75
   c. .66
   d. 7.5

3. ⅛ = _____
   a. 1.25
   b. .18
   c. .125
   d. 1.8

4. ⁵⁄₁₆ = _____
   a. .516
   b. 3.125
   c. .156
   d. .3125

5. ²⁰⁄₆₄ = _____
   a. .1235
   b. .3125
   c. 2.64
   d. 1.235

6.3.0 Converting Decimals to Fractions

Let's say you have .25 of a dollar. What fraction of a dollar is that? Follow these steps to find out:

Step 1 Say the decimal in words.

.25 is expressed as "twenty-five hundredths"

Step 2 Write the decimal as a fraction.

.25 is written as a fraction as ²⁵⁄₁₀₀

Step 3 Reduce it to its lowest terms.

$$\frac{25}{100} = \frac{25 \div 25}{100 \div 25} = \frac{1}{4}$$

Step 4 Read that .25 converted to a fraction is ¼. If you have .25 of a dollar, you have ¼ of a dollar.

Review Questions

Section 6.3.0

Find the answers to the following conversion problems and express them in lowest terms:

Fractions and Decimals

Remember that fractions and decimals are interchangeable. If it is hard to multiply a measurement using a fraction, convert it to decimal form. For example, it may be easier to multiply a number by .001 rather than by 1/1,000

When using a calculator, you will need to convert fractions into decimals to perform your calculations.

1. 0.5 = _____
   a. $\frac{5}{10}$
   b. $\frac{5}{100}$
   c. $\frac{1}{2}$
   d. $\frac{1}{20}$

2. 0.12 = _____
   a. $\frac{12}{10}$
   b. $\frac{6}{5}$
   c. $\frac{12}{100}$
   d. $\frac{3}{25}$

3. 0.125 = _____
   a. $\frac{12}{5}$
   b. $\frac{5}{12}$
   c. $\frac{1}{4}$
   d. $\frac{1}{8}$

Convert the following mixed decimals (whole numbers with decimals) to their equivalent improper fractions (expressed in lowest terms):

4. 2.8 = _____
   a. $\frac{14}{5}$
   b. $\frac{4}{5}$
   c. $\frac{28}{10}$
   d. $\frac{10}{8}$

5. 5.05 = _____
   a. $\frac{6}{20}$
   b. $\frac{55}{20}$
   c. $\frac{5}{100}$
   d. $\frac{101}{20}$

6.4.0 Converting Inches to Decimal Equivalents in Feet

What happens if you need to convert inches to their decimal equivalents in feet? For example: 3 inches equals what decimal equivalent in feet?

Here's a hint: First, express the inches as a fraction that has 12 as the denominator. You use 12 because there are 12 inches in a foot. Then reduce the fraction, and convert it to a decimal.

In this example, the fraction $\frac{3}{12}$ reduces to $\frac{1}{4}$.

You convert the fraction $\frac{1}{4}$ to a decimal by dividing the 4 into 1.00:

$$
\begin{array}{r}
.25 \\
4\,\overline{)\,1.00} \\
-0.8 \\
\hline
0.20 \\
-0.20 \\
\hline
0
\end{array}
$$

Thus, 3 inches converts to .25 feet.

Section 6.4.0

Find the answers to the following conversion problems, and round them to the nearest hundredth:

1. 9 inches = _____ feet
   a. .75
   b. .90
   c. .92
   d. .12

2. 10 inches = _____ feet
   a. .10
   b. .38
   c. .83
   d. .28

3. 2 inches = _____ feet
   a. .17
   b. .06
   c. .22
   d. .16

4. 4 inches = _____ feet
   a. .33
   b. 3.3
   c. .44
   d. .43

5. 8 inches = _____ feet
   a. 1.5
   b. .67
   c. .66
   d. .80

DID YOU KNOW?

HVAC Technicians

The heating, ventilating, and air conditioning (HVAC) trade is really many trades. It requires electrical, plumbing, carpentry, welding, and some insulation and sheetmetal work. HVAC technicians install, maintain, diagnose, and correct problems in heating and cooling systems. To do this they work with many mechanical, electrical, and electronic components. HVAC systems can involve electricity, chemicals and gases, oil, water, or coal. Technicians may specialize in new installations or maintenance of existing climate-control systems. HVAC technicians must have good mathematical skills so they can perform accurate installations and operate precision testing equipment to ensure that the systems function properly.

7.0.0 ◆ INTRODUCTION TO THE METRIC SYSTEM

The metric system is a system of measurement that uses a base-ten method of determining weight, length, volume, and temperature. That means that all measurements are counted in tens. Much of the world uses the metric system, and the company you work for may do business in places where the metric system is in use. So you need to learn the metric system in case the projects you are working on for your company rely on metric system measures.

You may be surprised to find you are already familiar with some of the common metric units. For example, have you purchased a 2-liter bottle of soda? Have you run a 10K (kilometer) race lately?

7.1.0 Units of Weight, Length, Volume, and Temperature

The name of each metric measurement (see *Figure 18*) tells you two things:

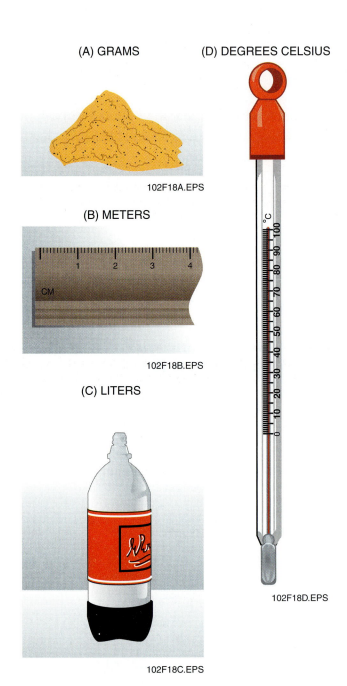

(A) GRAMS

102F18A.EPS

(B) METERS

CM

102F18B.EPS

(C) LITERS

102F18C.EPS

(D) DEGREES CELSIUS

°C

102F18D.EPS

Figure 18 ◆ What type of measurement is it?

- What type of measurement it is (the basic unit):

  Grams = Weight   Meters = Length
  Liters = Volume   Celsius = Temperature

- Its size (in relation to the basic unit, such as the meter):

  deka (da) = 10   deci (d)   = .1
  hecto (h) = 100   centi (c) = .01
  kilo (k)   = 1,000   milli (m) = .001
  mega (M)   = 1,000,000   micro (μ) = .000001

The most common prefixes are kilo, milli, and centi. Hecto is used mainly with meters in calculating land size. Mega and micro are used in scientific and engineering measurements.

Use *Tables 1, 2,* and *3* to learn the relationship between the names for different units of measurement.

Table 1   Weight Units

1 Kilogram	=	1,000 Grams
1 Hectogram	=	100 Grams
1 Dekagram	=	10 Grams
1 Gram	=	1 Gram
1 Decigram	=	0.1 Gram
1 Centigram	=	0.01 Gram
1 Milligram	=	0.001 Gram

Table 2   Length Units

1 Kilometer	=	1,000 Meters
1 Hectometer	=	100 Meters
1 Dekameter	=	10 Meters
1 Meter	=	1 Meter
1 Decimeter	=	0.1 Meter
1 Centimeter	=	0.01 Meter
1 Millimeter	=	0.001 Meter

Table 3   Volume Units

1 Kiloliter	=	1,000 Liters
1 Hectoliter	=	100 Liters
1 Dekaliter	=	10 Liters
1 Liter	=	1 Liter
1 Deciliter	=	0.1 Liter
1 Centiliter	=	0.01 Liter
1 Milliliter	=	0.001 Liter

Measurement Memory Tools

Here's a way to help you memorize the metric prefixes:

If you won $10, you'd buy a *deck of* cards.	deka = 10
If you won $100, you'd have a *heckuva* good time.	hecto = 100
If you won $1,000, you might *keel over*.	kilo = 1,000
If you won $1,000,000, you'd be *mega*-rich.	mega = 1,000,000

Make up your own memorizing tool if you like.
And a memory tool for the smaller units:

Desi sent Milli to *Micronesia*.

From small to smaller, that's

Desi	deci	=	.1	(1 tenth)
sent	centi	=	.01	(1 hundredth)
Milli	milli	=	.001	(1 thousandth)
Micronesia	micro	=	.000001	(1 millionth)

Here are a few more memory tools to help you look through the metric system (see *Figure 19*).

A gram is a little more than the weight of a paper clip.

A kilogram is a little more than two pounds (about 2.2 pounds).

Five milliliters make a teaspoon.

A liter is a little larger than a quart (about 1.06 quarts).

A millimeter is about the size of the diameter of a paper clip wire.

A centimeter is a little more than the width of a paper clip (about 0.4 inch).

A meter is a little longer than three feet (about 1.1 yards).

A kilometer is a little over a half of a mile (about 0.6 of a mile).

Review Questions

Section 7.1.0

1. A **deka**gram is _____ gram(s).
   a. .10
   b. 1
   c. 10
   d. 100

2. A **hecto**gram is _____ gram(s).
   a. 1.0
   b. 100
   c. .100
   d. 10,000

3. A **mega**gram is _____ grams.
   a. .10
   b. .00000001
   c. 100,000
   d. 1,000,000

4. A **deci**meter is _____ meter(s).
   a. .1
   b. .01
   c. 1.0
   d. .0001

5. A **milli**meter is _____ meter(s).
   a. 100,000
   b. 10
   c. .00001
   d. .001

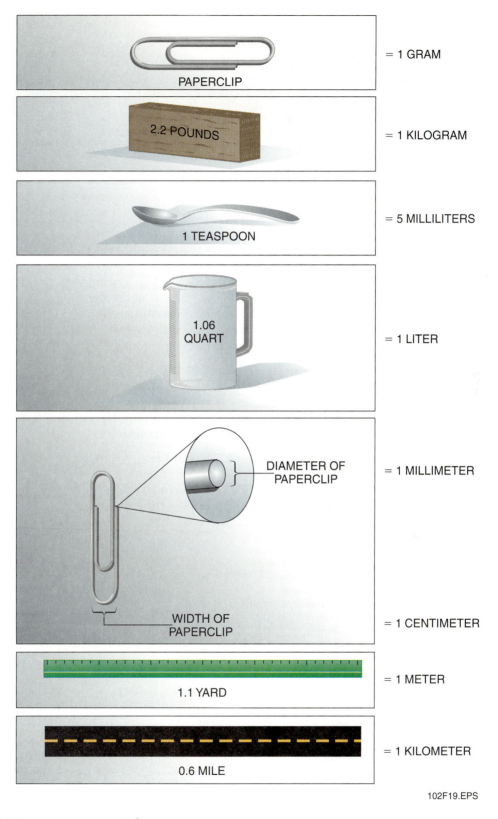

Figure 19 ◆ Metric measure memory tools.

102F19.EPS

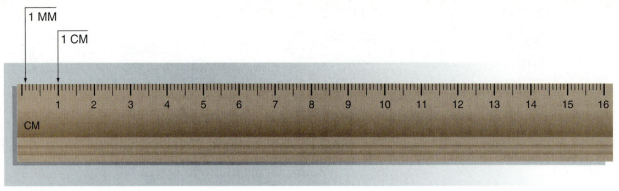

Figure 20 ◆ Metric ruler.

7.2.0 Using a Metric Ruler

Metric rulers are often used with blueprints, in which most measurements are given either in centimeters or in millimeters. In metalworking, for example, it is most common for you to take measurements in millimeters.

The metric ruler in *Figure 20* is divided into centimeters (cm) and millimeters (mm).

Review Questions

Section 7.2.0

Use the metric ruler in *Figure 21* to find the answers to the following questions.

1. P is at the _____ mark.
   a. 5 centimeter
   b. 55 centimeter
   c. 5.5 centimeter
   d. 50 centimeter

2. Q is at the _____ mark.
   a. 5 centimeter
   b. 55 centimeter
   c. 5 millimeter
   d. 55 millimeter

3. R is at the _____ mark.
   a. 25 centimeter
   b. 2.5 centimeter
   c. 250 millimeter
   d. 250 centimeter

4. S is at the _____ mark.
   a. 38 cm
   b. 3 cm and 8 mm
   c. 30 mm and 80 cm
   d. 83 mm

5. T is at the _____ mark.
   a. 1.3 mm
   b. 133 mm
   c. 13 cm
   d. 1.3 cm

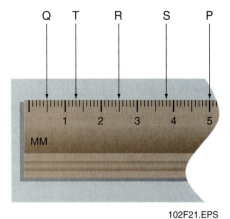

Figure 21 ◆ Review questions metric ruler.

7.3.0 Converting Measurements

Sometimes it may be necessary to change from one unit of measurement to another—say, from inches to yards or from centimeters to meters.

In the standard measurement system, also called the English system, this may involve several steps. If you want, for example, to change from inches to yards, you must first divide the number of inches by 12 (the number of inches in a foot) and then divide that number by three (the number of feet in a yard), or divide by 36 (the number of inches in a yard).

How many yards are in 72 inches? See *Figure 22*. There are 2 yards in 72 inches.

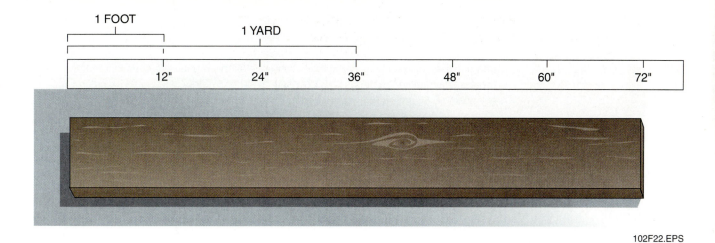

Figure 22 ◆ 72 inches of material.

The metric system makes the conversion much simpler. You would simply move the decimal point, because the system is built on multiples of 10.

How many meters are there in 72 centimeters? See *Figure 23*.

Because 1 centimeter = .01 meter (move decimal 2 places to the left), 72 centimeters = .72 meters.

There are .72 meters in 72 centimeters.

This problem is similar to working with money. If you had 72 cents in your pocket, how much of a dollar would you have? You'd have $.72, or 72 hundredths of a dollar.

Converting measurements from the English system to the metric system, and vice versa, is more complicated. At this stage in your training, you will not be responsible for making such conversions, but it is important that you at least be aware of them. Many dictionaries and other reference

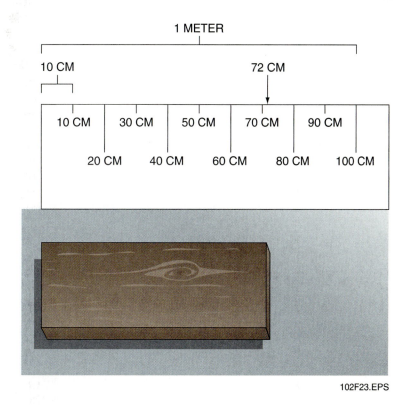

Figure 23 ◆ 72 centimeters of material.

Metrics on the Job

Imagine the problems you would have if you mistakenly measured yards instead of meters when cutting material. A yard is only 91.4 centimeters. You would be cutting everything 8.6 centimeters, or more than 3 inches, too short.

More and more companies are using the universally accepted metric system as their unit of measure. With the increasingly international character of large firms, employees must be able to rely on a standard system of measurement.

In 1999, a $1.25 million satellite sent to observe the planet Mars bounced off the Martian atmosphere and spun uselessly out into space because workers on Earth forgot to convert their measurements from feet and inches into meters before entering the data into the computer. The costly mistake has set back the scientific research of this planet by years.

books contain simple comparison charts that show some basic equivalents between English system measurements and metric system measurements. An example of such a chart appears in Appendix B at the end of this module.

Review Questions

Section 7.3.0

Find the answers to the following conversion problems:

1. .45 meter = _____ centimeter(s)
   a. 405
   b. 450
   c. 45
   d. .45

2. 3 yards = _____ inches
   a. 72
   b. 108
   c. 12
   d. 16

3. 36 inches = _____ yards
   a. 13
   b. 3
   c. 10
   d. 1

4. 90 inches = _____ yards
   a. 22
   b. 2.2
   c. 2½
   d. 200

5. 1 centimeter = _____ meters
   a. .1
   b. .01
   c. 10
   d. 100

 DID YOU KNOW?

Common Terms

The most common units of measure in the metric system are the meter, kilometer, millimeter, and centimeter. Unit names for larger measurements than a kilometer are not common. For example, the term hectometer (100 meters) is not often used. In the Olympics you hear about the 200 meter and the 400 meter races, not the 2 hectometer and the 4 hectometer races.

Measurements smaller than the millimeter (.001 meter) are usually used by scientists. The terms for units such as a micrometer (.000001 meter) or nanometer (.000000001) are simply not used in everyday measuring.

DID YOU KNOW?

Simple System

The metric system is a very easy measuring system to use. The basic unit of length is a meter. Anything longer than one meter is a multiple of the meter. For example, *kilo-* is a prefix meaning thousand. So a kilometer is 1,000 meters. The prefix *mega-* means one million. So a megameter is 1,000,000 meters (although you rarely hear this term used).

Anything shorter than a meter is a part of a meter and can be calculated using division. The most common smaller units of a meter used in measuring are the centimeter and the millimeter. The *centi-* prefix means one hundredth (1/100) and *milli-* means one thousandth (1/1,000). A centimeter is 1/100 of a meter. A millimeter is 1/1,000 of a meter.

All the units of measure in the metric system differ by multiples of 10. This makes it easy to convert measurements within the metric system.

Centimeter = 1 × .01
Millimeter = 1 × .001
Meter = 1
Kilometer = 1 × 1,000
Megameter = 1 × 1,000,000

To convert 1 kilometer to meters, you multiply by 1,000. Thus, 1 kilometer equals 1,000 meters (1 × 1000). To convert 1 meter to centimeters, you divide by .01. Thus, 1 meter equals 100 centimeters (1 ÷ .01 = 100).

DID YOU KNOW?

Millwrights

Millwrights install, repair, replace, and dismantle the machinery and heavy equipment used in almost every industry. They may be responsible for placement and installation of machines in a plant or shop. They use hoists, pulleys, jacks, and come-alongs to perform tasks. Other tools used vary from carpentry to masonry to mechanical trade hand tools such as micrometers and calipers.

Millwrights fit bearings, align gears and wheels, attach motors, and connect belts according to manufacturers' specifications. They may be in charge of preventive maintenance such as lubrication and fixing or replacing worn parts. Precision leveling and alignment are important in the assembly process; millwrights must have good mathematical skills so they can measure **angles,** material thickness, and small distances.

8.0.0 ◆ INTRODUCTION TO CONSTRUCTION GEOMETRY

Geometry might sound like something scary, but it is really made up of everyday stuff you already know—circles, triangles, squares, and rectangles! The construction industry exists in a world of measurements. You should recognize the basic shapes and measurements that make your work possible.

8.1.0 Angles

An angle is an important term in the construction trades. It is used by all building trades to describe the shape made by two straight lines that meet in a point or vertex. Angles are measured in degrees. To measure angles, you use a measurement tool called a protractor. The following are the typical angles (see *Figure 24*) you will measure in construction:

- **Acute angle:** an angle that measures between 0 and 90 degrees. The most common acute angles are 30, 45, and 60 degrees.

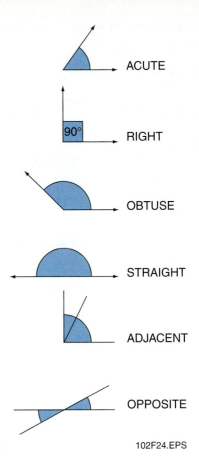

Figure 24 ◆ Typical angles.

- **Right angle:** an angle that measures 90 degrees. The two lines that form the right angle are perpendicular to each other. *This is the angle used most often in the construction trade.* A right angle is indicated in plans or drawings with this symbol: ⌐
- **Obtuse angle:** an angle that measures between 90 and 180 degrees.
- **Straight angle:** an angle that measures 180 degrees (a flat line).
- **Adjacent angles:** angles that have the same vertex and one side in common. Adjacent refers to objects that are next to each other.
- **Opposite angles:** angles formed by two straight lines that cross. Opposite angles are always equal.

8.2.0 Shapes

Common shapes (see *Figure 25*) you are already familiar with that are essential to your work in the trades include rectangles, squares, triangles, and circles.

DID YOU KNOW?

Rope Stretchers

The word *geometry* comes from two Greek words: *geos*, meaning land, and *metrein*, meaning to measure.

In ancient Egypt most farms were located beside the Nile River. Every year during the flood season, the Nile overflowed its banks and deposited mineral-rich silt over the farmland. The floodwaters destroyed the markers used to establish property lines between farms. When this happened, the farms had to be measured again. Men called rope stretchers remarked the property lines.

The rope stretchers calculated distances and directions using ropes that had equally spaced knots tied along the length of the rope. Stretching the ropes to measure distances on level land was easy. In many cases, however, the rope stretchers had to measure property lines from one side of a hill to the opposite side or across a pond.

The hills and ponds made this measurement difficult. To adjust to the uneven land or ponds that stood in the way, rope stretchers determined new ways of measuring. Such discoveries became the foundation of geometry.

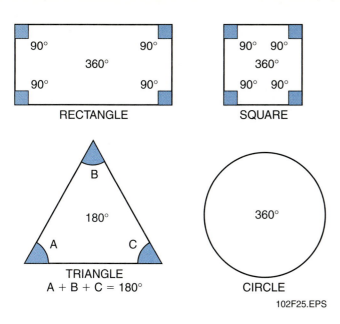

Figure 25 ◆ Common shapes.

8.2.1 Rectangle

A rectangle is a four-sided shape with four 90-degree angles. (The sum of all four angles in all rectangles is 360 degrees.) A rectangle has two pairs of equal sides that are parallel to each other. The **diagonals** of a rectangle are always equal. Diagonals are lines connecting opposite corners. If you cut a rectangle on the diagonal, you will have two **right triangles,** as shown in *Figure 26.*

8.2.2 Square

A square is a special type of rectangle with four equal sides and four 90-degree angles. (The sum of all four angles in all squares is 360 degrees.) If you cut a square on the diagonal, you will also have two right triangles. Each right triangle will have two 45-degree angles and one 90-degree angle, as shown in *Figure 27.*

When measuring the outside lines of a rectangle or a square, you are determining the **perimeter.** The perimeter is the sum of all four sides of a rectangle or square. It may be necessary to calculate the perimeter of a shape so that you can measure, mark, and cut the proper amount of material. For example, if you need to install shoe

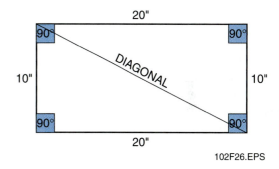

Figure 26 ◆ Cutting a rectangle on the diagonal produces two right triangles.

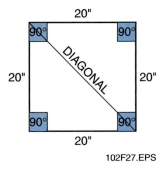

102F27.EPS

Figure 27 ◆ Cutting a square on the diagonal produces two right triangles.

molding along all four walls of a room, you must know the perimeter measurement. If the room is 14 feet by 12 feet, you would calculate: 14 + 12 + 14 + 12 = 52 feet of shoe molding. Another way to calculate would be: (2 × 14 feet) + (2 × 12 feet) = 52 feet.

8.2.3 Triangle

A triangle is a closed shape that has three sides and three angles. Although the angles in a triangle can vary, the sum of the three angles is always 180

 DID YOU KNOW?

A **formula** developed by a Greek mathematician named Pythagoras helps you find the side lengths for any right triangle. The Pythagorean theorem states that the sum of the squares of the two shorter sides is equal to the square of the longest side, or the **hypotenuse.** The hypotenuse is always opposite the 90-degree angle of the triangle. The mathematical equation for this theorem looks like this:

$$A^2 + B^2 = C^2$$

You know that the word *square* refers to the shape but, in mathematical terms, it also refers to the product of a number multiplied by itself. For example, 25 is the square of 5, and 16 is the square of 4. Another way to say this is that 25 is 5 squared, or 5 times itself, and 16 is 4 squared, or 4 times itself.

In a mathematical equation this might appear as $5^2 = 25$ or $4^2 = 16$. In these examples, the numbers 5 and 4 are called the square roots, because you have to square them—or multiply them by themselves—to arrive at the squares.

degrees (see *Figure 28*). The following are different types of triangles you will use in construction (see *Figure 28*):

- Right triangle—contains one 90-degree angle.
- Equilateral triangle—has three equal angles and three equal sides.
- Isosceles triangle—has two equal sides and two equal angles. A line that **bisects** (runs from the center of the base of the triangle to the highest point) an isosceles triangle creates two adjacent right angles.

8.2.4 Circle

A circle is a closed curved line around a center point. Every point on the curved line is exactly the same distance from the center point. A circle measures 360 degrees. The following measurements apply to circles (see *Figure 29*):

- **Circumference**—the length of the closed curved line that forms the circle. The formula for finding circumference is **pi** (3.14) × diameter.

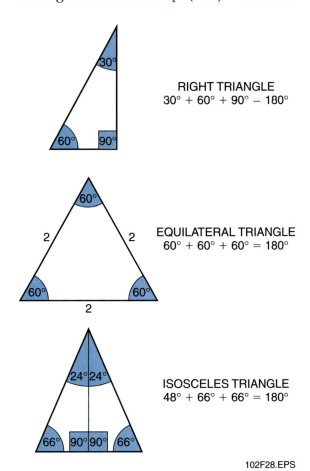

102F28.EPS

Figure 28 ◆ The sum of a triangle's 3 angles always equals 180 degrees.

- **Diameter**—the length of a straight line that crosses from one side of the circle through the center point to a point on the opposite side. The diameter is the longest straight line you can draw inside a circle.
- **pi or π**—a mathematical constant value of approximately 3.14 (or 22/7) used to determine the **area** and circumference of circles.
- **Radius**—the length of a straight line from the center point of the circle to any point on the closed curved line that forms the circle, and equal to half the diameter.

Review Questions

Section 8.2.0

1. You can use _____ to check if material is cut in a true rectangle.
   a. perimeters
   b. diameters
   c. diagonals
   d. right angles

2. A closed curved shape in which every point on the line is an equal distance from a center point is a _____.
   a. circumference
   b. diameter
   c. radius
   d. circle

3. An angle that measures between 0 and 90 degrees is called a(n) _____ angle.
   a. obtuse
   b. acute
   c. right
   d. lateral

4. A triangle with two equal sides and two equal angles is a(n) _____ triangle.
   a. equilateral
   b. isosceles
   c. right
   d. hypotenuse

5. The _____ is the sum of all four sides of a rectangle or a square.
   a. perimeter
   b. diagonal
   c. circumference
   d. square

8.3.0 Area of Shapes

Area is the measurement of the surface of an object. You must calculate the area of a shape, such

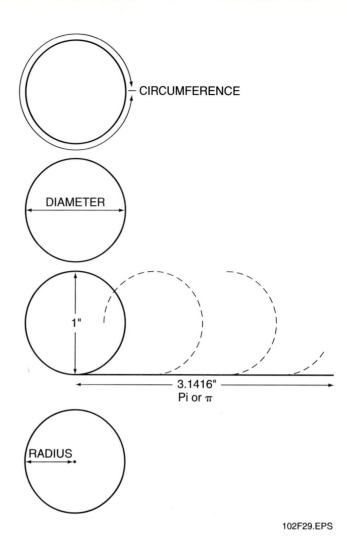

CIRCUMFERENCE

DIAMETER

1"

3.1416"
Pi or π

RADIUS

102F29.EPS

Figure 29 ◆ Measurements that apply to circles.

as a floor or a wall, to order the proper amount of material, such as carpeting or paint. Square units of measure describe the amount of surface area. Measurements are in square inches (sq in), square feet (sq ft), and square yards (sq yd):

- Square inch = 1 inch × 1 inch (inch^2)
- Square foot = 1 foot × 1 foot (foot^2)
- Square yard = 1 yard × 1 yard (yard^2)

You must be able to calculate the area of the shapes discussed before. There are mathematical formulas that make this very easy to do. In Appendix C you will find a list of the formulas for calculating the areas of these shapes. You will need to be familiar with these formulas at this stage in your training.

- Rectangle—area = length × width. For example, you have to paint a wall that is 20 feet long and 8 feet high. The calculation of area is 20 ft × 8 ft = 160 sq ft.

- Square—area = length × width (but remember that all sides of a square are equal, so the formula can also be area = side × side, or side^2). For example, you have to tile a 12-foot square room. The area is 12 ft × 12 ft = 144 sq ft (or 12 ft^2 = 144 sq ft).
- Circle—area = pi × radius^2. In this formula, you must use the mathematical constant pi, which has an approximate value of 3.14. You multiply pi by the radius of the circle squared. For example, to find the area of a circular driveway to be sealed, you must first find the radius, which is 20 feet. The calculation is 3.14 × (20 ft)^2 or 3.14 × 400 sq ft = 1,256 sq ft.
- Triangle—area = 0.5 × base × height. The base is the side the triangle sits on. The height is the length of the triangle from its base to the highest point. For example, you have to install siding on a triangular section of a building. You find the triangle has a base of 2 feet and a height of 4 feet. The calculation is 0.5 × 2 ft × 4 ft = 4 sq ft.

Review Questions

Section 8.3.0

1. The area of a rectangle that is 8 feet long and 4 feet wide is ___.
   a. 12 ft
   b. 32 ft
   c. 32 sq ft
   d. 12 sq ft

2. The area of a 16-inch square is ____.
   a. 256 sq in
   b. 32 in
   c. 256 in
   d. 32 sq in

3. The area of a circle with a 14-foot diameter is ____.
   a. 153 ft
   b. 43.96 sq ft
   c. 153.9 sq ft
   d. 196 sq ft

4. The area of a triangle with a base of 4 feet and a height of 6 feet is ____.
   a. 12 sq ft
   b. 24 sq ft
   c. 10 sq ft
   d. 5 sq ft

5. The area of a rectangle that is 14 feet long and 5 feet wide is _____.
   a. 145 sq ft
   b. 38 sq ft
   c. 70 sq ft
   d. 19 sq ft

8.4.0 Volume of Shapes

Volume is the amount of space occupied in three dimensions. To measure volume, you must use three measurements: length, width, and height (depth or thickness). **Cubic** units of measure describe the volume of different spaces. Measurements are in cubic

ON-SITE

CONSTRUCTION

Calculating the Area of a Cylinder

Let's say you have to paint the outside of a storage tank that is 10 feet in diameter and 20 feet high. How much total area will you have to paint? To know this, you have to calculate the area of a cylinder (see *Figure 30*).

First, the top of the storage tank is a circle, so you can use the formula to calculate the area of a circle, (pi)r^2, to find the area of the top. If the diameter is 10 feet, the radius is half of that, or 5 feet. The calculation is 3.14 × 5^2 or 3.14 × 25 = 78.5 square feet.

Then, what do you do about the sides? Imagine that you could unroll the tank—you would see a rectangle shape! You know the height of the tank is 20 feet. To calculate the length (remember, although you're visualizing a rectangle, it's still a circle), you must find the circumference of the top (pi, or 3.14, × diameter). Therefore, the length is 31.4 feet. You now know both the length and the width. To find the area (area = length × width), calculate 20 ft × 31.4 ft = 628 sq ft.

Now add the two areas together to find out how much area you must paint: 78.5 sq ft + 628 sq ft = 706.5 sq ft of tank surface.

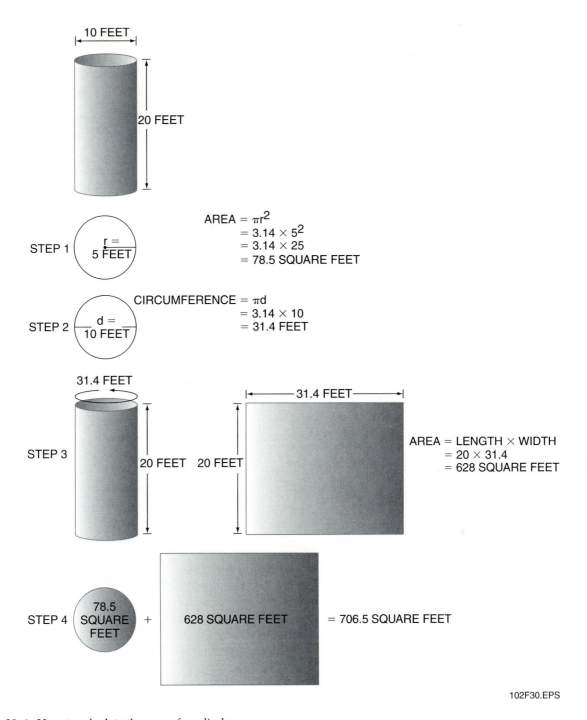

STEP 1

AREA = πr^2
= 3.14 × 5²
= 3.14 × 25
= 78.5 SQUARE FEET

r =
5 FEET

STEP 2

CIRCUMFERENCE = πd
= 3.14 × 10
= 31.4 FEET

d =
10 FEET

STEP 3

AREA = LENGTH × WIDTH
= 20 × 31.4
= 628 SQUARE FEET

STEP 4

78.5 SQUARE FEET + 628 SQUARE FEET = 706.5 SQUARE FEET

102F30.EPS

Figure 30 ◆ How to calculate the area of a cylinder.

inches (cu in), cubic feet (cu ft), and cubic yards (cu yd):

- Cubic inch = 1 inch × 1 inch × 1 inch (or inch³)
- Cubic foot = 1 foot × 1 foot × 1 foot (or foot³)
- Cubic yard = 1 yard × 1 yard × 1 yard (or yard³)

You must be able to calculate the volume of the shapes discussed before. The following are mathematical formulas that make this very easy to do.

In Appendix C you will find a list of the formulas for calculating the volumes of these shapes. You will need to be familiar with these formulas at this stage in your training. Remember to convert dimensions before multiplying (see On-Site: Unit Conversion).

- Rectangle—volume = length × width × depth. For example, you have to order the right amount of cubic yards of concrete for a slab that is 20 feet long and 8 feet wide and 4 inches thick (see *Figure 31*). You must know the total volume of the slab. To calculate that, perform the following steps:

Step 1 Convert inches to feet.

$$20 \text{ ft} \times 8 \text{ ft} \times (4 \text{ in} \div 12) =$$

Step 2 Multiply length × width × depth.

$$20 \text{ ft} \times 8 \text{ ft} \times .33 \text{ ft} = 52.8 \text{ cu ft}$$

Step 3 Convert cubic feet to cubic yards.

$$52.8 \text{ cu ft} \div 27 \text{ (cu ft per cu yd)} =$$
$$1.95 \text{ cu yd of concrete}$$

- Square—volume = length × width × depth. For example, you have to order more concrete for a slab that is 12 feet squared and 5 inches thick. The calculation for volume is:

Step 1 Convert inches to feet.

$$12 \text{ ft} \times 12 \text{ ft} \times (5 \text{ in} \div 12) =$$

Step 2 Multiply length × width × depth.

$$12 \text{ ft} \times 12 \text{ ft} \times .42 \text{ ft} = 60.5 \text{ cu ft}$$

Step 3 Convert cubic feet to cubic yards.

$$60.5 \text{ cu ft} \div 27 \text{ (cu ft per cu yd)} =$$
$$2.24 \text{ cu yd of concrete}$$

- Cube (square)—A cube is a special type of three-dimensional rectangle in that its length, width, and height are equal (see *Figure 32*). To find the volume of a cube, you can cube one dimension (multiply the number by itself three times). Perform the following steps to see how you order concrete for a cube to be used as a support member of a structure:

Step 1 Determine the volume of a cube that is 8 feet cubed.

$$8 \text{ ft} \times 8 \text{ ft} \times 8 \text{ ft} = 512 \text{ cu ft}$$

Step 2 Convert cubic feet to cubic yards.

$$512 \text{ cu ft} \div 27 = 18.96 \text{ cu yd of concrete}$$

- Cylinder (circle)—volume = pi × radius² × height. In this formula, you can use a shortcut: the area of a circle × height. For example, you must fill a cylinder that is 22 feet in diameter and 10 feet high (see *Figure 33*):

Step 1 Calculate the area of the circle. [(pi)r²].

$$3.14 \times 11^2 = 379.94 \text{ sq ft}$$

Step 2 Calculate volume (area × height).

$$379.94 \text{ sq ft} \times 10 \text{ ft} = 3,799.4 \text{ cu ft to fill}$$

- Triangle—volume = 0.5 × base × height × depth (thickness). In this formula you can use a shortcut: the area of a triangle × depth. For example, you must fill a triangular shape that

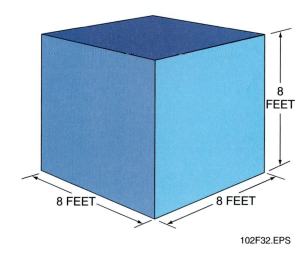

Figure 32 ◆ Volume of a cube.

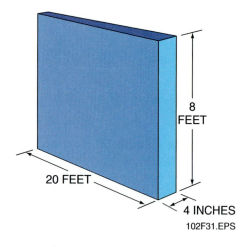

Figure 31 ◆ Volume of a rectangle.

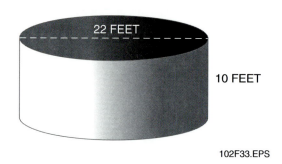

Figure 33 ◆ Volume of a cylinder

Unit Conversion

When calculating areas or volume of shapes, you must often work with different units of measure (inch, foot, yard) for dimensions. You cannot use different units together in a calculation. Before doing your calculation, you must convert the measurements to the same units. Use the following conversion table to change dimensions to all inches, feet, or yards.

Unit Conversion Table

Inches	Feet	Yards
1	$\frac{1}{12}$	$\frac{1}{36}$
12	1	$\frac{1}{3}$
36	3	1

For example, you may have lumber that is sized 2 inches × 4 inches × 12 feet. You cannot multiply these different units together to determine the number of **board feet** in this piece of lumber. You cannot multiply 2 inches by 4 inches by 12 feet and get a correct answer. You must convert all the dimensions to inches using the conversion table:

2 inches × 1 = 2 inches (no change)

4 inches × 1 = 4 inches (no change)

12 feet × 12 = 144 inches (changed feet to inches)

Now you can calculate the total cubic inches in this piece of lumber: 2" × 4" × 144" = 1152 cubic inches. You must now perform a final conversion to determine how many board feet this is. There are 144 cubic inches in 1 board foot. Use this number to convert the cubic inches to board feet.

1152 cu in ÷ 144 cu in = 8 board feet.

In another example, you must determine the area of a hallway that is 30 feet long and 36 inches wide. First, convert the dimensions to the same units. In this case, you calculate: 36 inches ÷ 12 (inches in 1 ft) = 3 feet. Now perform the calculation for area (length × width = area): 30 ft × 3 ft = 90 sq ft. The area of the hallway is 90 square feet.

---

has a base of 6 inches, a height of 12 inches, and a depth of 11 inches:

Step 1 Calculate the area of the triangle.

0.5 × 6 in × 12 in = 36 sq in

Step 2 Calculate volume.

36 sq in × 11 in = 396 cu in to fill

Review Questions

Section 8.4.0

1. The volume of a rectangular shape 5 feet high, 6 feet thick, and 13 feet long is _____.
   a. 24 cu ft
   b. 390 cu ft
   c. 43 ft
   d. 95 ft

2. The volume of a 3-foot cube is _____.
   a. 27 cu ft
   b. 9 cu ft
   c. 12 cu ft
   d. 27 cu yd

3. The volume of a triangular shape that has a 6-inch base, a 2-inch height, and a 4-inch depth is _____.
   a. 12 cu in
   b. 24 cu in
   c. 48 cu in
   d. 36 cu in

4. The volume of a cylinder that is six feet wide and 12 feet high is _____.
   a. 28.26 sq ft
   b. 18 sq ft
   c. 339 cu ft
   d. 360 cu ft

5. If a square measures 17 feet square and is 6 inches thick, its volume is _____ cubic yards.
   a. 144.5
   b. 5.35
   c. 102
   d. 3.77

Summary

Mathematics is not just something you need when you are in school. The construction site requires math every day to get the work done. Whether you are cutting stock, painting a wall, or installing cable, you need your math skills on the job. Basic operations such as addition, subtraction, multiplication, and division are the keys to completing your tasks. Knowing how to measure, mark, and use materials and supplies increases your value to your employer. Be sure you have math in your brain's tool belt when you start work each day! You have completed the learning and practice portion of this module, *Introduction to Construction Math*. Your instructor will now provide you with a module examination to test your comprehension of this course.

Matthew B. Barnett, BE&K Construction Company

Welder
BE&K Construction Company
Birmingham, Alabama

Matthew Barnett was born in South Korea and grew up on a farm in Alabama. After high school, he attended a two-year technical college, where he was named best welder in his class. He is a manual and orbital welder for BE&K Construction in Birmingham. He is also a technical assistant on metal inert gas welding machines, and a technician using various specialized TRI-tool prep equipment.

What made you decide to be a welder?
I was influenced by being raised in a rural environment, watching my dad weld on his tractors, bush hog, and sawmill equipment. The bright arc fascinated me and made me want to learn more about it, which I did in high school. It seemed like the perfect career for me, doing hands-on work that was primarily outside, and something that could be a source of pride to me. My sisters went to college, but I wanted to jump right into a construction career, one that would let me travel around the United States and to other countries.

How did you pursue your goal?
After graduation, I accepted a scholarship to attend Reid State Technical College in Evergreen, Alabama. At the end of the two-year program, I applied for a job with BE&K, mainly because I had read about their accomplishments in the construction field and their policy of offering higher pay for more education. Once I was hired, I began to travel and to use my education to make my career dream a reality.

Through the construction jobs I worked on, I started learning more ways to weld with specialized machines, and I realized that I could expand my career and create better job opportunities for myself.

Welding isn't just a job; it's an exciting career that can lead you to the other side of the world, to the depths of the ocean, or to the top of the tallest skyscraper. And there is always the possibility of taking your skills to more advanced levels.

What subject in school has turned out to be most helpful?
Mathematics was an important part of my studies, and now I find myself using math in all areas of my life: at work, at home, and in sports and other recreational activities. Math is a primary tool to design and build any construction project, of any size. In my work, I use math to establish and test welding procedures for Procedures Qualification Reports, and to make calculations of heat, travel, position of filler metal, length of ground and electrode cable, direction of current, and distance of arc gap.

How important is safety?
Safety is extremely important. Respect your coworkers, and work safely and carefully. There are three things that cause an accident: "I didn't see, I didn't know, and I was in too much of a hurry." Don't take anything for granted, and follow the rules on each job site. If you focus on having pride in your work, safety will be part of that.

What do you like best about the job?
Many welders say they like the chance to travel. I do like to travel, but most of all I like the appreciation and respect that I receive from coworkers, my employer, and clients. I believe that if you work hard at your job, you probably will be rewarded with bigger paychecks and opportunities for advancement. But the biggest reward is the respect you earn from others and from yourself.

Do you have any advice for someone entering the trades?
I'd say, if welding—or another trade—interests you, "just do it" by getting into some type of educational program or training. To be successful in a construction/welding career, you have to set high goals for yourself and take pride in your work. Always try to do the job right the first time, but, as my instructor told me, "Don't accept what you just did as your best, because you can always do better." Also, I've learned that when someone gives me advice, I should listen. But most of all, open yourself to learning. Go out and get the education and training that you need, and keep learning on the job. Only you can decide what your "perfect job" is and what you need to do to get there.

Trade Terms Introduced in This Module

Acute angle: Any angle between 0 degrees and 90 degrees.

Adjacent angles: Angles that have the same vertex and one side in common.

Angle: The shape made by two straight lines coming together at a point; the space between those two lines is measured in degrees.

Area: The surface or amount of space occupied by a two-dimensional object such as a rectangle, circle, or square. To calculate the area for rectangles and squares, multiply the length and width. To calculate the area for circles, multiply the square of the radius and pi.

Bisect: To divide into equal parts.

Board feet: The basic unit of measure for lumber. One board foot is equal to a board that is 1 inch thick, 12 inches long, and 1 inch wide.

Borrowing: Process of moving numbers from one value column (such as the tens column) to another value column (such as units) to perform subtraction problems.

Circle: A closed curved line around a central point. A circle measures 360 degrees.

Circumference: The distance around the curved line that forms a circle.

Convert: To change from one unit or expression to another. For example, convert a decimal to a percent: .25 converts to 25%; or, convert a fraction to an equivalent: ¾ to ⅝.

Cubic: Measurement found by multiplying a number against itself three times; it describes volume.

Decimal: Part of a number represented by digits to the right of a point, called a decimal point.

For example, in the number 1.25, .25 is the decimal part of the number.

Degree: A unit of measurement for angles. For example, a right angle is 90 degrees, an acute angle measures between 0 and 90 degrees, and an obtuse angle measures between 90 and 180 degrees.

Denominator: The part of a fraction below the dividing line. For example, the 2 in ½ is the denominator.

Diagonal: Line drawn from one corner of a rectangle or square to the opposite corner.

Diameter: The length of a straight line that crosses from one side of a circle, through the center point, to a point on the opposite side. The diameter is the longest straight line you can draw inside a circle.

Difference: The result you get when you subtract one number from another. For example, in the problem $8 - 3 = 5$, the number 5 is the difference.

Digit: Any of the numerical symbols 0 to 9.

Dividend: The number to be divided in a division problem. For example, in the problem $10 \div 5 = 2$, 10 is the dividend.

Divisor: The number by which a dividend is divided. For example, in the problem $10 \div 5 = 2$, 5 is the divisor.

English ruler: Instrument that measures English measurements; the standard ruler. For example, units of English measure are inches, feet, and yards.

Equilateral triangle: A triangle that has three equal sides and three equal angles.

Equivalent fractions: Fractions having different numerators and denominators, but equal values, such as ½ and ¾.

Formula: A mathematical process used to solve a problem. For example, the formula for finding the area of a rectangle is side A times side B = Area, or A × B = Area.

Fraction: A number represented by a numerator and a denominator, such as ½.

Hypotenuse: The longest side of a right triangle. It is opposite the right or 90-degree angle.

Improper fraction: A fraction whose numerator is larger than its denominator. For example, ¾ and ⅗ are improper fractions.

Invert: To reverse the order or position of numbers. In fractions, to turn upside down, such as ¾ to ⅓.

Isosceles triangle: A triangle that has two equal sides and two equal angles.

Long division: Process of writing out each step of a division problem until you reach the answer and identify any remainder that can no longer be divided by the divisor.

Machinist's rule: A ruler that is marked so that the inches are divided into 10 equal parts, or tenths.

Meter: The base unit of length in the metric system; approximately 39.37 inches.

Metric ruler: Instrument that measures metric lengths. Units of measure can include millimeters, centimeters, and meters.

Micrometer: A precision tool that can measure to the nearest 1/1000 (or 0.001) of an inch.

Mixed number: A combination of a whole number with a fraction or decimal. For example, mixed numbers are 3⁷⁄₁₆, 5.75, and 1¼.

Negative numbers: Numbers less than zero. For example, −1, −2, and −3 are negative numbers.

Numerator: The part of a fraction above the dividing line. For example, the 1 in ½ is the numerator.

Obtuse angle: Any angle between 90 degrees and 180 degrees.

Opposite angles: Two angles that are formed by two straight lines crossing. They are always equal.

Percent: Of or out of one hundred.

Perimeter: The distance around the outside of any closed shape, such as a rectangle, circle, or square.

Pi: A mathematical value of approximately 3.14 (or 22/7) used to determine the area and circumference of circles. It is sometimes symbolized by π.

Place value: The exact quantity of a digit, determined by its place within the whole number or by its relationship to the decimal point.

Positive numbers: Numbers greater than zero. For example, 1, 2, and 3 are positive numbers.

Radius: The distance from a center point of a circle to any point on the curved line, or half the width (diameter) of a circle.

Rectangle: A four-sided shape with four 90-degree angles. Opposite sides of a rectangle are always parallel and the same length.

Remainder: The leftover amount in a division problem. For example, in the problem 34 ÷ 8, 8 goes into 34 four times (8 × 4 = 32) and 2 is left over, or, in other words, it is the remainder.

Right angle: An angle that measures 90 degrees. The two lines that form a right angle are perpendicular to each other. This is the angle used most in the trades.

Right triangle: A triangle that includes one 90-degree angle.

Shim: A thin, tapered piece of material, such as wood, used to level or plumb a surface. The shim is placed between two structures, such as a wall and a cabinet, to fill in uneven areas and make a level surface.

Square: (1) A special type of rectangle with four equal sides and four 90-degree angles. (2) The product of a number multiplied by itself. For example, 25 is the square of 5; 16 is the square of 4.

Standard ruler: An instrument that measures English lengths (inches, feet, and yards). See *English ruler*.

Straight angle: A 180-degree angle or flat line.

Sum: The total in an addition problem. For example, in the problem 7 + 8 = 15, 15 is the sum.

Triangle: A closed shape that has three sides and three angles.

Vertex: A point at which two or more lines or curves come together.

Volume: The amount of space occupied in three dimensions (length, width, and height/depth/thickness).

Whole numbers: Complete units without fractions or decimals.

Multiplication Table

Trace across and down from the numbers that you want to multiply, and find the answer. In the example highlighted in yellow, 6 × 7 = 42.

	2	3	4	5	6	7	8	9	10	11	12
2	4	6	8	10	12	14	16	18	20	22	24
3	6	9	12	15	18	21	24	27	30	33	36
4	8	12	16	20	24	28	32	36	40	44	48
5	10	15	20	25	30	35	40	45	50	55	60
6	12	18	24	30	36	42	48	54	60	66	72
7	14	21	28	35	42	49	56	63	70	77	84
8	16	24	32	40	48	56	64	72	80	88	96
9	18	27	36	45	54	63	72	81	90	99	108
10	20	30	40	50	60	70	80	90	100	110	120
11	22	33	44	55	66	77	88	99	110	121	132
12	24	36	48	60	72	84	96	108	120	132	144

Conversion Tables

How to Convert Units of Volume

Metric to English			English to Metric		
From	**Multiply By**	**To Obtain**	**From**	**Multiply By**	**To Obtain**
Liters	1.0567	Quarts	Quarts	0.946	Liters
Liters	2.1134	Pints	Pints	0.473	Liters
Liters	0.2642	Gallons	Gallons	3.785	Liters

How to Convert Units of Weight

Metric to English			English to Metric		
From	**Multiply By**	**To Obtain**	**From**	**Multiply By**	**To Obtain**
Grams	0.0353	Ounces	Pounds	0.4536	Kilograms
Grams	15.4321	Grains	Pounds	453.6	Grams
Kilograms	2.2046	Pounds	Ounces	28.35	Grams
Kilograms	0.0011	Tons (short)	Grains	0.0648	Grams
Tons (metric)	1.1023	Tons (short)	Tons (short)	0.9072	Tons (metric)

How to Convert Units of Length

Metric to English			English to Metric		
From	**Multiply By**	**To Obtain**	**From**	**Multiply By**	**To Obtain**
Meters	39.37	Inches	Inches	2.54	Centimeters
Meters	3.2808	Feet	Inches	0.0254	Meters
Meters	1.0936	Yards	Inches	25.4	Millimeters
			Miles	1,609,344	Millimeters
Centimeters	0.3937	Inches	Feet	0.3048	Meters
Millimeters	0.03937	Inches	Feet	30.48	Centimeters
Kilometers	0.6214	Miles	Yards	0.9144	Meters
			Yards	91.44	Centimeters
			Miles	1.6093	Kilometers

Area and Volume Formulas

Area Formulas

Rectangle: area = length × width
Square: area = length × width or side^2
Circle: area = pi × radius^2
Triangle: area = 0.5 × base × height

Volume Formulas

Rectangle: volume = length × width × depth
Square: volume = length × width × depth
Cube (square):   volume = side^3
Cylinder (circle): volume = pi × radius^2 × height or area of a circle × height
Triangle: volume = 0.5 × base × height × depth (thickness) or area of a triangle × depth

Answers to Review Questions

Section 2.1.0
1. b
2. b
3. d
4. c
5. d

Section 2.2.1
1. b
2. b
3. d
4. b
5. c

Section 2.2.2
1. b
2. d
3. b
4. b
5. d

Section 2.3.0
1. b
2. b
3. c
4. c
5. b

Section 2.4.0
1. a
2. b
3. d
4. b
5. c

Section 2.4.1
1. a
2. c

3. d
4. b
5. d

Section 2.5.0
1. b
2. a
3. d
4. d
5. b

Section 2.5.1
1. a
2. b
3. c
4. b
5. b

Section 2.6.2
1. a
2. a
3. c
4. b
5. d

Section 2.6.3
1. d
2. a
3. c
4. c
5. b

Section 2.6.4
1. a
2. b
3. a
4. b
5. b

Section 2.6.6
1. a
2. c
3. b
4. a
5. a

Section 3.1.0
1. c
2. a
3. b
4. c
5. c

Section 4.1.0
1. b
2. c
3. d
4. a
5. c

Section 4.2.0
1. c
2. d
3. d
4. a
5. b

Section 4.3.0
1. c
2. b
3. d
4. b
5. d

Section 4.4.0
1. b
2. b

3. a
4. d
5. c

Section 4.5.0
1. b
2. d
3. a
4. a
5. b

Section 4.5.1
1. a
2. d
3. d
4. b
5. c

Section 4.6.0
1. b
2. a
3. d
4. b
5. c

Section 4.7.0
1. c
2. d
3. c
4. a
5. d

Section 5.0.0
1. a
2. c
3. b
4. b
5. c

Section 5.1.0
1. b
2. c
3. a
4. b
5. b

Section 5.2.0
1. c
2. d
3. a
4. b
5. b

Section 5.3.0
1. a
2. b
3. a
4. b
5. b

Section 5.4.0
1. c
2. d
3. b
4. a
5. b

Section 5.5.1
1. c
2. b
3. a
4. b
5. c

Section 5.5.2
1. c
2. a
3. c
4. b
5. d

Section 5.5.3
1. a
2. d
3. c
4. b
5. a

Section 5.6.0
1. b
2. a
3. d
4. c
5. b

Section 5.7.0
1. d
2. b
3. d
4. a
5. c

Section 6.1.0
1. b
2. d
3. c
4. b
5. a

Section 6.2.0
1. a
2. b
3. c
4. d
5. b

Section 6.3.0
1. c
2. d
3. d
4. a
5. d

Section 6.4.0
1. a
2. c
3. a
4. a
5. b

Section 7.1.0
1. c
2. b
3. d
4. a
5. d

Section 7.2.0
1. a
2. c
3. b
4. b
5. d

Section 7.3.0
1. c
2. b
3. d
4. c
5. b

Section 8.2.0
1. c
2. d
3. b
4. b
5. a

Section 8.3.0
1. c
2. a
3. c
4. a
5. c

Section 8.4.0
1. b
2. a
3. b
4. c
5. b

Additional Resources

This module is intended to present thorough resources for task training. The following reference works are suggested for further study. These are optional materials for continued education rather than for task training.

All the Math You'll Ever Need, 1999 Edition. Stephen Slavin. New York: John Wiley & Sons.

Calculator Math for Job and Personal Use, 1992 Edition. William Pasewark and Merle Wood. Phoenix, AZ: South-Western.

Conquering Math Phobia: A Painless Primer, 1991 Edition. Calvin Clawson. New York: John Wiley & Sons.

Math for the Industrial Shop, 1989 Edition. Diane Cheatham. Detroit, MI: Chatfield College.

Math to Build On, A Book for Those Who Build, 1993 Edition. Johnny and Margaret Hamilton. Clinton, NC: Construction Trades Press.

Mathematics for the Million, 1993 Edition. Lancelot Thomas Hogben. New York: W. W. Norton & Company.

Practical Problems in Mathematics, 1980 Edition. John E. Ball. Albany, NY: Delmar Publishers.

The NCCER makes every effort to keep these textbooks up-to-date and free of technical errors. We appreciate your help in this process. If you have an idea for improving this textbook, or if you find an error, a typographical mistake, or an inaccuracy in NCCER's Contren™ textbooks, please write us, using this form or a photocopy. Be sure to include the exact module number, page number, a detailed description, and the correction, if applicable. Your input will be brought to the attention of the Technical Review Committee. Thank you for your assistance.

Instructors – If you found that additional materials were necessary in order to teach this module effectively, please let us know so that we may include them in the Equipment/Materials list in the Instructor's Guide.

Write: Curriculum Revision and Development Department
National Center for Construction Education and Research
P.O. Box 141104, Gainesville, FL 32614-1104

Fax: 352-334-0932

E-mail: curriculum@nccer.org

Craft _____ Module Name _____

Copyright Date _____ Module Number _____ Page Number(s) _____

Description _____

_____

_____

_____

(Optional) Correction _____

_____

_____

(Optional) Your Name and Address _____

_____

_____

Welding Safety

Course Map

This course map shows all of the modules in the AWS Entry Level Welder – Phase 1 curriculum. The suggested training order begins at the bottom and proceeds up. Skill levels increase as you advance on the course map. The local Training Program Sponsor may adjust the training order.

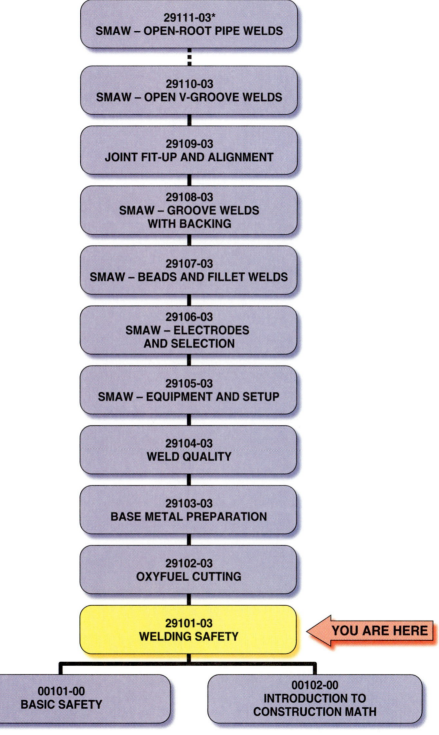

AWS ENTRY LEVEL WELDER—PHASE 1

29111-03*
SMAW – OPEN-ROOT PIPE WELDS

29110-03
SMAW – OPEN V-GROOVE WELDS

29109-03
JOINT FIT-UP AND ALIGNMENT

29108-03
SMAW – GROOVE WELDS
WITH BACKING

29107-03
SMAW – BEADS AND FILLET WELDS

29106-03
SMAW – ELECTRODES
AND SELECTION

29105-03
SMAW – EQUIPMENT AND SETUP

29104-03
WELD QUALITY

29103-03
BASE METAL PREPARATION

29102-03
OXYFUEL CUTTING

29101-03
WELDING SAFETY

YOU ARE HERE

00101-00
BASIC SAFETY

00102-00
INTRODUCTION TO
CONSTRUCTION MATH

*Module 29111-03 is an elective for those students progressing through the AWS Entry Level Welder program.

29101CMAP.EPS

Figures

Table

Welding Safety

Objectives

When you have completed this module, you will be able to do the following:

1. Identify some common hazards in welding.
2. Explain and identify proper personal protection used in welding.
3. Demonstrate how to avoid welding fumes.
4. Explain some of the causes of accidents.
5. Identify and explain uses for material safety data sheets.
6. Demonstrate safety techniques for storing and handling cylinders.
7. Explain how to avoid electric shock when welding.
8. Demonstrate proper material handling methods.

Prerequisites

Before you begin this module, it is recommended that you successfully complete Modules 00100-00 and 00102-00.

Required Trainee Materials

1. Pencil and paper
2. Appropriate personal protective equipment

1.0.0 ◆ INTRODUCTION

Many hazards are associated with welding, cutting, and related processes. Some of these are electric shock, infrared rays, ultraviolet (UV) rays, hot metal, slag, sparks, and associated welding fumes.

It is necessary that trainees understand the hazards of welding and develop the proper attitude toward safety. This begins with an understanding of the need for proper clothing, eye protection, and face protection, along with general shop safety. *Figure 1* shows a welder utilizing proper protective equipment.

The industry is fortunate to have a well-developed standard for safety: *American National Standards Institute (ANSI) Z49.1-1999, Safety in Welding, Cutting, and Allied Processes.* Everyone involved with welding, cutting, and related processes should be familiar with this document. The document discusses personal protective equipment, ventilation, and fire prevention as well as welding in confined spaces. It contains precautionary information for welding processes that is important to welders, supervisors, and managers. The document also details specific welding process safety procedures.

This module is an extension of the *Core Curriculum* safety module. Its purpose is to identify general safety considerations that apply to

101F01.EPS

Figure 1 ◆ Welder utilizing proper protective equipment.

various aspects of the field of welding and the steps needed to avoid job-related deaths and injuries. Specific equipment or process hazards are covered in more detail in modules that deal with the use of the equipment or process in a particular application.

2.0.0 ◆ JOB-SITE ACCIDENTS

There are many causes of accidents. They can usually be divided into two broad categories: personal and physical. It is essential that you be aware of the factors that can cause accidents so you can understand the consequences of your actions on the job site.

2.1.0 Personal Factors That Cause Accidents

Accidents can often be traced to personal factors such as poor health, lack of experience, and the improper use of alcohol and medications.

People who are ill or injured may not be able to concentrate on their work. In addition, they may be physically weakened and unable to handle strenuous work. A person doing work that is inherently dangerous is more likely to cause an accident when he or she is sick or injured. Mental stress can also play a role in accidents. Again, concentration is the issue. An employee who is worried about a serious personal problem is likely to be distracted and unable to focus on work. Workers need to realize that they endanger themselves and those around them when they are not 100% effective. As a worker, you need to be able to recognize when others may not be up to par and take appropriate precautions.

Age and inexperience often play a role in accidents. Insurance company studies show that a person who lacks experience is more likely to take risks that cause accidents. Sometimes, accidents occur just because someone hasn't had enough experience to learn how to avoid them. In fact, insurance company statistics show that more accidents occur in the under-18 age group than in any other. An inexperienced person often is not able to foresee the outcome of an action or just lacks the knowledge to know what works and what doesn't.

People who consume alcohol or use illegal drugs while working risk their lives and the lives of their co-workers. A drug or alcohol hangover is nearly as bad. Legal prescription and over-the-counter drugs can also cause problems. Some cold remedies and cough medicines contain alcohol or other substances that will make people drowsy. Alcohol and drugs affect coordination, alertness, and decision-making ability. Never use these sub-stances while you are working, and don't work with people who use them on the job.

2.2.0 Physical Factors That Cause Accidents

Accidents are often caused by conditions at the job site. For example, foot and vehicle traffic congestion increases at starting and quitting times. As quitting time approaches, people are more likely to hurry, possibly taking risks they might not otherwise take. People often slow down right after a meal, so workers may be less attentive after lunch than at other times.

People who learned to put their toys away when they were young probably keep a neat, safe work site as adults. The others are likely to have tools, equipment, scrap material, and other stuff lying around their work site that will trip people. If you keep a messy work site, eventually you or someone else will be injured because of it. Put away your tools, dispose of scrap materials, and secure your equipment and cables when not in use. If you have flammable or hazardous material at your work site, keep it properly contained and covered. When you're done with it, return it to the designated location for disposal or storage.

Damaged or defective welding equipment is dangerous. For example, a frayed or damaged welding cable could give someone a lethal shock or start a fire. Any damaged or defective equipment should be so tagged and either repaired or replaced. All equipment should be periodically inspected for damage. Make sure that periodic maintenance procedures recommended by the manufacturer are performed on schedule.

3.0.0 ◆ APPROPRIATE PERSONAL PROTECTIVE EQUIPMENT

Each person in the shop wears general work clothing. Extra protection is needed for each person in direct contact with hot materials. Depending on the specific job and conditions, protective equipment can include:

- Body protection
- Foot protection
- Hand protection
- Ear protection
- Eye, face, and head protection

3.1.0 Body Protection

Basic clothing should offer protection from flying sparks, heat, and ultraviolet radiation from an electric arc. Shirts should be made of a tight-weave fabric, have long sleeves and pocket flaps, and be worn with the collar buttoned. Pants must not have cuffs and should fit so they hang straight down the leg, touching the shoe tops without creases. Cuffs and creases can catch sparks, which can cause fires. Never wear polyester or other synthetic fibers; sparks will melt these materials, causing serious burns. Wear wool and cotton, which are more resistant to sparks, instead of synthetic fibers.

For cutting operations, out-of-position or overhead welding/cutting, or welding operations that produce spattering molten metal and sparks or large amounts of heat, additional protective coverings are required. *Figure 2* shows a welder in protective clothing and equipment. Protective coverings, usually made of leather because of its durability, are often just referred to as *leathers*.

However, because leather is heavy and hot to wear, lightweight fire- and heat-resistant clothing is also available. *Figure 3* shows a variety of leather and fire-resistant items that can be worn in combinations to provide various degrees of covering. Full jackets offer the most protection. A cape is cooler but offers only shoulder, chest, and arm protection. Leather pants, in combination with a jacket, will protect a welder's lap and legs. If the weather is hot and full leather pants are uncomfortable, leggings, sometimes called chaps, are available. They are strapped to the legs, leaving the back open. However, an apron alone is cooler and will protect a welder's lap and most of the leg area when squatting, sitting, or bending over a table. In some cases, if the welding is at or below the waist level, only full- or half-sleeve arm covers are required in combination with an apron.

SAFETY GLASSES

WELDING CAP
(VISOR TURNED BACK)

EAR PLUGS

COTTON OR
WOOL OUTER
GARMENTS

CLEAR OR
SHADED FACE
SHIELD

GAUNTLET-
TYPE WELDING
GLOVES

ALTERNATE
HEAD AND FACE
PROTECTION

PANT LEG EXTENDS
ALL THE WAY TO
THE INSTEP OF THE
BOOT (NO CUFF)

HIGH-TOP
LEATHER BOOTS

101F02.EPS

Figure 2 ◆ Typical personal protective equipment.

LEATHER JACKET

LEATHER AND FIRE-RESISTANT
CLOTH JACKET

LEATHER
APRON

FIRE-RESISTANT
ARM COVER

LEATHER
ARM COVER

LEATHER
BOOT OR SHOE
PROTECTION
(SPATS)

CAPE (FASTENS TO
AN APRON)

LEATHER CHAPS

101F03.EPS

Figure 3 ◆ Leather and fire-resistant protective coverings.

3.2.0 Foot Protection

The Occupational Safety and Health Administration (OSHA) requires that protective footwear be worn when working where falling, rolling, or sharp objects pose a danger of foot injuries and where feet are exposed to electrical hazards. Footwear with leather-reinforced soles or inner-soles of flexible metal are recommended. High-top safety shoes or boots 8" or taller should be worn. Ensure that the pant leg covers the tongue and lace area of the footwear. If the tongue and lace area is exposed, wear leather spats (*Figure 3*) under the pants or leggings and over the front top of the footwear to protect them from sparks or falling molten metal. Spats will prevent sparks from burning through the front of lace-up boots. Protective footwear must comply with *ANSI Z49.1-1999*. Sneakers, tennis shoes, and similar types of footwear must never be worn on the job site.

3.3.0 Hand Protection

Gloves are the primary type of hand protection. Gloves must be selected on the basis of the hazards involved in doing the work. Gauntlet-type welding gloves must be worn when welding or cutting to protect against UV rays from an electric arc and heat from any thermal welding/cutting process. The most common glove worn is the standard leather welding glove shown in *Figure 4*. For heavy-duty welding or cutting, special heat-resistant or heat-reflective gloves are also available.

HEAVY-DUTY INSULATED
FIRE-RESISTANT
KEVLAR GLOVES

INSULATING
COTTON LINERS
FOR GLOVES

HEAVY-DUTY
HEAT-REFLECTIVE
GLOVES

STANDARD
LEATHER
GLOVES

101F04.EPS

Figure 4 ◆ Gauntlet-type welding gloves.

Lightweight Leather Gloves

Soft, flexible, lightweight leather gloves are used for light-duty work and for operations involving gas metal arc welding (GMAW), gas tungsten arc welding (GTAW), brazing, soldering, and oxyfuel welding where free hand movement is required.

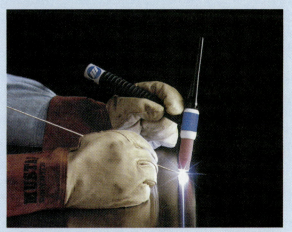

101P0101.EPS

3.4.0 Ear Protection

Welding areas can be very noisy. In addition, if overhead work is being performed, hot sparks can cause burns to the ears and ear canals unless a leather hood is used. For maximum protection, earmuff-type hearing protectors (*Figure 5*) should also be used. They are available in varying degrees of protection from all noise, including low frequencies. Most earmuffs have adjustable headbands that can be worn over the head, behind the neck, or under the chin. To use earmuffs, adjust the tension on the headband and ear cushion pads to obtain the best possible seal. Check the earmuff shell for cracks and the ear cushion pads for tears before each use. Any damaged, cracked, or torn part must be repaired or replaced. As minimal protection, earplugs (*Figure 5*) can be used. Disposable earplugs are the most common form of hearing protection used in the industry. These devices usually have an outer layer of pliable foam and a core layer of acoustical fiber that filters out harmful noise yet allows you to hear normal conversation. To use disposable earplugs, simply roll each plug into a cone and insert the tapered end into the ear canal while pulling up on the upper portion of your ear. The earplugs will expand, filling the ear canal and creating a proper fit. Reusable earplugs that can be cleaned and worn repeatedly are also commonly used. These are typically cleaned with boiling water or alcohol. Plain cotton placed in the ear is not an acceptable protective device.

3.5.0 Eye, Face, and Head Protection

The heat and light produced by cutting or welding operations can damage the skin and eyes. Injury to the eyes may result in permanent loss of vision. Oxyfuel cutting and welding can cause eye fatigue and mild burns to the skin because of the infrared heat radiated by the process. Welding or cutting operations involving an electric arc of any kind produce UV radiation, which can cause severe burns to the eyes and exposed skin and permanent damage to the retina. A flash burn can harm unprotected eyes in just seconds. Welders should never view an electric arc directly or indirectly without wearing a properly tinted lens designed for electric arc use. If electric arc operations are occurring in the vicinity, safety goggles with a tinted lens (shades 3 to 5) and tinted side shields must be worn at all times.

For oxyfuel welding and cutting, wear tinted welding goggles (shades 4 to 6) over safety glasses, and wear a clear face shield. Clear safety glasses and goggles with a tinted face shield can also be used. Most oxyfuel welders prefer the lat-

Figure 5 ◆ Typical ear protection.

ter combination because, for clear vision, only the face shield has to be flipped up. For overhead oxyfuel operations, a leather hood may be used in place of the face shield to obtain protection from sparks and molten metal (*Figure 6*).

For electric arc operations, a leather hood or welding helmet with a properly tinted lens (shades 9 to 14) must be worn over safety goggles to provide proper protection. Many varieties of helmets are available; typical styles are shown in *Figure 7*. Some of the helmets are available with additional side-view lenses in a lighter tint so that welders can sense, by peripheral vision, any activities occurring beside them.

Most welding and cutting tasks require the use of safety goggles, chemical-resistant goggles, dust goggles, or face shields. Always check the material safety data sheet (MSDS) for the welding or cutting product being used to find out what type of eye protection is needed.

4.0.0 ◆ VENTILATION

Adequate mechanical ventilation should be provided to remove fumes that are produced by welding or cutting processes. *ANSI Z49.1-1999* on welding safety covers such ventilation procedures. The gases, dust, and fumes caused by welding or cutting operations can be hazardous if the appropriate safety precautions are not observed. The following general rules can be used to determine if there is adequate ventilation:

- The welding area must contain at least 10,000 cubic feet of air for each welder.
- There must be air circulation.
- Partitions, structural barriers, or equipment must not block air circulation.

TINTED HEADBAND
WELDING GOGGLES
(SPRING-LOADED)

CLEAR
ELASTIC-STRAP
SAFETY GOGGLES

TINTED
ELASTIC-STRAP
WELDING GOGGLES

TINTED HEADBAND
WELDING
FACE SHIELD

CLEAR HEADBAND
SAFETY
FACE SHIELD

101F06.EPS

Figure 6 ◆ Oxyfuel welding/cutting goggles and face shield combinations.

STANDARD SIZE
FLIP-LENS
FACEPLATE

STANDARD SIZE
FIXED-LENS FACEPLATE
(STANDARD HELMET)

LARGE-LENS FACEPLATE

101F07.EPS

Figure 7 ◆ Typical electric arc welding helmets.

Even when there is adequate ventilation, the welder should try to avoid inhaling welding or cutting fumes and smoke. The heated fumes and smoke generally rise straight up. Observe the column of smoke and position yourself to avoid it. A small fan may also be used to divert the smoke, but take care to keep the fan from blowing directly on the work area; the fumes and gases must be present at an electric arc in order to protect the molten metal from the air.

4.1.0 Fume Hazards

Welding or cutting processes create fumes. Fumes are solid particles consisting of the base metal, electrodes or welding wire, and any coatings applied to them. Most fumes are not considered dangerous as long as there is adequate ventilation. If ventilation is questionable, use air sampling to determine the need for corrective measures. Adequate ventilation can be a problem in tight or cramped working quarters. To ensure adequate room ventilation, local exhaust ventilation should be used to capture fumes *(Figure 8)*.

The exhaust hood should be kept four to six inches away from the source of the fumes. Welders should recognize that fumes of any type, regardless of their source, should not be inhaled. The best way to avoid problems is to provide adequate ventilation. If this is not possible, breathing protection must be used. Protective devices for use in poorly ventilated or confined spaces are shown in *Figures 9* and *10*.

If respirators are used, your employer should offer worker training on respirator fitting and usage. Medical screenings should also be given.

WARNING!

Fumes and gases can be dangerous. Overexposure can cause nausea, headaches, dizziness, metal fume fever, and severe toxic effects that can be fatal. Studies have demonstrated irritation to eyes, skin, and the respiratory system and even more severe complications. In confined spaces, fumes and gases may cause asphyxiation.

Auto-Darkening Helmets

Many new helmets are available with an adjustable auto-darkening lens that changes from a shade 3 or 4 to a shade 10, 11, or 12 the instant an electric arc starts to occur. A sensor just above the lens detects the start of an electric arc and electronically causes the shade of the lens to change in less than $\frac{1}{25,000}$ of a second. The use of this type of lens eliminates the need for the welder to raise the helmet or flip up the faceplate to view the work before striking an arc. The helmet is expensive, and most require periodic battery replacement and testing to ensure proper operation.

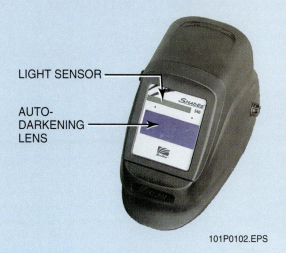

LIGHT SENSOR

AUTO-DARKENING LENS

101P0102.EPS

101F08.EPS

Figure 8 ◆ A flexible exhaust pickup.

101F09.EPS

Figure 9 ◆ Typical respirator.

4.2.0 Respirators

Special metals require the use of respirators to protect welders from harmful fumes. Respirators are grouped into three main types based on how they work to protect the wearer from contaminants. The types are:

- Air-purifying respirators
- Supplied-air respirators (SARs)
- Self-contained breathing apparatus (SCBA)

101F10.EPS

Figure 10 ◆ A belt-mounted respirator.

4.2.1 Air-Purifying Respirators

Air-purifying respirators provide the lowest level of protection. They are made for use only in atmospheres that have enough oxygen to sustain life (at least 19.5%). Air-purifying respirators use spe-

cial filters and cartridges to remove specific gases, vapors, and particles from the air. The respirator cartridges contain charcoal, which absorbs certain toxic vapors and gases. When the wearer detects any taste or smell, the charcoal's absorption capacity has been reached and the cartridge can no longer remove the contaminant. The respirator filters remove particles such as dust, mists, and metal fumes by trapping them within the filter material. Filters should be changed when it becomes difficult to breathe. Depending on the contaminants, cartridges can be used alone or in combination with a filter/pre-filter and filter cover. Air-purifying respirators should be used for protection only against the types of contaminants listed on the filters and cartridges and on the National Institute for Occupational Safety and Health (NIOSH) approval label affixed to each respirator carton and replacement filter/cartridge carton. Respirator manufacturers typically classify air-purifying respirators into four groups:

- No maintenance
- Low maintenance
- Reusable
- Powered air-purifying respirators (PAPRs)

No-maintenance and low-maintenance respirators are typically used for residential or light commercial work that does not call for constant and heavy respirator use. No-maintenance respirators are typically half-mask respirators with permanently attached cartridges or filters. The entire

Know What You're Cutting or Welding

Cutting or welding operations involving materials, coatings, or electrodes containing cadmium, mercury, lead, zinc, chromium, and beryllium result in toxic fumes. For cutting or welding such materials, always use proper area ventilation and wear an approved full-face, supplied-air respirator (SAR) that uses breathing air from outside the work area. For occasional, very short-term exposure to fumes from zinc- or copper-coated materials, a high-efficiency particulate arresting (HEPA)-rated or metal-fume filter may be used on a standard respirator.

Ventilation

Areas that have less than 10,000 cubic feet (283 cubic meters) per welder or that have ceilings less than 16' (4.9 meters) high require forced ventilation. This is the equivalent of a room 25' by 25' with a ceiling of 16'. What is the cubic area of a 90' long by 25' wide room with 8' ceilings?

respirator is discarded when the cartridges or filters are spent. Low-maintenance respirators generally are also half-mask respirators that use replaceable cartridges and filters. However, they are not designed for constant use.

Reusable respirators (*Figure 11*) are made in half-mask and full facepiece styles. These respirators require the replacement of cartridges, filters, and respirator parts. Their use also requires a complete respirator maintenance program. Air respirator maintenance is discussed later.

Powered air-purifying respirators (PAPRs) are made in half-mask, full facepiece, and hood styles. They use battery-operated blowers to pull outside air through the cartridges and filters attached to the respirator. The blower motors can be either mask- or belt-mounted. Depending on the cartridges used, they can filter particulates, dusts, fumes, and mists along with certain gases and vapors. PAPRs like the one shown in *Figure 12* have a belt-mounted, powered air-purifier unit

connected to the mask by a breathing tube. Many models also have an audible and visual alarm that is activated when airflow falls below the required minimum level. This feature gives an immediate indication of a loaded filter or low battery charge condition. Units with the blower mounted in the mask do not use a belt-mounted powered air purifier connected to a breathing tube.

4.2.2 Supplied-Air Respirators

Supplied-air respirators (*Figure 13*) provide a supply of air for extended periods of time via a high-

101F12.EPS

Figure 12 ◆ Typical powered air-purifying respirator (PAPR).

101F11.EPS

Figure 11 ◆ Reusable half-mask air-purifying respirator.

101F13.EPS

Figure 13 ◆ Supplied-air respirator.

pressure hose that is connected to an external source of air, such as a compressor, compressed-air cylinder, or pump. They provide a higher level of protection in atmospheres where air-purifying respirators are not adequate. Supplied-air respirators are typically used in toxic atmospheres. Some can be used in atmospheres that are immediately dangerous to life and health (IDLH) as long as they are equipped with an air cylinder for emergency escape. An atmosphere is considered IDLH if it poses an immediate hazard to life or produces immediate, irreversible, and debilitating effects on health. There are two types of supplied-air respirators: continuous-flow and pressure-demand.

The continuous-flow supplied-air respirator provides air to the user in a constant stream. One or two hoses are used to deliver the air from the air source to the facepiece. Unless the compressor or pump is especially designed to filter the air or a portable air-filtering system is used, the unit must be located where there is breathable air (grade D or better as described in *Compressed Gas Association [CGA] Commodity Specification G-7.1*). Continuous-flow respirators are made with tight-fitting half-masks or full facepieces. They are also made with hoods. The flow of air to the user may be adjusted either at the air source (fixed flow) or on the unit's regulator (adjustable flow). Pressure-demand supplied-air respirators are similar to the continuous-flow type except that they supply air to the user's facepiece via a pressure-demand valve as the user inhales and fresh air is required. They typically have a two-position exhalation valve that allows the worker to switch between pressure-demand and negative-pressure modes to facilitate entry into, movement within, and exit from a work area.

4.2.3 Self-Contained Breathing Apparatus (SCBA)

SCBAs (*Figure 14*) provide the highest level of respiratory protection. They can be used in oxygen-deficient atmospheres (below 19.5% oxygen), in poorly ventilated or confined spaces, and in IDLH atmospheres. These respirators provide a supply of air for 30 to 60 minutes from a compressed-air cylinder worn on the user's back. Note that the emergency escape breathing apparatus (EEBA) is a smaller version of an SCBA cylinder. EEBA units are used for escape from hazardous environments and generally provide a five- to ten-minute supply of air.

4.3.0 Respiratory Program

Local and OSHA procedures must be followed when selecting the proper type of respirator for

101F14.EPS

Figure 14 ◆ Self-contained breathing apparatus.

BEFORE USING A RESPIRATOR YOU MUST DETERMINE THE FOLLOWING:

1. THE TYPE OF CONTAMINANT(S) FOR WHICH THE RESPIRATOR IS BEING SELECTED
2. THE CONCENTRATION LEVEL OF THAT CONTAMINANT
3. WHETHER THE RESPIRATOR CAN BE PROPERLY FITTED ON THE WEARER'S FACE

YOU MUST READ AND UNDERSTAND ALL RESPIRATOR INSTRUCTIONS, WARNINGS, AND USE LIMITATIONS CONTAINED ON EACH PACKAGE BEFORE USE.

101F15.EPS

Figure 15 ◆ Use the right respirator for the job.

a particular job (*Figure 15*). A respirator must be properly selected (based on the contaminant present and its concentration level), properly fitted, and used in accordance with the manufacturer's instructions. It must be worn during all times of exposure. Regardless of the kind of respirator needed, OSHA regulations require employers to have a respirator protection program consisting of:

- Standard operating procedures for selection and use
- Employee training
- Regular cleaning and disinfecting
- Sanitary storage
- Regular inspection
- Annual fit testing
- Pulmonary function testing

As an employee, you are responsible for wearing respiratory protection when needed. When it comes to vapors or fumes, both can be eliminated in certain concentrations by the use of air-purifying devices as long as oxygen levels are acceptable. Examples of fumes are smoke billowing from a fire or the fumes generated when welding. Always check the cartridge on your respirator to make sure it is the correct type to use for the air conditions and contaminants found on the job site.

When selecting a respirator to wear while working with specific materials, you must first determine the hazardous ingredients contained in the material and their exposure levels, then choose the proper respirator to protect yourself at these levels. Always read the product's MSDS. It identifies the hazardous ingredients and should list the type of respirator and cartridge recommended for use with the product.

Limitations that apply to all half-mask (air-purifying) respirators are as follows:

- These respirators do not completely eliminate exposure to contaminants, but they will reduce the level of exposure to below hazardous levels.
- These respirators do not supply oxygen and must not be used in areas where the oxygen level is below 19.5%.
- These respirators must not be used in areas where chemicals have poor warning signs, such as no taste or odor.

If your breathing becomes difficult, if you become dizzy or nauseated, if you smell or taste the chemical, or if you have other noticeable effects, leave the area immediately, return to a fresh air area, and seek any necessary assistance.

4.3.1 Positive and Negative Fit Checks

All respirators are useless unless properly fit-tested to each individual. To obtain the best protection from your respirator, you must perform positive and negative fit checks each time you wear it. These fit checks must be done until you have obtained a good face seal.

To perform the positive fit check, do the following:

Step 1 Adjust the facepiece for the best fit, then adjust the head and neck straps to ensure good fit and comfort.

> **WARNING!**
> Do not overtighten the head and neck straps. Tighten them only enough to stop leakage. Overtightening can cause facepiece distortion and dangerous leaks.

Step 2 Block the exhalation valve with your hand or other material.

Step 3 Breathe out into the mask.

Step 4 Check for air leakage around the edges of the facepiece.

Step 5 If the facepiece puffs out slightly for a few seconds, a good face seal has been obtained.

To perform a negative fit check, do the following:

Step 1 Block the inhalation valve with your hand or other material.

Step 2 Attempt to inhale.

Step 3 Check for air leakage around the edges of the facepiece.

Step 4 If the facepiece caves in slightly for a few seconds, a good face seal has been obtained.

5.0.0 ◆ CONFINED SPACE PERMITS

A confined space refers to a relatively small or restricted space, such as a storage tank, boiler, or pressure vessel or small compartments, such as underground utility vaults, small rooms, or the unventilated corners of a room.

OSHA 29 CFR 1910.146 defines a confined space (*Figure 16*) as a space that:

- Is large enough and so configured that an employee can bodily enter and perform assigned work
- Has a limited or restricted means of entry or exit; for example, tanks, vessels, silos, storage bins, hoppers, vaults, and pits
- Is not designed for continuous employee occupancy

Respirator Inspection, Care, and Maintenance

A respirator must be clean and in good condition, and all of its parts must be in place for it to give you proper protection. Respirators must be cleaned every day. Failure to do so will limit their effectiveness and offer little or no protection. For example, suppose you wore the respirator yesterday and did not clean it. The bacteria from breathing into the respirator, plus the airborne contaminants that managed to enter the facepiece, will have made the inside of your respirator very unsanitary. Continued use may cause you more harm than good. Remember, only a clean and complete respirator will provide you with the necessary protection. Follow these guidelines:

- Inspect the condition of your respirator before and after each use.

- Do not wear a respirator if the facepiece is distorted or if it is worn and cracked. You will not be able to get a proper face seal.

- Do not wear a respirator if any part of it is missing. Replace worn straps or missing parts before using.

- Do not expose respirators to excessive heat or cold, chemicals, or sunlight.

- Clean and wash your respirator after every time you use it. Remove the cartridge and filter, hand wash the respirator using mild soap and a soft brush, and let it air dry overnight.

- Sanitize your respirator each week. Remove the cartridge and filter, then soak the respirator in a sanitizing solution for at least two minutes. Thoroughly rinse with warm water and let it air dry overnight.

- Store the clean and sanitized respirator in its resealable plastic bag. Do not store the respirator face down. This will cause distortion of the facepiece.

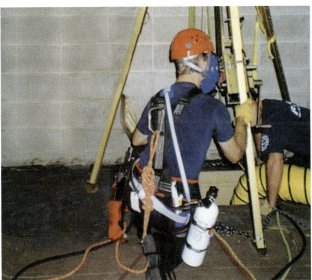

101F16.EPS

Figure 16 ◆ Worker entering a confined space with a restricted opening for entry and exit.

OSHA 29 CFR 1910.146 further defines a permit-required confined space as a space that:

- Contains or has the potential to contain a hazardous atmosphere
- Contains a material that has the potential for engulfing an entrant
- Has an internal configuration such that an entrant could be trapped or asphyxiated by inwardly converging walls or by a floor that slopes downward and tapers to a smaller cross section
- Contains any other recognized serious safety or health hazard

For safe working conditions, the oxygen level in a confined space atmosphere must range between 19.5% and 21.5% by volume, with 21% being considered the normal level. Oxygen concentrations below 19.5% by volume are considered deficient; those above 23.5% by volume are considered enriched (*Table 1*).

When welding or cutting is being performed in any confined space, the gas cylinders and welding machines are left on the outside. Before operations are started, the wheels for heavy portable

Confined Space Precautions

Table 1 indicates the effect of increases and decreases in oxygen levels in a confined space. If too much oxygen is ventilated into a confined space, it can be absorbed by the welder's clothing and ignite. If too little oxygen is present, it can lead to the welder's death in minutes. For this reason, the following precautions apply:

- Make sure confined spaces are ventilated properly for cutting or welding purposes.
- Never use oxygen in confined spaces for ventilation purposes.

Table 1 Effects of an Increase or Decrease in Oxygen Levels

Oxygen Level	Effects
> 21.5%	Easy ignition of flammable material such as clothes
19.5% – 21.5%	Normal
17%	Deterioration of night vision, increased breathing volume, accelerated heartbeat
14% – 16%	Very poor muscular coordination, rapid fatigue, intermittent respiration
6% – 10%	Nausea, vomiting, inability to perform, unconsciousness
< 6%	Spasmodic breathing, convulsive movements, and death in minutes

equipment are securely blocked to prevent accidental movement. Where a welder must enter a confined space through a manhole or other opening, all means are provided for quickly removing the worker in case of emergency. When safety belts and lifelines are used for this purpose, they are attached to the welder's body so that he or she cannot be jammed in a small exit opening. An attendant with a pre-planned rescue procedure is stationed outside to observe the welder at all times and must be capable of putting the rescue operations into effect.

When welding or cutting operations are suspended for any substantial period of time, such as during lunch or overnight, all electrodes are removed from the holders, and the holders are carefully located so that accidental contact cannot occur. The welding machines are also disconnected from the power source.

In order to eliminate the possibility of gas escaping through leaks or improperly closed valves when gas welding or cutting, the gas and oxygen supply valves must be closed, the regulators released, the gas and oxygen lines bled, and

the valves on the torch shut off when the equipment will not be used for a substantial period of time. Where practical, the torch and hose are also removed from the confined space. After welding operations are completed, the welder must mark the hot metal or provide some other means of warning other workers.

6.0.0 ◆ AREA SAFETY

An important factor in area safety is good housekeeping. The work area should be picked up and swept clean. The floors and workbenches should be free of dirt, scrap metal, grease, oil, and anything that is not essential to accomplishing the given task. Collections of steel, welding electrode studs, wire, hoses, and cables are difficult to work around and easy to trip over. An electrode caddy can be used to hold the electrodes and studs. Hooks can be made to hold hoses and cables, and scrap steel should be thrown into scrap bins.

The ideal welding shop should have bare concrete floors and bare metal walls and ceilings to reduce the possibility of fire. Never weld or cut over wood floors, as this increases the possibility of fire.

It is important to keep flammable liquids as well as rags, wood scraps, piles of paper, and other combustibles out of the welding area.

If you must weld in an enclosed building, make every effort to eliminate anything that could trap a spark. Sparks can smolder for hours and then burst into flames. Regardless of where you're welding, be sure to have a fire extinguisher nearby. Also, keep a five-gallon bucket of water handy to cool off hot metal and quickly douse small fires. If a piece of hot metal must be left unattended, use soapstone to write the word *HOT* on it before leaving. This procedure can also be used to warn people of hot tables, vises, firebricks, and tools.

Never use a cutting torch inside your workshop unless a proper cutting area is available. Take whatever you're going to cut outside, away from

101F17.EPS

Figure 17 ◆ A typical welding screen.

flammables. Also be aware that welding sparks can ignite gasoline fumes in a confined space.

Whenever welding must be done outside a welding booth, use portable screens to protect other personnel from the arc or reflected glare (see *Figure 17*). The portable screen also prevents drafts of air from interfering with the stability of the arc.

The most common welding accident is burned hands and arms. Keep first-aid equipment nearby to treat burns in the work area. Eye injuries can also occur if you are careless. Post emergency phone numbers in a prominent location.

The following are some work-area reminders:

- Eliminate tripping hazards by coiling cables and keeping clamps and other tools off the floor.
- Don't get entangled in cables, loose wires, or clothing while you work. This allows you to move freely, especially should your clothing ignite or some other accident occur.
- Clean up oil, grease, or other agents that may ignite and splatter off surfaces while welding.
- Shut off the welder and disconnect the power plug before performing any service or maintenance.
- Keep the floor free of electrodes once you begin to weld. They could cause a slip or fall.
- Work in a dry area, booth, or other shielded area whenever possible.
- Make sure there are no open doors or windows through which sparks may travel to flammable materials.

7.0.0 ◆ HOT WORK PERMITS AND FIRE WATCHES

A hot work permit (*Figure 18*) is an official authorization from the site manager to perform work that may pose a fire hazard. The permit includes information such as the time, location, and type of work being done. The hot work permit system promotes the development of standard fire safety guidelines. Permits also help managers keep records of who is working where and at what time. This information is essential in the event of an emergency or at other times when personnel need to be evacuated.

During a fire watch, a person other than the welder or cutting operator must constantly scan the work area for fires. Fire watch personnel must have ready access to fire extinguishers and alarms and know how to use them. Cutting operations must never be performed without a fire watch. Whenever oxyfuel cutting equipment is used, there is a great danger of fire. Hot work permits and fire watches are used to minimize this danger. Most sites require the use of hot work permits and fire watches. When they are violated, severe penalties are imposed.

 WARNING!

Never perform any type of heating, cutting, or welding until you have obtained a hot work permit and established a fire watch. If you are unsure of the procedure, check with your supervisor. Violation of hot work permit and fire watch procedures can result in serious injury or death.

8.0.0 ◆ OXYFUEL GAS WELDING AND CUTTING SAFETY

Most welding environment fires occur during oxyfuel gas welding or cutting. Welders should be well trained in the function and operation of each part of an oxyfuel gas welding or cutting station. Welders should also be trained and tested in the correct methods of starting, testing for leaks, and shutting down an oxyfuel gas welding or cutting station. They must be aware of the hazards involved in the use of fuel gas, oxygen, and shielding gas cylinders and how these cylinders are stored safely.

It is recommended that at least one fire watch be posted with an extinguisher to watch for possible fires.

9.0.0 ◆ CUTTING CONTAINERS

Cutting and welding activities present unique hazards depending upon the material being cut or welded and the fuel used to power the equipment. All cutting and welding should be done in designated areas of the shop if possible. These areas should be made safe for welding and cutting

HOT WORK PERMIT

For Cutting, Welding, or Soldering with Portable Gas or ARC Equipment

(References: 1997 Uniform Fire Code Article 49 & National Fire Protection Association Standard NFPA 51B.)

Job Date _____ Start Date _____ Expiration _____ WO# _____

Name of Applicant _____ Company _____ Phone _____

Supervisor _____ Phone _____

Location / Description of work _____

_____

_____

IS FIRE WATCH REQUIRED?

1._____ (yes or no)   Are combustible materials in building construction closer than 35 feet to the point of operation?

2._____ (yes or no)   Are combustibles more than 35 feet away but would be easily ignited by sparks?

3._____ (yes or no)   Are wall or floor openings within a 35 foot radius exposing combustible material in adjacent areas, including concealed spaces in floors or walls?

4._____ (yes or no)   Are combustible materials adjacent to the other side of metal partitions, walls, ceilings, or roofs which could be ignited by conduction or radiation?

5._____ (yes or no)   Does the work necessitate disabling a fire detection, suppression, or alarm system component?

YES to any of the above indicates that a qualified fire watch is required.

Fire Watcher Names(s)_____ Phone_____

NOTIFICATIONS

Notify the following groups at least 72 hours prior to work and 30 minutes after work is completed.
Write in names of persons contacted.
Notify in person OR by phone ONLY if question #5 above is answered "yes":

• Facilities Management Fire Alarm Supervisor _____

Notify by phone or in person: (If by phone, write down name of person and send them a completed copy of this permit.)

• Facilities Management Fire Protection Group _____

• Environmental Health & Safety Industrial Hygiene Group _____

SIGNATURES REQUIRED

University Project Manager _____ Date _____ Phone _____

I understand and will abide by the conditions described in this permit. I will implement the necessary precautions which are outlined on both sides of this permit form. Thirty minutes after each hot work session, I will reinspect work areas and adjacent areas to which spark and heat might have spread to verify that they are fire safe, and contact Facilities Management Alarm Technicians to have any disabled fire protection systems reactivated.

_____   _____ Date _____ Phone _____
   Permit Applicant   Company or Department
1/17/03

Figure 18 ◆ Hot work permit.

Oxyfuel Gas Welding Safety Precautions

- Always light the oxyfuel gas torch flame using an approved torch lighter to avoid burning your fingers.

- Never point the torch tip at anyone when lighting it or using it.

- Never point the torch at the cylinders, regulators, hoses, or anything else that may be damaged and cause a fire or explosion.

- Never lay a lighted torch down on the bench or work piece, and do not hang it up while it is lighted. If the torch is not in the welder's hands, it must be off.

- To prevent possible fires and explosions, check valves and flashback arrestors must be installed in all oxyfuel gas welding and cutting outfits.

- When cutting with oxyfuel gas equipment, clear the area of all combustible materials.

- Skin contact with liquid oxygen can cause frostbite. Be careful when handling liquid oxygen.

- Never use oxygen as a substitute for compressed air.

- All oxygen cylinders with leaky valves or safety fuse plugs and discs should be set aside and marked for the attention of the supplier. Do not tamper with or attempt to repair oxygen cylinder valves. Do not use a hammer or wrench to open the valves.

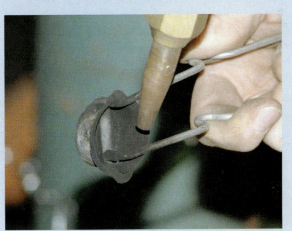

101P0103.EPS

operations with concrete floors, arc filter screens, protective drapes, curtains or blankets (*Figure 19*), and fire extinguishers. No combustibles should be stored nearby.

Containers must be cleaned by steam cleaning, flushing with water, or washing with detergent until all traces of the material have been removed.

WARNING!

Welding or cutting must never be performed on drums, barrels, tanks, vessels, or other containers until they have been emptied and cleaned thoroughly, eliminating all flammable materials and all substances (such as detergents, solvents, greases, tars, or acids) that might produce flammable, toxic, or explosive vapors when heated.

WARNING!

Clean containers only in well-ventilated areas. Vapors can accumulate during cleaning, causing explosions or injury.

After cleaning the container (*Figure 20*), fill it with water or an inert gas such as argon or carbon dioxide (CO_2) for additional safety. Air, which contains oxygen, is displaced from inside the container by the water or inert gas. Without oxygen,

3,000°F INTERMITTENT, 1,500°F CONTINUOUS
SILICON DIOXIDE CLOTH

101F19.EPS

Figure 19 ◆ A typical welding blanket.

combustion cannot take place. When using water, position the container to minimize the air space. When using an inert gas, provide a vent hole so the inert gas can purge the air.

 WARNING!

Do not assume that a container that has held combustibles is clean and safe until proven so by proper tests. Do not weld in places where dust or other combustible particles are suspended in air or where explosive vapors are present. Removal of flammable materials from vessels/containers may be done by steaming or boiling.

Proper procedures for cutting or welding hazardous containers are described in the *American Welding Society (AWS) F4.1-1999, Recommended Safe Practices for the Preparation for Welding and Cutting of Containers and Piping.* You should also consult with the local fire marshal before welding or cutting such containers.

10.0.0 ◆ CYLINDER STORAGE AND HANDLING

Oxygen and fuel gas cylinders or other flammable materials must be stored separately. The storage areas must be separated by 20' or by a wall 5' high with at least a 30-minute burn rating. The purpose of the distance or wall is to keep the heat of a small fire from causing the oxygen cylinder safety valve to release. If the safety valve releases the oxygen, a small fire would become a raging inferno.

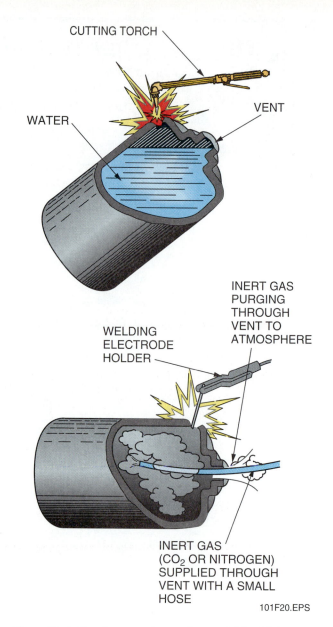

101F20.EPS

Figure 20 ◆ Purging containers of potential health hazards.

Inert gas cylinders may be stored separately or with either fuel cylinders or oxygen cylinders. Empty cylinders must be stored separately from full cylinders, although they may be stored in the same room or area. All cylinders must be stored vertically and have the protective caps screwed on firmly.

10.1.0 Securing Gas Cylinders

Cylinders must be secured with a chain or other device so that they cannot be knocked over accidentally. Even though they are more stable, cylinders attached to a manifold should be chained, as should cylinders stored in a special room used only for cylinder storage.

10.2.0 Storage Areas

Cylinder storage areas must be located away from halls, stairwells, and exits so that in case of an emergency they will not block an escape route. Storage areas should also be located away from heat, radiators, furnaces, and welding sparks. The location of storage areas should be where unauthorized people cannot tamper with the cylinders. A warning sign that reads *Danger—No Smoking, Matches, or Open Lights,* or similar wording, should be posted in the storage area.

10.3.0 Cylinders with Valve Protection Caps

Cylinders equipped with a valve protection cap must have the cap in place unless the cylinder is in use. The protection cap prevents the valve from being broken off if the cylinder is knocked over. If the valve of a full high-pressure cylinder (argon, oxygen, CO_2, or mixed gases) is broken off, the cylinder can fly around the shop like a missile if it has not been secured properly. Never lift a cylinder by the safety cap or valve. The valve can easily break off or be damaged. When moving cylinders, the valve protection cap must be replaced (*Figure 21*), especially if the cylinders are mounted on a truck or trailer. Cylinders must never be dropped or handled roughly.

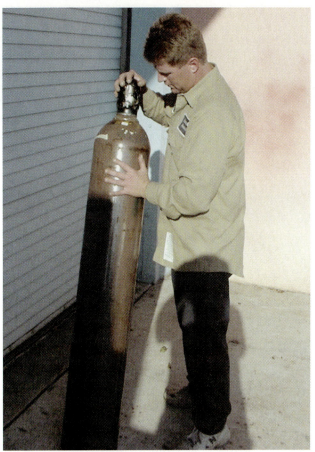

101F21.EPS

Figure 21 ◆ Handle bottles with care.

10.4.0 General Precautions

Use warm water (not boiling) to loosen cylinders that are frozen to the ground. Any cylinder that leaks, has a bad valve, or has gas-damaged threads must be identified and reported to the supplier. Use a piece of soapstone to write the problem on the cylinder. If closing the cylinder valve cannot stop the leak, move the cylinder outdoors to a safe location, away from any source of ignition, and notify the supplier. Post a warning sign, then slowly release the pressure.

In its gaseous form, acetylene is extremely unstable and explodes easily. For this reason it must remain at pressures below 15 pounds per square inch (psi). If an acetylene cylinder is tipped over, stand it upright and wait at least 30 minutes before using it. If liquid acetone is withdrawn from a cylinder, it will gum up the safety check valves and regulators and decrease the stability of the acetylene stored in the cylinder. For this reason, acetylene must never be withdrawn at a per-hour rate that exceeds 10% of the volume of the cylinder(s) in use. Acetylene cylinders in use should be opened no more than one and one-half

turns and, preferably, no more than three-fourths of a turn.

Other precautions include:

- Use only compressed gas cylinders containing the correct gas for the process used and properly operating regulators designed for the gas and pressure used.
- Make sure all hoses, fittings, and other parts are suitable and maintained in good condition.
- Keep cylinders in the upright position and securely chained to an undercarriage or fixed support.
- Keep combustible cylinders in one area of the building for safety. Cylinders must be at a safe distance from arc welding or cutting operations and any other source of heat, sparks, or flame.
- Never allow the electrode, electrode holder, or any other electrically hot parts to come in contact with the cylinder.
- When opening a cylinder valve, keep your head and face clear of the valve outlet.
- Always use valve protection caps on cylinders when they are not in use or are being transported.

Power Tool Safety Precautions

- Wear heavy protective clothing, a hard hat, goggles, and heavy work gloves to guard against injury from flying debris.

- Use adequate hearing protection.

- Use the appropriate respiratory protection when dust hazards are produced.

- To avoid shock, ensure that electrically powered tools are adequately grounded and used in a dry environment. For added protection, use tools that are double insulated and equipped with ground fault circuit interrupters.

- Do not use power tools in confined spaces where sparks could cause explosions.

- Use nonsparking tools around combustible and flammable materials.

- Do not exceed the rated speed of the tool.

- Use recommended tool guards.

- Inspect equipment regularly, and repair or replace it as necessary.

11.0.0 ◆ POWER TOOL SAFETY

All power tools must be properly grounded to prevent accidental electric shock. If you feel even a slight tingle while using a power tool, stop and have the tool checked by a technician. Power tools should never be used with force or allowed to overheat from excessive or incorrect use. If an extension cord is used, it should have a large enough current rating to carry the load. An extension cord that is too small will cause the tool and the cord to overheat.

12.0.0 ◆ ELECTRICAL SAFETY

Electric shock from welding and cutting equipment can kill or cause severe burns by coming in contact with bare skin. Serious injury can also result if a fall occurs because of the shock. The amount of current that passes through the human body determines the outcome of an electric shock. The higher the voltage, the greater the chance for a fatal shock. Electric current flows along the path of least resistance to return to its source. If you come in contact with a live conductor, you become a load. *Figure 22* shows how much resistance the human body presents under various circumstances and how this converts into amps or milliamps when the voltage is 110V. Note that the potential for shock increases dramatically if the skin is damp. A cut will also reduce resistance and increase shock potential. Currents of less than 1 amp can severely injure and even kill a person.

Each welder and operator of the equipment must be trained and able to recognize the dangers associated with each particular type of equipment to avoid injuries, fatalities, and other electrical accidents.

It is a good idea for all involved to know first-aid procedures, such as cardiopulmonary resuscitation (CPR) for treating electric shock. Treat electrical burns like any other type of burn by applying ice or a cold compress over the burned area. Prevent infection and cover with a clean, dry dressing. Seek medical attention if necessary.

Electrical Safety Precautions

- Never operate arc-welding equipment on a wet or damp floor.

- Never touch an electrode to a grounded surface because these surfaces will become electrically live. The electrode and work circuit is electrically energized when the output is on.

- Never place electrode holders in contact with a grounded metal surface because it could short circuit the welding machine.

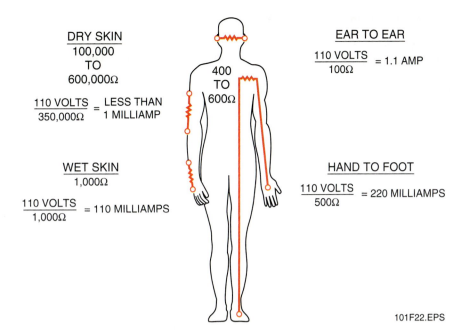

Figure 22 ◆ Typical body resistances and current.

DRY SKIN
100,000
TO
600,000Ω

$$\frac{110\ VOLTS}{350,000\Omega} = \text{LESS THAN 1 MILLIAMP}$$

WET SKIN
1,000Ω

$$\frac{110\ VOLTS}{1,000\Omega} = 110\ MILLIAMPS$$

400
TO
600Ω

EAR TO EAR

$$\frac{110\ VOLTS}{100\Omega} = 1.1\ AMP$$

HAND TO FOOT

$$\frac{110\ VOLTS}{500\Omega} = 220\ MILLIAMPS$$

101F22.EPS

Shock Prevention

- Floors must be dry at all times. Use wooden platforms or rubberized carpet/floor coverings or other insulated material.

- Only experienced electricians may work on electric arc welding machine power connections.

- Workbenches must be grounded.

- With the power off, cable connections must be checked for cracked insulators, loose contacts, and worn or cut hoses.

- A machine must never be operated above its rated capacity.

13.0.0 ◆ MATERIAL SAFETY DATA SHEETS

All manufacturers of potentially hazardous materials must provide detailed information regarding possible hazards resulting from the use of their product in the form of a material safety data sheet (MSDS) to the users of their products. An MSDS must be provided to anyone using the product or anyone working in the area where the product is in use. Often, companies will post these sheets on a bulletin board or put them in a convenient place near the work area.

All MSDSs must contain the same basic information in at least nine required sections. How-ever, some MSDSs may have additional sections covering transportation, environmental issues, and other subjects, as deemed necessary by the manufacturer. Depending on the manufacturer, MSDSs are often formatted in different ways. *Figure 23* shows a portion of an MSDS produced by one major manufacturer.

14.0.0 ◆ MATERIAL HANDLING

Proper lifting, moving, and handling of large, heavy, welded assemblies is important to the safety of the workers and the weldment. Improper work habits can cause serious personal injury and/or damage to equipment and materials.

HAYNES
International

MATERIAL SAFETY DATA SHEET

SAFETY DEPARTMENT
1020 WEST PARK AVENUE
P.O. BOX 9013
KOKOMO, INDIANA 46904-9013
INFORMATION: 765-456-6614

HAYNES INTERNATIONAL, INC.
Welding Consumables

MSDS IDENTIFICATION NUMBER **H1072-4** This replaces H1072-3	PREVIOUS REVISION DATE 01/07/97 DATE REVISED 07/31/99	EMERGENCY PHONE NUMBERS HAYNES: 765-456-6894 CHEMTREC: 800-424-9300 (24-hour contact for Health & Transportation Emergencies)

This Material Safety Data Sheet (MSDS) provides information on a specific group of manufactured metal products. Since these metal products share a common physical nature and constituents, the data presented are applicable to all alloys identified. This document was prepared to meet the requirements of OSHA's Hazard Communication Standard, 29 CFR 1910.1200 and the Superfund Amendments and Reauthorization Act of 1986 Public Law 99-949.

15. REGULATORY INFORMATION	
U.S. FEDERAL REGULATIONS	**OSHA:** Listed as an air contaminant (29 CFR 1910.1000). Hazardous by definition of Hazard Communication Standard (29 CFR 1910.1200).
	TSCA (Toxic Substance Control Act): Components of this material are listed on the TSCA inventory.
	CERCLA: Hazardous Substance (40 CFR 302.4): Chromium (as a powder), Copper, Nickel (as a powder or dust) Extremely Hazardous Substance (40 CFR 355): Not Listed
	SARA HAZARD CATEGORY: Listed below are the hazard categories for Sections 311 and 312 of the Superfund Amendment and Reauthorization Act of 1986 (SARA Title III):
	Immediate Hazard: X Delayed Hazard: X Fire Hazard: – Pressure Hazard: – Reactivity Hazard: –

101F23.EPS

Figure 23 ◆ Sample MSDS.

14.1.0 Lifting

When lifting a heavy object, distribute the weight of the object evenly between both hands. Use your legs to lift, not your back. Do not try to lift a large or bulky object without help if the object is heavier than you can lift with one hand.

14.2.0 Hoists or Cranes

Hoists or cranes can be overloaded with welded assemblies. Check the capacity of the equipment before trying to lift a load. Keep any load as close to the ground as possible while it is being moved. Pushing a load on a crane is better than pulling a load. If it is necessary to pull a load, use a rope. When moving or lifting a load with ropes, chains, and cables, stand to one side of the load to prevent injury if the equipment or material holding the load should break or snap back.

15.0.0 ◆ SAFETY PLANNING AND EMERGENCY ACTION PLANS

Planning is critical to safety. Accidents occur if you let them, but if you anticipate and plan ways to avoid them, accidents can be reduced or eliminated. Hazards in welding can arise from conditions at the work site and from the tools, equipment, and materials being used. Before work begins, the site should be surveyed for hazardous materials and unsafe conditions. Regardless of the size of the job, the site should always be evaluated to reveal any hazard such as confined spaces, unguarded openings, possible electrical hazards, and environmental hazards.

Once all potential hazards have been identified, a comprehensive emergency action plan should be developed to deal with all emergencies. Depending on prevailing OSHA standards and the number of people involved, the plan may be required to be in writing, or it may be communicated orally. As appropriate, an emergency action plan must include the following elements:

- Emergency escape procedures and emergency route assignments
- Procedures to be followed by employees who remain to take corrective actions
- Procedures that account for all employees after any emergency evacuation has been completed
- Rescue and medical duties, along with the necessary training for those individuals who are to perform them

- The preferred means for reporting fires and other emergencies, including all phone numbers
- Names and/or job titles of persons or departments to be contacted for further information or explanation of duties under the plan

Summary

Safety is everyone's responsibility. Proper clothing, footwear, and eye protection are essential for safe welding and cutting. Workers who fail to comply with safety rules are subject to dismissal. All welding shops must have established plans for dealing with accidents. Take the time to learn the proper procedures for accident response and reporting before you need to respond to an emergency.

Accidents are very harmful for both employees and employers, and they are often caused by poor behavior and unsafe conditions. Most accidents can be prevented. By knowing and avoiding the behaviors that cause accidents and by keeping working conditions safe, it may be possible to avoid injuries and reduce hazards.

Trying to bluff your way through a job you do not understand is asking for trouble. Even if you think you know the correct procedures, a review may bring out an important part of the job that you may have forgotten. Don't be afraid to ask questions. The responses you receive may help new or less experienced co-workers get answers to questions they may be too bashful to ask.

Materials, fumes, welding radiation exposure, and storing and handling cylinders present particular hazards for the welder. Developing an attitude of safety is an excellent way for every worker to avoid or reduce all of these hazards. Practicing good safety attitudes means that you:

- Report all unsafe conditions and behaviors immediately.
- Keep work areas clean and orderly at all times.
- Immediately report all accidents and injuries, no matter how minor.
- Be certain you completely understand the instructions given before starting work.
- Know how and where needed medical help may be obtained.
- Wear the required protective devices when working in a hazardous operation area.
- Do not use alcohol or drugs. If you are ill and must take prescribed medication, notify your supervisor immediately.

Review Questions

1. Accidents are more likely for employees in the age group _____.
   a. 18 and under
   b. 19 to 25
   c. 26 to 45
   d. 60 and over

2. Body protection is best provided by wearing _____.
   a. pants with cuffs
   b. polyester clothing
   c. cotton clothing
   d. loose-fitting shirts

3. To prevent fume hazards there must be at least _____ cubic feet of air for each welder.
   a. 2,000
   b. 4,000
   c. 5,000
   d. 10,000

4. The type of respirator that provides the most protection when you are working in an area with either a lack of oxygen or high levels of toxic materials in the air is the _____ type.
   a. half-mask
   b. full-mask
   c. supplied-air
   d. SCBA

5. To make sure your respirator provides proper protection, you must _____ before using it.
   a. perform a positive fit check
   b. perform both positive and negative fit checks
   c. perform a negative fit check
   d. tighten the respirator head and neck straps as much as possible

6. A confined space is one that _____.
   a. has a flammable atmosphere
   b. has unrestricted means of entry or exit
   c. is designed for continuous employee occupancy
   d. is large enough that an employee can bodily enter and perform tasks

7. A good area safety practice is _____.
   a. welding near an open window
   b. welding over wooden floors
   c. to use cardboard to deflect welding/grinding sparks away from others
   d. writing *HOT* on hot metal before leaving it unattended

8. A hot work permit _____.
   a. authorizes the performance of work potentially posing a fire hazard
   b. promotes development of standard fire safety guidelines
   c. records unsafe conditions at a job site
   d. helps the manager keep records of hazardous spaces

9. Cutting operations should never be performed without a _____ in the area.
   a. bucket of sand
   b. bucket of water
   c. fire watch
   d. fire hose

10. Before performing oxyfuel cutting on a cleaned container, fill the container with an inert gas or _____ and position the container to displace as much air as possible from inside the container.
    a. air
    b. sand
    c. water
    d. solvent

11. Because of its instability, the acetylene gas used in oxyfuel cutting must remain at pressures less than _____ psi when in its gaseous form.
    a. 10
    b. 15
    c. 25
    d. 35

12. Because of the liquid acetone inside an acetylene cylinder, acetylene cylinders must always be used in a(n) _____ position.
    a. upright
    b. 45°
    c. horizontal
    d. 60°

13. The potential for electric shock _____.
    a. decreases when the skin is damp
    b. remains the same when the skin is damp
    c. increases when the skin is damp
    d. decreases when the skin is cut

14. An MSDS is a form used to _____.
    a. file a worker's compensation claim with an insurance company
    b. list the contents, hazards, and precautions that pertain to a chemical or material
    c. record unsafe conditions that exist at a job site
    d. recommend changes to your employer's safety program

15. When moving a load with a crane, _____.
  a. keep the load as close to the ground as possible
  b. pull it instead of pushing it
  c. hoist the load as high as possible
  d. stand in the path of the load when directing the crane operator

Additional Resources

This module is intended to present thorough resources for task training. The following reference works are suggested for further study. These are optional materials for continued education rather than for task training.

Modern Welding, A. D. Althouse, C. H. Turnquist, W. A. Bowditch, and K. E. Bowditch, 2000. Tinley Park, IL: The Goodheart-Willcox Company, Inc.

Safety in Welding, Cutting, and Allied Processes, ANSI Z49.1-99, 1999. Miami, FL: American Welding Society.

Welder's Handbook, Richard Finch, 1997. New York, NY: The Berkley Publishing Group, Inc.

The NCCER makes every effort to keep these textbooks up-to-date and free of technical errors. We appreciate your help in this process. If you have an idea for improving this textbook, or if you find an error, a typographical mistake, or an inaccuracy in NCCER's Contren™ textbooks, please write us, using this form or a photocopy. Be sure to include the exact module number, page number, a detailed description, and the correction, if applicable. Your input will be brought to the attention of the Technical Review Committee. Thank you for your assistance.

Instructors—If you found that additional materials were necessary in order to teach this module effectively, please let us know so that we may include them in the Equipment/Materials list in the Instructor's Guide.

Write: Curriculum Revision and Development Department
National Center for Construction Education and Research
P.O. Box 141104, Gainesville, FL 32614-1104

Fax: 352-334-0932

E-mail: curriculum@nccer.org

Craft _____ Module Name _____

Copyright Date _____ Module Number _____ Page Number(s) _____

Description _____

_____

_____

_____

(Optional) Correction _____

_____

_____

(Optional) Your Name and Address _____

_____

_____

Oxyfuel Cutting

Course Map

This course map shows all of the modules in the AWS Entry Level Welder – Phase 1 curriculum. The suggested training order begins at the bottom and proceeds up. Skill levels increase as you advance on the course map. The local Training Program Sponsor may adjust the training order.

AWS ENTRY LEVEL WELDER—PHASE 1

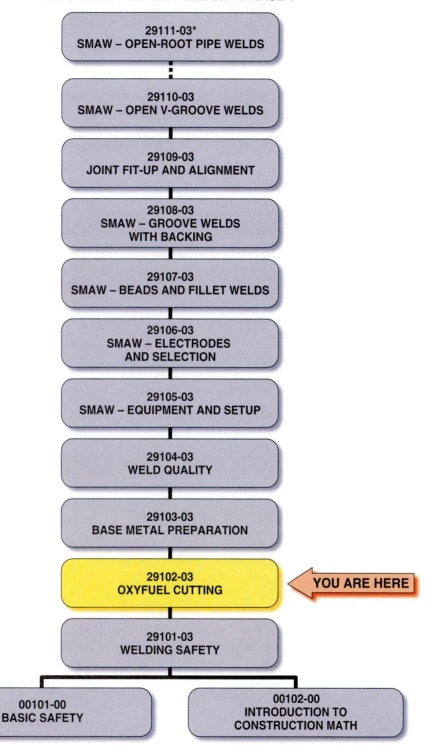

29111-03*
SMAW – OPEN-ROOT PIPE WELDS

29110-03
SMAW – OPEN V-GROOVE WELDS

29109-03
JOINT FIT-UP AND ALIGNMENT

29108-03
SMAW – GROOVE WELDS
WITH BACKING

29107-03
SMAW – BEADS AND FILLET WELDS

29106-03
SMAW – ELECTRODES
AND SELECTION

29105-03
SMAW – EQUIPMENT AND SETUP

29104-03
WELD QUALITY

29103-03
BASE METAL PREPARATION

29102-03
OXYFUEL CUTTING   ← YOU ARE HERE

29101-03
WELDING SAFETY

00101-00
BASIC SAFETY

00102-00
INTRODUCTION TO
CONSTRUCTION MATH

*Module 29111-03 is an elective for those students progressing through the AWS Entry Level Welder program.

29102CMAP.EPS

Figures

Tables

AWS ENTRY LEVEL WELDER – TRAINEE MODULE 29102-03

Oxyfuel Cutting

Objectives

When you have completed this module, you will be able to do the following:

1. Identify and explain the use of oxyfuel cutting equipment.
2. Set up oxyfuel equipment.
3. Light and adjust an oxyfuel torch.
4. Shut down oxyfuel cutting equipment.
5. Disassemble oxyfuel equipment.
6. Change empty cylinders.
7. Perform oxyfuel cutting:
   • Straight line and square shapes
   • Piercing and slot cutting
   • Bevels
   • Washing
   • Gouging
8. Operate a motorized, portable oxyfuel gas cutting machine.

Prerequisites

Before you begin this module, it is recommended that you successfully complete Modules 00100-00 through 29101-03.

Required Trainee Materials

1. Pencil and paper
2. Appropriate personal protective equipment

1.0.0 ◆ INTRODUCTION

Oxyfuel cutting (OFC), also called flame cutting or burning, is a process that uses the flame and oxygen from a cutting torch to cut **ferrous metals.** The flame is produced by burning a fuel gas mixed with pure oxygen. The flame heats the metal to be cut to the kindling temperature (a cherry-red color); then a stream of high-pressure pure oxygen is directed from the torch at the metal's surface. This causes the metal to instantaneously oxidize or burn. The cutting process results in oxides that mix with molten iron and produce **slag,** which is blown from the cut by the jet of cutting oxygen. This oxidation process, which takes place during the cutting operation, is similar to a greatly sped-up rusting process. *Figure 1* shows oxyfuel cutting.

The oxyfuel cutting process is usually used only on ferrous metals such as straight carbon steels, which oxidize rapidly. This process can be used to quickly cut, trim, and shape ferrous metals, including the hardest steel.

Oxyfuel cutting can be used for certain metal alloys, such as stainless steel; however, the process requires higher preheat temperatures (white heat) and about 20% more oxygen for cutting. In addition, sacrificial steel plate or rod may have to be placed on top of the cut to help maintain the burning process. Other methods, such as carbon arc cutting, powder cutting, inert gas cutting, and plasma arc cutting, are much more practical for cutting steel alloys and nonferrous metals. These other methods will be covered in a later module.

102F01.EPS

Figure 1 ◆ Oxyfuel cutting.

2.0.0 ◆ OXYFUEL SAFETY SUMMARY

The proper safety equipment and precautions must be used when working with oxyfuel equipment because of the potential danger from the high-pressure flammable gases and high temperatures used. The following is a summary of safety procedures and practices that must be observed while cutting or welding. Keep in mind that this is just a summary. Complete safety coverage is provided in the Level One module, *Welding Safety*. If you have not completed that module, do so before continuing. Above all, be sure to wear appropriate protective clothing and equipment when welding or cutting.

2.1.0 Protective Clothing and Equipment

- Always use safety goggles with a full face shield or a helmet. The goggles, face shield, or helmet lens must have the proper light-reducing tint for the type of welding or cutting to be performed. Never directly or indirectly view an electric arc without using a properly tinted lens *(Figure 2)*.
- Wear proper protective leather and/or flame retardant clothing along with welding gloves that will protect you from flying sparks and molten metal, as well as heat.
- Wear 8" or taller high-top safety shoes or boots. Make sure that the tongue and lace area of the footwear will be covered by a pant leg. If the tongue and lace area is exposed or the footwear must be protected from burn marks, wear leather spats under the pants or chaps and over the top of the footwear.
- Wear a solid material (nonmesh) hat with a bill pointing to the rear or, if much overhead cutting or welding is required, a full leather hood with a welding face plate and the correctly tinted lens. If a hard hat is required, use a hard hat that allows the attachment of rear deflector material and a face shield.
- If a full leather hood is not worn, wear a face shield and snugly fitting welding goggles over safety glasses for gas welding or cutting. Either the face shield or the lenses of the welding gog-

EAR PLUGS

TINTED VISOR

CLEAR GOGGLES OVER SAFETY GLASSES

LEATHER CAP (VISOR TURNED BACK)

LEATHER JACKET

ALTERNATE HEAD AND FACE PROTECTION

GAUNTLET-TYPE WELDING GLOVES

LEATHER CHAPS OVER CUFFLESS PANTS (OR LEATHER PANTS)

HIGH-TOP LEATHER BOOTS

102F02.EPS

Figure 2 ◆ Typical personal protective equipment.

gles must be an approved shade 5 or 6 filter. Depending on the method used for electric arc cutting, wear safety goggles and a welding hood with the correctly tinted lens (shade 5 to 14).

• If a full leather hood is not worn, wear ear-muffs, or at least earplugs, to protect your ear canals from sparks.

2.2.0 Fire/Explosion Prevention

• Never carry matches or gas-filled lighters in your pockets. Sparks can cause the matches to ignite or the lighter to explode, resulting in serious injury.

• Never perform any type of heating, cutting, or welding until a hot work permit is obtained and an approved fire watch is established. Most work site fires caused by these types of operations are started by cutting torches.

• Never use oxygen to blow off clothing. The oxygen can remain trapped in the fabric for a time. If a spark hits the clothing during this time, the clothing can burn rapidly and violently out of control.

• Make sure that any flammable material in the work area is moved or shielded by a fire-resistant covering. Approved fire extinguishers must be available before attempting any heating, welding, or cutting operations.

• Always comply with any site requirement for a hot-work permit and/or a fire watch.

• Never release a large amount of oxygen or use oxygen as compressed air. Its presence around flammable materials or sparks can cause rapid and uncontrolled combustion. Keep oxygen away from oil, grease, and other petroleum products.

• Never release a large amount of fuel gas, especially acetylene. Methane and propane tend to concentrate in and along low areas and can ignite at a considerable distance from the release point. Acetylene is lighter than air but is even more dangerous. When mixed with air or oxygen, it will explode at much lower concentrations than any other fuel.

• To prevent fires, maintain a neat and clean work area and make sure that any metal scrap or slag is cold before disposal.

• Before cutting or welding containers such as tanks or barrels, check to see if they have contained any explosive, hazardous, or flammable materials, including petroleum products, citrus products, or chemicals that decompose into toxic fumes when heated. As a standard practice, always clean and then fill any tanks or barrels with water, or purge them with a flow of inert gas to displace any oxygen (*Figure 3*).

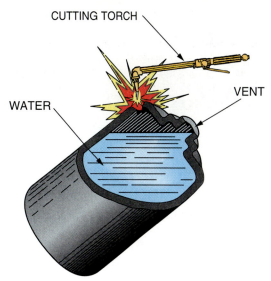

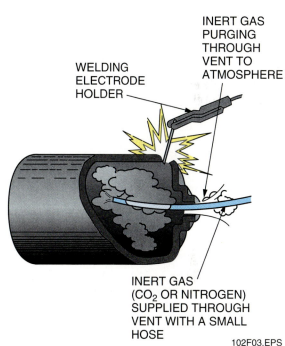

Figure 3 ◆ Eliminating/minimizing oxygen in a container.

2.3.0 Work Area Ventilation

• Make sure confined-space procedures are followed before conducting any welding or cutting in the confined space.

• Always perform cutting or welding operations in a well-ventilated area. Cutting or welding operations involving materials, coatings, or electrodes containing cadmium, mercury, lead, zinc, chromium, and beryllium result in toxic fumes. For long-term cutting or welding of such materials, always wear an approved full face, supplied-air respirator (SAR) that uses breathing air supplied externally of the work

area. For occasional, very short-term exposure to zinc or copper fumes, a high-efficiency particulate arresting (HEPA)-rated or metal-fume filter may be used on a standard respirator.

- Make sure confined spaces are ventilated properly for cutting or welding purposes.
- Never use oxygen in confined spaces for ventilation purposes.

3.0.0 ◆ OXYFUEL CUTTING EQUIPMENT

The equipment used to perform oxyfuel cutting includes oxygen and fuel gas cylinders, oxygen and fuel gas regulators, hoses, and a cutting torch. A typical movable oxyfuel (oxyacetylene) cutting outfit is shown in *Figure 4*.

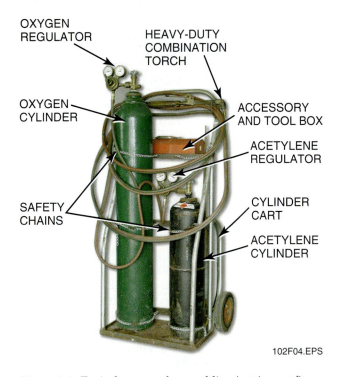

OXYGEN REGULATOR

HEAVY-DUTY COMBINATION TORCH

OXYGEN CYLINDER

ACCESSORY AND TOOL BOX

ACETYLENE REGULATOR

SAFETY CHAINS

CYLINDER CART

ACETYLENE CYLINDER

102F04.EPS

Figure 4 ◆ Typical oxyacetylene welding/cutting outfit.

3.1.0 Oxygen

Oxygen (O_2) is a colorless, odorless, tasteless gas that supports combustion. Combined with burning material, pure oxygen causes a fire to flare and burn out of control. When mixed with fuel gases, oxygen produces the high-temperature flame required in order to flame cut metals.

3.1.1 Oxygen Cylinders

Oxygen is stored at more than 2,000 pounds per square inch (psi) in hollow steel cylinders. The cylinders come in a variety of sizes based on the cubic feet of oxygen they hold. The smallest standard cylinder holds about 85 cubic feet of oxygen, and the largest ultra-high-pressure cylinder holds about 485 cubic feet. The most common size oxygen cylinder used for welding and cutting operations is the 227 cubic foot cylinder. It is more than 4' tall and 9" in diameter. The shoulder of the oxygen cylinder has the name of the gas and/or the supplier stamped, labeled, or stenciled on it. *Figure 5* shows standard high-pressure oxygen cylinder markings and sizes. The cylinders must be tested every 10 years.

Oxygen cylinders have bronze cylinder valves on top *(Figure 6)*. Turning the cylinder valve handwheel controls the flow of oxygen out of the cylinder. A safety plug on the side of the cylinder valve allows oxygen in the cylinder to escape if the pressure in the cylinder rises too high. Oxygen cylinders are usually equipped with Compressed Gas Association (CGA) 540 valves for service up to 3,000 per square inch gauge (psig). Some cylinders are equipped with CGA 577 valves for up to 4,000 psig service or CGA 701 valves for up to 5,500 psig service. Use care when handling oxygen cylinders because oxygen is stored at such high pressures. When it is not in use, always cover the cylinder valve with the protective steel safety cap *(Figure 7)*.

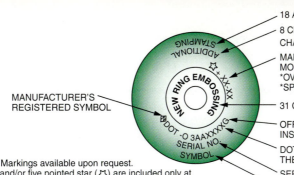

18 ADDITIONAL CHARACTERS-5/16"

8 CHARACTERS-1/2" OR 12 CHARACTERS- 5/16"

MANUFACTURING TEST DATE: MONTH-YEAR
*OVERFILL MARK "+"
*SPECIAL 10-YEAR RETEST MARK-"☆"

31 CHARACTERS-7/16"

OFFICIAL MARK OF INDEPENDENT INSPECTOR-"G"

DOT SPECIFICATIONS TO WHICH THE CYLINDER WAS MANUFACTURED

SERIAL NUMBER

PURCHASER'S USER MARK (UP TO 11 CHARACTERS-1/2")

MANUFACTURER'S REGISTERED SYMBOL

Transport Canada Markings available upon request.
*The plus sign (+) and/or five pointed star (☆) are included only at customer's request, and indicate compliance with applicable requirements of the Code of Federal Regulations, Title 49, Transportation.

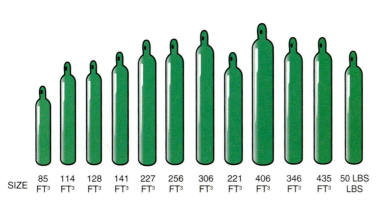

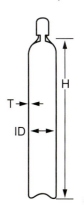

| SIZE | 85 FT³ | 114 FT³ | 128 FT³ | 141 FT³ | 227 FT³ | 256 FT³ | 306 FT³ | 221 FT³ | 406 FT³ | 346 FT³ | 435 FT³ | 50 LBS LBS |

HIGH PRESSURE CYLINDER MARKINGS

DOT SPECIFICATIONS	O₂ CAPACITY (FT³)		WATER CAPACITY (IN³)		NOMINAL DIMENSIONS (IN)			NOMINAL WEIGHT (LB)	PRESSURE (PSI)	
	AT RATED SERVICE PRESSURE	AT 10% OVERCHARGE	MINIMUM	MAXIMUM	AVG. INSIDE DIAMETER "ID"	HEIGHT "H"	MINIMUM WALL "T"		SERVICE	TEST

STANDARD HIGH PRESSURE CYLINDERS[1]

DOT SPECIFICATIONS	AT RATED SERVICE PRESSURE	AT 10% OVERCHARGE	MINIMUM	MAXIMUM	AVG. INSIDE DIAMETER "ID"	HEIGHT "H"	MINIMUM WALL "T"	NOMINAL WEIGHT (LB)	SERVICE	TEST
3AA2015	85	93	960	1040	6.625	32.50	0.144	48	2015	3360
3AA2015	114	125	1320	1355	6.625	43.00	0.144	61	2015	3360
3AA2265	128	140	1320	1355	6.625	43.00	0.162	62	2265	3775
3AA2015	141	155	1630	1690	7.000	46.00	0.150	70	2015	3360
3AA2015	227	250	2640	2710	8.625	51.00	0.184	116	2015	3360
3AA2265	256	281	2640	2710	8.625	51.00	0.208	117	2265	3775
3AA2400	306	336	2995	3060	8.813	55.00	0.226	140	2400	4000
3AA2400	405	444	3960	4040	10.060	56.00	0.258	181	2400	4000

ULTRALIGHT® HIGH PRESSURE CYLINDERS[1]

DOT SPECIFICATIONS	AT RATED SERVICE PRESSURE	AT 10% OVERCHARGE	MINIMUM	MAXIMUM	AVG. INSIDE DIAMETER "ID"	HEIGHT "H"	MINIMUM WALL "T"	NOMINAL WEIGHT (LB)	SERVICE	TEST
E-9370-3280	365	NA	2640	2710	8.625	51.00	0.211	122	3280	4920
E-9370-3330	442	NA	3181	3220	8.813	57.50	0.219	147	3330	4995

ULTRA HIGH PRESSURE CYLINDERS[2]

DOT SPECIFICATIONS	AT RATED SERVICE PRESSURE	AT 10% OVERCHARGE	MINIMUM	MAXIMUM	AVG. INSIDE DIAMETER "ID"	HEIGHT "H"	MINIMUM WALL "T"	NOMINAL WEIGHT (LB)	SERVICE	TEST
3AA3600	347[3]	374	2640	2690	8.500	51.00	0.336	170	3600	6000
3AA6000	434[3]	458	2285	2360	8.147	51.00	0.568	267	6000	10000
E-10869-4500	435[3]	NA	2750	2890	8.813	51.00	0.260	148	4500	6750
E-10869-4500	485[3]	NA	3058	3210	8.813	56.00	0.260	158	4500	6750

1. Regulators normally permit filling these cylinders with 10% overcharge, provided other requirements are met.
2. Under no circumstances are these cylinders to be filled to a pressure exceeding the marked service pressure at 70°F.
3. Nitrogen capacity at 70°F.

All cylinders normally furnished with 3/4" NGT internal threads, unless otherwise specified. Nominal weights include neck ring but exclude valve and cap, add 2 lbs. (.91 kg) for cap and 1 1/2 lb. (.8 kg) for valve.
Cap adds approximately 5 in. (127 mm) to height.
Cylinder capacities are approximately 5 in. (127 mm) to height.
Cylinder capacities are approximately at 70°F. (21°C).

102F05.EPS

Figure 5 ◆ High-pressure oxygen cylinder markings and sizes.

HANDWHEEL

SAFETY PLUG

102F06.EPS

Figure 6 ◆ Oxygen cylinder valve.

 WARNING!

Do not remove the protective cap unless the cylinder is secured. If the cylinder falls over and the nozzle breaks off, the cylinder will be propelled like a rocket, causing severe injury or death to anyone in its way.

3.2.0 Acetylene

Acetylene gas (C_2H_2), a compound of carbon and hydrogen, is lighter than air. It is formed by dissolving calcium carbide in water. It has a strong, distinctive, garlic-like odor. In its gaseous form, acetylene is extremely unstable and explodes easily. Because of this instability, it cannot be compressed at pressures of more than 15 psi when in its gaseous form. At higher pressures, acetylene gas breaks down chemically, producing heat and pressure that could result in a violent explosion.

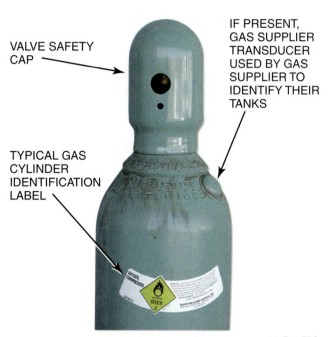

VALVE SAFETY CAP

IF PRESENT, GAS SUPPLIER TRANSDUCER USED BY GAS SUPPLIER TO IDENTIFY THEIR TANKS

TYPICAL GAS CYLINDER IDENTIFICATION LABEL

102F07.EPS

Figure 7 ◆ Oxygen cylinder with standard safety cap.

When combined with oxygen, acetylene creates a flame that burns hotter than 5,500°F, one of the hottest gas flames. Acetylene can be used for flame cutting, welding, heating, flame hardening, and stress relieving.

3.2.1 Acetylene Cylinders

Because of the explosive nature of acetylene gas, it cannot be stored above 15 psi in a hollow cylinder. To solve this problem, acetylene cylinders are specially constructed to store acetylene at higher pressures. The acetylene cylinder is filled with a porous material that creates a solid, instead of a hollow, cylinder. The porous material is soaked with acetone, which absorbs the

Alternate High-Pressure Cylinder Valve Cap

High-pressure cylinders can also be equipped with a clamshell cap that can be closed to protect the cylinder valve with or without a regulator installed on the valve. This enables safe movement of the cylinder after the cylinder valve is closed. This type of cap is usually secured to the cylinder body cap threads when it is installed so that it cannot be removed. When the clamshell is closed, it can also be padlocked to prevent unauthorized operation of the cylinder valve.

CLAMSHELL OPEN TO ALLOW
CYLINDER VALVE OPERATION

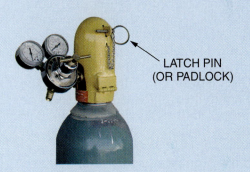

LATCH PIN
(OR PADLOCK)

CLAMSHELL CLOSED FOR MOVEMENT OR PADLOCKED
TO PREVENT OPERATION OF CYLINDER VALVE

CLAMSHELL CLOSED FOR TRANSPORT

102P0201.EPS

acetylene, stabilizing it for storage at higher pressures. Because of the liquid acetone inside the cylinder, acetylene cylinders must always be used in an upright position. If the cylinder is tipped over, stand the cylinder upright and wait at least 30 minutes before using it. If liquid acetone is withdrawn from a cylinder, it will gum up the safety check valves and regulators. Always take care to withdraw acetylene gas from a cylinder at pressures less than 15 psig and at hourly rates that do not exceed one-tenth of the cylinder capacity. Higher rates may cause liquid acetone to be withdrawn along with the acetylene.

Acetylene cylinders have safety fuse plugs in the top and bottom of the cylinder that melt at 220°F (Figure 8). In the event of a fire, the fuse plugs will release the acetylene gas, preventing the cylinder from exploding.

Acetylene cylinders are available in a variety of sizes based on the cubic feet of acetylene that they can hold. The smallest standard cylinder holds about 10 cubic feet of gas. The largest standard cylinder holds about 420 cubic feet of gas. A cylinder that holds about 850 cubic feet is also available. Figure 9 shows standard acetylene cylinder markings and sizes. Like oxygen cylinders, acetylene cylinders must be tested every 10 years.

Acetylene cylinders are usually equipped with a standard CGA 510 brass cylinder valve (see Figure 9). The handwheel of the valve controls the flow of acetylene from the cylinder to a regulator. Some acetylene cylinders are equipped with an alternate standard CGA 300 valve. Some obsolete valves still in use require a special long-handled wrench with a square socket end to operate the valve.

The smallest standard acetylene cylinder, which holds 10 cubic feet, is equipped with a CGA 200 small series valve, and 40 cubic foot cylinders use a CGA 520 small series valve. As with oxygen cylinders, place a protective valve cap on the acetylene cylinders during transport (Figure 10).

WARNING!

Do not remove the protective cap unless the cylinder is secured. If the cylinder falls over and the nozzle breaks off, the cylinder will release highly explosive gas.

3.3.0 Liquefied Fuel Gases

Many fuel gases other than acetylene are used for cutting. They include natural gas and liquefied fuel gases such as methylacetylene propadiene

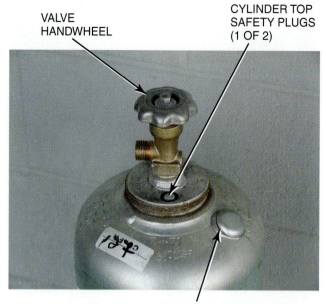

VALVE HANDWHEEL

CYLINDER TOP SAFETY PLUGS (1 OF 2)

IF PRESENT, GAS SUPPLIER TRANSDUCER FOR CYLINDER IDENTIFICATION

CYLINDER BOTTOM SAFETY PLUGS

102F08.EPS

Figure 8 ◆ Standard acetylene cylinder valve and safety plugs.

(MAPP®), propylene, and propane. Their flames are not as hot as acetylene, but they have higher British thermal unit (Btu) ratings and are cheaper and safer to use. The supervisor at your job site will determine which fuel gas to use.

Table 1 compares the flame temperatures of oxygen mixed with various fuel gases.

MAPP® is a Dow Chemical Company product that is a chemical combination of acetylene and propane gases. MAPP® gas burns at temperatures

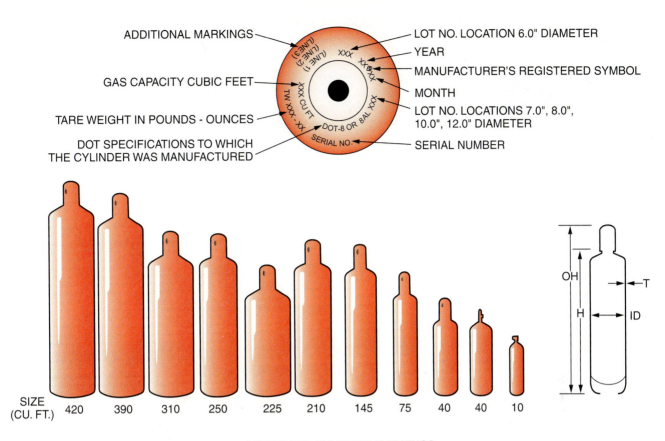

ACETYLENE CYLINDER MARKINGS

DOT SPECIFICATIONS	CAPACITY			NOMINAL DIMENSIONS (IN)				ACETONE (LB - OZ)	APPROXIATE TARE WEIGHT WITH VALVE WITHOUT CAP (LB)
	ACETYLENE	MIN. WATER							
	(FT³)	(IN³)	(LB.)	AVG. INSIDE DIAMETER "ID"	HEIGHT W/OUT VALVE OR CAP "H"	HEIGHT W/VALVE AND CAP "OH"	MINIMUM WALL "T"		
8 AL[1]	10	125	4.5	3.83	13.1375	14.75	0.0650	1-6	8
8[1]	40	466	16.8	6.00	19.8000	23.31	0.0870	5-7	25
8[2]	40	466	16.8	6.00	19.8000	28.30	0.0870	5-7	28
8[3]	75	855	30.8	7.00	25.5000	31.25	0.0890	9-8	45
8	100	1055	38.0	7.00	30.7500	36.50	0.0890	12-2	55
8	145	1527	55.0	8.00	34.2500	40.00	0.1020	18-10	76
8	210	2194	79.0	10.00	32.2500	38.00	0.0940	25-13	105
8AL	225	2630	94.7	12.00	27.5000	32.75	0.1280	29-6	110
8	250	2606	93.8	10.00	38.0000	43.75	0.0940	30-12	115
8AL	310	3240	116.7	12.00	32.7500	38.50	0.1120	39-5	140
8AL	390	4151	150.0	12.00	41.0000	46.75	0.1120	49-14	170
8AL	420	4375	157.5	12.00	43.2500	49.00	0.1120	51-14	187
8	60	666	24.0	7.00	25.79 OH		0.0890	7-11	40
8	130	1480	53.3	8.00	36.00 OH		0.1020	17-2	75
8AL	390	4215	151.8	12.00	46.00 OH		0.1120	49-14	180

1. Tapped for 3/8" valve but are not equipped with valve protection caps.
2. Includes valve protection cap.
3. Can be tared to hold 60 ft³ (1.7 m³) of acetylene gas.
   Standard tapping (except cylinders tapped for 3/8") 3/4"-14 NGT.

Weight includes saturation gas, filler, paint, solvent, valve, fuse plugs.
Does not include cap of 2 lb. (.91 kg.)
Cylinder capacities are based upon commercially pure acetylene gas at 250 psi (17.5 kg/cm²), and 70°F (15°C).

102F09.EPS

Figure 9 ◆ Acetylene cylinder markings and sizes.

Alternate Acetylene Cylinder Safety Cap

Acetylene cylinders can be equipped with a ring guard cap that protects the cylinder valve with or without a regulator installed on the valve. This enables safe movement of the cylinder after the cylinder valve is closed. This type of cap is usually secured to the cylinder body cap threads when it is installed so that it cannot be removed.

102P0202.EPS

102F10.EPS

Figure 10 ◆ Acetylene cylinder with standard valve safety cap.

Table 1 Flame Temperatures of Oxygen with Various Fuel Gases

Type of Gas	Flame Temperature
Acetylene	More than 5,500°F
MAPP®	5,300°F
Propylene	5,190°F
Natural gas	4,600°F
Propane	4,580°F

almost as high as acetylene and has the stability of propane. Because of this stability, it can be used at pressures over 15 psi and is not as likely as acetylene to **backfire** or **flashback**. MAPP® also has an offensive odor that can be detected easily. MAPP® gas can be used for flame cutting, heating, stress relieving, brazing, soldering, and scarfing (cleaning cutting slag or other material from the workpiece).

Propylene mixtures are hydrocarbon-based gases that are stable and shock-resistant, making them relatively safe to use. They are purchased under trade names such as *High Purity Gas (HPG™)*, *Apachi™*, and *Prestolene™*. These gases and others have distinctive odors to make leak detection easier. They burn at temperatures around 5,193°F, hotter than natural gas and propane. Propylene gases are used for flame cutting, scarfing, heating, stress relieving, brazing, and soldering.

Propane is also known as liquefied petroleum (LP) gas. It is stable and shock-resistant, and it has a distinctive odor for easy leak detection. It burns at 4,580°F, which is the lowest temperature of any fuel gas. It has a slight tendency toward backfire and flashback and is used quite extensively for cutting procedures.

Natural gas is delivered by pipeline rather than by cylinders. It burns at about 4,600°F. Natural gas is relatively stable and shock-resistant and has a slight tendency toward backfire and flashback. Because of its recognizable odor, leaks are easily detectable. Natural gas is used primarily for cutting on job sites with permanent cutting stations.

3.3.1 Liquefied Fuel Gas Cylinders

Liquefied fuel gases are shipped in hollow steel cylinders *(Figure 11)*. When empty, they are much lighter than acetylene cylinders.

The liquefied fuel gas is stored in hollow steel cylinders of various sizes. They can hold from 30 to 225 pounds of fuel gas. As the cylinder valve is opened, the vaporized gas is withdrawn from the cylinder. The remaining liquefied gas absorbs heat and releases additional vaporized gas. The pressure of the vaporized gas varies with the outside temperature. The colder the outside temperature, the lower the vaporized gas pressure will be. If high volumes of gas are removed from a liquefied fuel gas cylinder, the pressure drops, and the temperature of the cylinder will also drop. A ring of frost can form around the base of the cylinder. If high withdrawal rates continue, the regulator may also start to ice up. If high withdrawal rates are required, special regulators with electric heaters should be used.

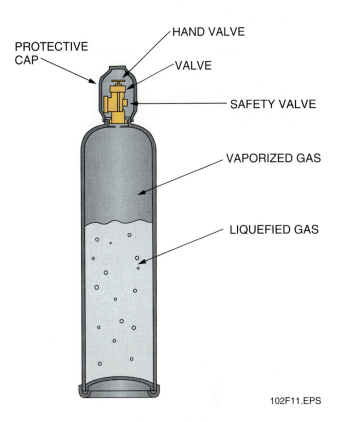

Figure 11 ◆ Liquefied fuel gas cylinder.

  WARNING!
Never apply heat directly to a cylinder or regulator. This can cause excessive pressures, resulting in an explosion.

The pressure inside a liquefied fuel gas cylinder is not an indicator of how full or empty the cylinder is. The weight of a cylinder determines how much liquefied gas is left. Liquefied fuel gas cylinders are equipped with CGA 510, 350, or 695 valves, depending on the fuel and storage pressures.

  WARNING!
Do not remove the protective cap on liquefied fuel gas cylinders unless the cylinder is secured. If the cylinder falls over and the nozzle breaks off, the cylinder will release highly explosive gas. Cylinders containing a liquid such as propane must be kept in an upright position. If the valve is broken or is opened with the cylinder horizontal, the fuel can emerge as a liquid that will shoot a long distance before it vaporizes. If it is ignited, it will act like a flamethrower.

Handling and Storing Liquefied Gas Cylinders

Liquefied fuel gas cylinders have a safety valve built into the valve at the top of the cylinder. The safety valve releases gas if the pressure begins to rise. Use care when handling fuel gas cylinders because the gas in cylinders is stored at such high pressures. Cylinders should never be dropped or hit with heavy objects, and they should always be stored in an upright position. When not in use, the cylinder valve must always be covered with the protective steel cap.

3.4.0 Regulators

Regulators *(Figure 12)* are attached to the cylinder valve. They reduce the high cylinder pressures to the required lower working pressures and maintain a steady flow of gas from the cylinder.

The pressure adjusting screw controls the gas pressure. Turned clockwise, it increases the flow of gas. Turned counterclockwise, it reduces the flow of gas. When turned counterclockwise until loose (released), it stops the flow of gas.

Most regulators contain two gauges. The high-pressure or cylinder-pressure gauge indicates the actual cylinder pressure; the low-pressure or working-pressure gauge indicates the pressure of the gas leaving the regulator.

Oxygen regulators differ from fuel gas regulators. Oxygen regulators are often painted green and always have right-hand threads on all connections. The oxygen regulator's high-pressure gauge generally reads up to 3,000 psi and includes a second scale that shows the amount of oxygen in the cylinder in terms of cubic feet. The low-pressure or working-pressure gauge may read 100 psi or higher.

Fuel gas regulators are often painted red and always have left-hand threads on all the connections. As a reminder that the regulator has left-hand threads, a V-notch may be cut around the nut. The fuel gas regulator's high-pressure gauge usually reads up to 400 psi. The low-pressure or working-pressure gauge may read up to 40 psi. Acetylene gauges, however, are always red-lined at 15 psi as a reminder that acetylene pressure should not be increased over 15 psi.

There are two types of regulators: single-stage and two-stage.

3.4.1 Single-Stage Regulators

Single-stage, spring-compensated regulators reduce pressure in one step. As gas is drawn from the cylinder, the internal pressure of the cylinder decreases. A single-stage, spring-compensated regulator is unable to automatically adjust for this

WARNING!

To prevent injury and damage to regulators, always follow these guidelines:

- Never subject regulators to jarring or shaking, as this can damage the equipment beyond repair.
- Always check that the adjusting screw is fully released before the cylinder valve is turned on and when the welding has been completed.
- Always open cylinder valves slowly and stand on the side of the cylinder opposite the regulator.
- Never use oil to lubricate a regulator. This can result in an explosion when the regulator is in use.
- Never use fuel gas regulators on oxygen cylinders or oxygen regulators on fuel gas cylinders.
- Never work with a defective regulator. If it is not working properly, shut off the gas supply and have the regulator repaired by someone who is qualified to work on it.
- Never use large wrenches, pipe wrenches, pliers, or slipjoint pliers to install or remove regulators.

decrease in internal cylinder pressure. Therefore, it becomes necessary to adjust the spring pressure to periodically raise the output gas pressure as the gas in the cylinder is consumed. These regulators are the most commonly used because of their low cost and high flow rates.

3.4.2 Two-Stage Regulators

The two-stage, pressure-compensated regulator reduces pressure in two steps. It first reduces the input pressure from the cylinder to a predetermined intermediate pressure. The intermediate pressure is then adjusted by the pressure-adjusting screw. With this type of regulator, the delivery pressure to the torch remains constant,

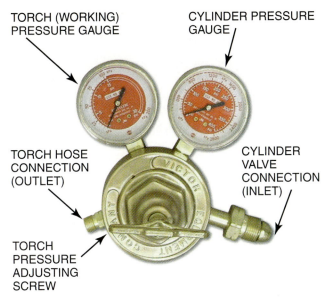

TORCH (WORKING) PRESSURE GAUGE

CYLINDER PRESSURE GAUGE

TORCH HOSE CONNECTION (OUTLET)

CYLINDER VALVE CONNECTION (INLET)

TORCH PRESSURE ADJUSTING SCREW

FUEL GAS REGULATOR

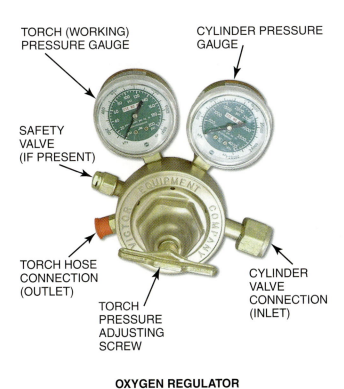

TORCH (WORKING) PRESSURE GAUGE

CYLINDER PRESSURE GAUGE

SAFETY VALVE (IF PRESENT)

TORCH HOSE CONNECTION (OUTLET)

CYLINDER VALVE CONNECTION (INLET)

TORCH PRESSURE ADJUSTING SCREW

OXYGEN REGULATOR

102F12.EPS

Figure 12 ◆ Oxygen and acetylene regulators.

and no readjustment is necessary as the gas in the cylinder is consumed. Standard two-stage regulators are more expensive than single-stage regulators and have lower flow rates. There are also heavy-duty types with higher flow rates that are usually preferred for thick material and/or continuous-duty cutting operations.

3.4.3 Check Valves and Flashback Arrestors

Check valves and flashback arrestors (*Figure 13*) are safety devices for regulators, hoses, and torches. Check valves allow gas to flow in one direction only. Flashback arrestors stop fire.

Check valves consist of a ball and spring that open inside a cylinder. The valve allows gas to move in one direction but closes if the gas attempts to flow in the opposite direction. When a torch is first pressurized or when it is being shut off, back-pressure check valves prevent the entry and mixing of acetylene with oxygen in the oxygen hose or the entry and mixing of oxygen with acetylene in the acetylene hose.

Flashback arrestors prevent flashbacks from reaching the hoses and/or regulator. They have a flame-retarding filter that will allow heat, but not flames, to pass through. Most flashback arrestors also contain a check valve.

Add-on check valves and flashback arrestors are designed to be attached either to the torch handle connections or to the regulator outlets. As a minimum, flashback arrestors with check valves should be attached to the torch handle connections. Both devices have arrows on them to

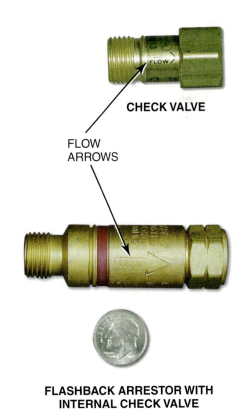

CHECK VALVE

FLOW ARROWS

FLASHBACK ARRESTOR WITH INTERNAL CHECK VALVE

102F13.EPS

Figure 13 ◆ Torch handle add-on check valve or flashback arrestor.

Torch Wrenches

Only a torch wrench, sometimes called a gang wrench, should be used to install regulators, hose connections, check valves, flashback arrestors, torches, and torch tips. The universal torch wrench shown is equipped with various size wrench cutouts for use with a variety of equipment and standard CGA components. The fittings for oxyfuel equipment are brass or bronze, and certain components are often fitted with soft, flexible, O-ring seals. The seal surfaces of the fittings or O-rings can be easily damaged by overtightening with standard wrenches. The length of a torch wrench is limited to reduce the chances of damage to fittings because of excessive torque. In some cases, manufacturers specify only hand-tightening for certain fitting connections of a torch set (tips or cutting/welding attachments). In any event, follow the manufacturer's specific instructions when connecting the components of a torch set.

102P0203.EPS

Flashback Arrestors

It is highly recommended that add-on flashback arrestors with check valves, rather than just check valves, be installed at both the torch handle and regulator connections to prevent serious injury or property damage. If the flame front of a flashback, sometimes called the backburn, gets by the check valves into the hoses or regulators, the hoses could burn through, and a large, very dangerous, uncontrolled fire could start. Newer torches are available with check valves and flashback arrestors built into the torch handle.

---

indicate flow direction. When installing add-on check valves and flashback arrestors, be sure the arrow matches the desired gas flow direction.

3.5.0 Hoses

Hoses transport gases from the regulators to the torch. Oxygen hoses are usually green or black with right-hand threaded connections. Hoses for fuel gas are usually red and have left-hand threaded connections. The fuel gas connections may also be grooved as a reminder that they have left-hand threads.

Proper care and maintenance of the hose is important for maintaining a safe, efficient work area. Remember the following guidelines for hoses:

- Protect the hose from molten slag or sparks, which will burn the exterior. Although some hoses are flame retardant, they will burn.

- Remove the hoses from under the metal being cut. If the hot metal falls on the hose, the hose will be damaged.
- Frequently inspect and replace hoses that show signs of cuts, burns, worn areas, cracks, or damaged fittings.
- Never use pipe-fitting compounds or lubricants around hose connections. These compounds often contain oil or grease, which ignite and burn or explode in the presence of oxygen.

3.6.0 Cutting Torches

Cutting torches mix oxygen and fuel gas for the torch flame and control the stream of oxygen necessary for the cutting jet. Depending on the job site, you may use either a one-piece or a combination cutting torch.

3.6.1 One-Piece Hand Cutting Torch

The one-piece hand cutting torch, sometimes called a demolition torch, contains the fuel gas and oxygen valves that allow the gases to enter the chambers and then flow into the tip where they are mixed. The main body of the torch is called the handle. The torch valves control the fuel gas and oxygen used for preheating the metal to be cut. The cutting oxygen lever, which is spring-loaded, controls the jet of cutting oxygen. Hose connections are located at the end of the torch body behind the valves. *Figure 14* shows a three-tube one-piece positive-pressure hand cutting torch in which the preheat fuel and oxygen are mixed in the tip. These torches are designed for heavy-duty cutting. They have long supply tubes from the torch handle to the torch head to reduce radiated heat to the operator's hands. The torches are generally available in capacities for cutting steel up to 12" thick. Larger-capacity torches, with the ability to cut steel up to 36" thick, can also be obtained.

Two different types of oxyfuel cutting torches are in general use. The positive-pressure torch is designed for use with fuel supplied through a regulator from pressurized fuel storage cylinders. The injector torch is designed to use a vacuum created by oxygen flow to draw the necessary amount of fuel from a very low-pressure fuel source, such as a natural gas line or acetylene generator. The injector torch, when used, is most often found in continuous-duty high-volume manufac-

turing applications. Both types may employ one of two different fuel-mixing methods:

- Torch-handle or supply-tube mixing
- Torch-head or tip mixing

The two methods can normally be distinguished by the number of supply tubes from the torch handle to the torch head. Torches that use three tubes (see *Figure 14*) from the handle to the head mix the preheat fuel and oxygen at the torch head or tip. This method tends to help eliminate any flashback damage to the torch head supply tubes and torch handle. Torches with two tubes usually mix the preheat fuel and oxygen in a mixing chamber in the torch body or in one of the supply tubes. Injector torches usually have the injector located in one of the supply tubes, and the mixing occurs in that tube from the injector to the torch head. Some older torches that have only two visible tubes are actually three-tube torches that mix the preheat fuel and oxygen in the torch head or tip. This is accomplished by using a separate preheat fuel tube inside a larger preheat oxygen tube.

3.6.2 Combination Torch

The combination torch consists of a cutting torch attachment that fits onto a welding torch handle. These torches are normally used in light-duty or medium-duty applications. Fuel gas and oxygen valves are on the torch handle. The cutting attachment has a cutting oxygen lever and

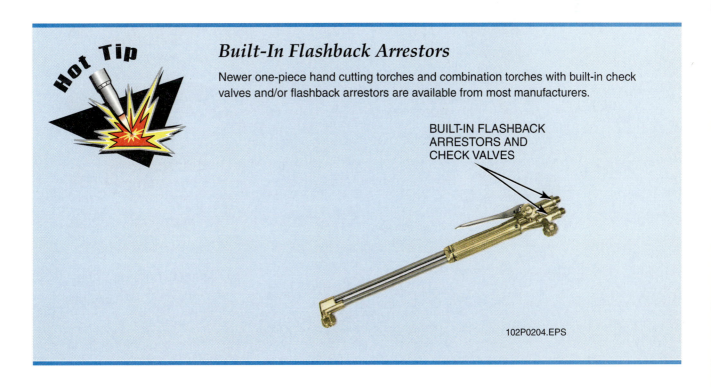

Hot Tip

Built-In Flashback Arrestors

Newer one-piece hand cutting torches and combination torches with built-in check valves and/or flashback arrestors are available from most manufacturers.

BUILT-IN FLASHBACK
ARRESTORS AND
CHECK VALVES

102P0204.EPS

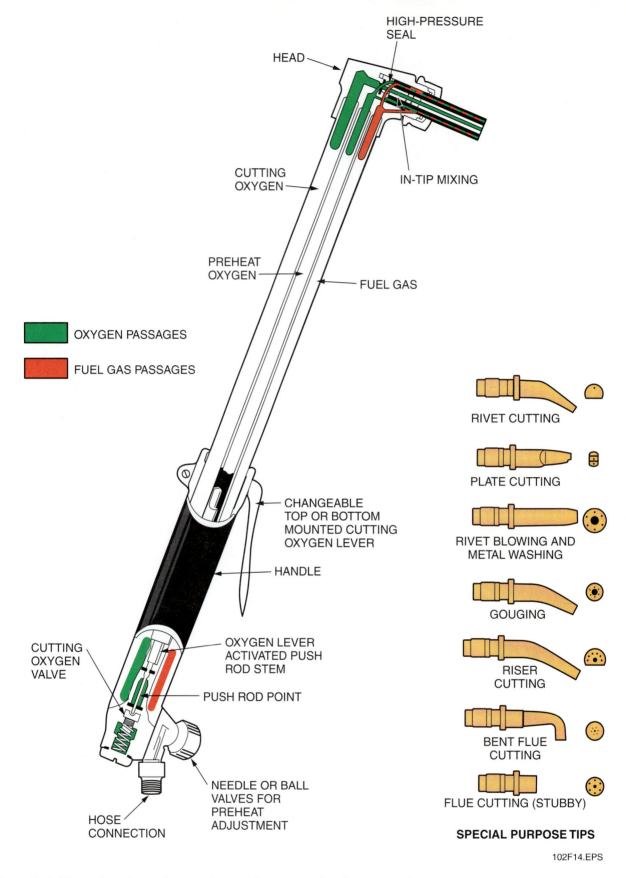

HIGH-PRESSURE SEAL

HEAD

IN-TIP MIXING

CUTTING OXYGEN

PREHEAT OXYGEN

FUEL GAS

OXYGEN PASSAGES

FUEL GAS PASSAGES

CHANGEABLE TOP OR BOTTOM MOUNTED CUTTING OXYGEN LEVER

HANDLE

CUTTING OXYGEN VALVE

OXYGEN LEVER ACTIVATED PUSH ROD STEM

PUSH ROD POINT

HOSE CONNECTION

NEEDLE OR BALL VALVES FOR PREHEAT ADJUSTMENT

RIVET CUTTING

PLATE CUTTING

RIVET BLOWING AND METAL WASHING

GOUGING

RISER CUTTING

BENT FLUE CUTTING

FLUE CUTTING (STUBBY)

SPECIAL PURPOSE TIPS

102F14.EPS

Figure 14 ◆ Heavy-duty three-tube one-piece positive-pressure hand cutting torch.

another oxygen valve to control the preheat flame. When the cutting attachment is screwed onto the torch handle, the torch handle oxygen valve is opened all the way, and the preheat oxygen is controlled by an oxygen valve on the cutting attachment. When the cutting attachment is removed, welding and heating tips can be screwed onto the torch handle. *Figure 15* shows a two-tube combination torch in which the preheat mixing is accomplished in a supply tube. These torches are usually positive-pressure torches with mixing occurring in the attachment body, supply tube, head, or tip. These torches are also equipped with built-in flashback arrestors and check valves.

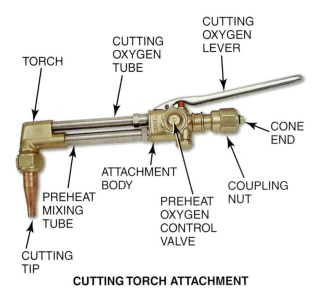

CUTTING TORCH ATTACHMENT

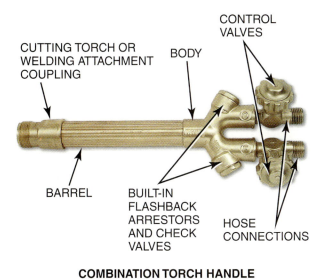

COMBINATION TORCH HANDLE

102F15.EPS

Figure 15 ◆ Typical combination torch.

3.7.0 Cutting Torch Tips

Cutting torch tips, or nozzles, fit into the cutting torch and are either screwed in or secured with a tip nut. There are one- and two-piece cutting tips (*Figure 16*).

One-piece cutting tips are made from a solid piece of copper. Two-piece cutting tips have a separate external sleeve and internal section.

Torch manufacturers supply literature explaining the appropriate torch tips and gas pressures to be used for various applications. *Table 2* shows a sample cutting tip chart that lists recommended tip sizes and gas pressures for use with acetylene fuel gas and a specific manufacturer's torch and tips.

> **WARNING!**
>
> Do not use the cutting tip chart from one manufacturer for the cutting tips of another manufacturer. The gas flow rate of the tips may be different, resulting in excessive flow rates. Different gas pressures may also be required.

The cutting torch tip to be used depends on the base metal thickness and fuel gas being used. Special-purpose tips are also available for use in such operations as gouging and grooving.

3.7.1 Cutting Tips for Acetylene

One-piece torch tips are always used with acetylene cutting because of the high temperatures involved. They can have four, six, or eight preheat holes in addition to the single cutting hole. *Figure 17* shows typical acetylene torch cutting tips.

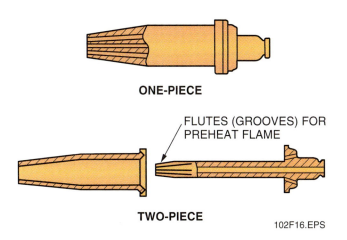

ONE-PIECE

FLUTES (GROOVES) FOR PREHEAT FLAME

TWO-PIECE

102F16.EPS

Figure 16 ◆ One- and two-piece cutting tips.

Torch Tip Styles

Nearly all manufacturers use different tip-to-torch mounting designs, sealing surfaces, and diameters. In addition, tip sizes and flow rates are usually not the same between manufacturers even though the number designations may be the same. This makes it impossible to safely interchange cutting tips between torches from different manufacturers. Even though some tips from different manufacturers may appear to be the same, do not interchange them. The sealing surfaces are very precise, and serious leaks may occur that could result in a dangerous fire or flashback.

102P0205.EPS

Table 2 Sample Acetylene Cutting Tip Chart

Cutting Tip Series 1-101, 3-101, and 5-101						
Metal Thickness	Tip Size	Cutting Oxygen Pressure (psig)	Preheat Oxygen (psig)	Acetylene Pressure (psig)	Speed (in./min.)	Kerf Width
⅛"	000	20/25	3/5	3/5	20/30	.04
¼"	00	20/25	3/5	3/5	20/28	.05
⅜"	0	25/30	3/5	3/5	18/26	.06
½"	0	30/35	3/6	3/5	16/22	.06
¾"	1	30/35	4/7	3/5	15/20	.07
1"	2	35/40	4/8	3/6	13/18	.09
2"	3	40/45	5/10	4/8	10/12	.11
3"	4	40/50	5/10	5/11	8/10	.12
4"	5	45/55	6/12	6/13	6/9	.15
6"	6	45/55	6/15	8/14	4/7	.15
10"	7	45/55	6/20	10/15	3/5	.34
12"	8	45/55	7/25	10/15	3/4	.41

Acetylene Flow Rates

Manufacturers provide listings of the maximum fuel flow rate for each acetylene tip size in addition to recommended acetylene pressures. When selecting a tip, make sure that its maximum flow rate (in cubic feet per hour) does not exceed one-tenth of the total fuel capacity (in cubic feet) for the acetylene cylinder in use. Multiple cylinders must be manifolded together if the flow rate exceeds the cylinder(s) in use in order to prevent withdrawal of acetone along with acetylene.

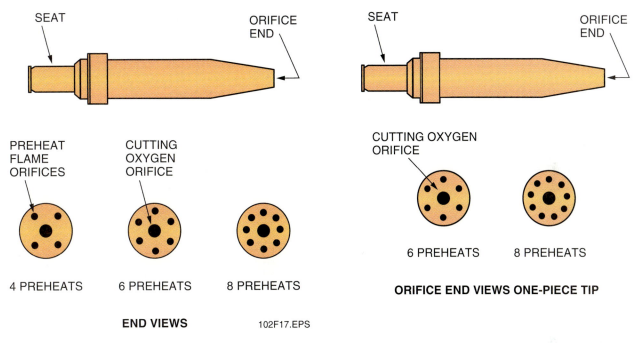

Figure 17 ◆ Typical acetylene torch cutting tips.

3.7.2 Cutting Tips for Liquefied Fuel Gases

Tips used with liquefied fuel gases must have at least six preheat holes. Because fuel gases burn at lower temperatures than acetylene, more holes are necessary for preheating. Tips used with liquefied fuel gases can be one- or two-piece cutting tips. *Figure 18* shows typical cutting tips used with liquefied fuel gases.

3.7.3 Special-Purpose Cutting Tips

Special-purpose tips are available for special cutting jobs. These jobs include cutting sheet metal, rivets, risers, and flues, as well as washing and gouging. *Figure 19* shows special-purpose torch cutting tips.

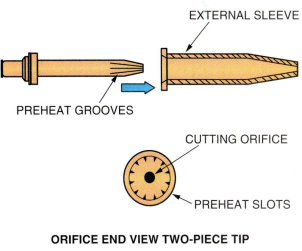

ORIFICE END VIEW TWO-PIECE TIP

102F18.EPS

Figure 18 ◆ Typical cutting tips for liquefied fuel gases.

Alternate Design Solid Copper Tip

A manufacturer of replacement torch tips for various brands of torches has a two-piece design for use in place of a standard solid copper tip. It consists of a single tip-to-head sealing adapter and a number of small interchangeable tip sizes that screw onto the adapter. The small screw-on tips are faster to change and somewhat less costly than a full standard tip.

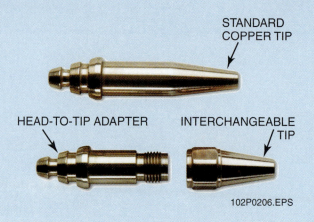

STANDARD
COPPER TIP

HEAD-TO-TIP ADAPTER

INTERCHANGEABLE
TIP

102P0206.EPS

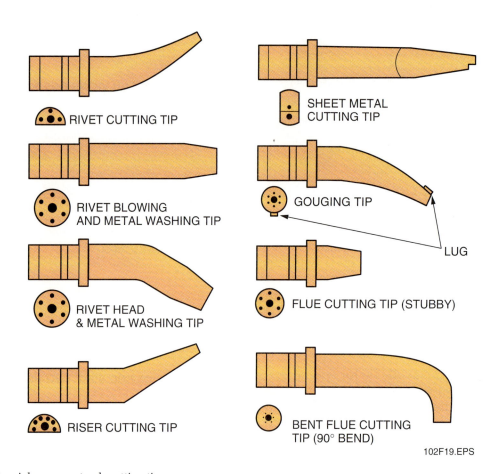

RIVET CUTTING TIP

SHEET METAL
CUTTING TIP

RIVET BLOWING
AND METAL WASHING TIP

GOUGING TIP

LUG

RIVET HEAD
& METAL WASHING TIP

FLUE CUTTING TIP (STUBBY)

RISER CUTTING TIP

BENT FLUE CUTTING
TIP (90° BEND)

102F19.EPS

Figure 19 ◆ Special-purpose torch cutting tips.

- The sheet metal cutting tip has only one preheat hole. This minimizes the heat and prevents distortion in the sheet metal. These tips are normally used with a motorized carriage but can also be used for hand cutting.
- Rivet cutting tips are used to cut off rivet heads, bolt heads, and nuts.
- Riser cutting tips are similar to rivet cutting tips and can also be used to cut off rivet heads, bolt heads, and nuts. They have extra preheat holes to cut risers, flanges, or angle legs faster. They can be used for any operation that requires a cut close to and parallel to another surface, such as in removing a metal backing.
- Rivet blowing and metal washing tips are heavy-duty tips designed to withstand high heat. They are used for coarse cutting and for removing such items as clips, angles, and brackets.
- Gouging tips are used to groove metal in preparation for welding.
- Flue cutting tips are designed to cut flues inside boilers. They also can be used for any cutting operation in tight quarters where it is difficult to get a conventional tip into position.

3.8.0 Tip Cleaners and Tip Drills

With use, cutting tips become dirty. Carbon and other impurities build up inside the holes, and molten metal often sprays and sticks onto the surface of the tip. A dirty tip will result in a poor-quality cut with an uneven kerf and excessive slag buildup. To ensure good cuts with straight kerfs and minimal slag buildup, clean cutting tips with tip cleaners or tip drills *(Figure 20)*.

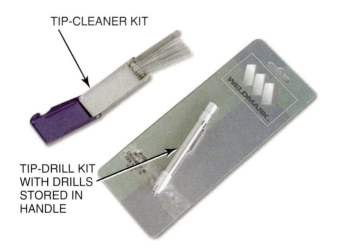

TIP-CLEANER KIT

TIP-DRILL KIT WITH DRILLS STORED IN HANDLE

102F20.EPS

Figure 20 ◆ Tip cleaner and drill kits.

Tip cleaners are small round files. They usually come in a set with files to match the diameters of the various tip holes. In addition, each set usually includes a file that can be used to lightly recondition the face of the cutting tip. Tip cleaners are inserted into the tip hole and moved back and forth a few times to remove deposits from the hole.

Tip drills are used for major cleaning and for holes that are plugged. Tip drills are tiny drill bits that are sized to match the diameters of tip holes. The drill fits into a drill handle for use. The handle is held, and the drill bit is turned carefully inside the hole to remove debris. They are more brittle than tip cleaners, making them more difficult to use.

CAUTION

Tip cleaners and tip drills are brittle. If you are not careful, they may break off inside a hole. Broken tip cleaners are difficult to remove. Improper use of tip cleaners or tip drills can enlarge the tip, causing improper burning of gases. If this occurs, tips must be discarded. If the end of the tip has been partially melted or deeply gouged, do not attempt to cut it off or file it flat. The tip should be discarded and replaced with a new tip. This is because some tips have tapered preheat holes, and if a significant amount of metal is removed from the end of the tip, the preheat holes will become too large.

3.9.0 Friction Lighters

Always use a friction lighter *(Figure 21)*, also known as a striker or spark-lighter, to ignite the cutting torch. The friction lighter works by rubbing a piece of flint on a steel surface to create sparks.

WARNING!

Do not use a match or a gas-filled lighter to light a torch. This could result in severe burns and/or could cause the lighter to explode.

3.10.0 Cylinder Cart

The cylinder cart, or bottle cart, is a modified hand truck that has been equipped with seats and chains to hold cylinders firmly in place. Bottle carts help ensure the safe transportation of gas

Cup-Type Striker

When using a cup-type striker to ignite a welding torch, hold the cup of the striker slightly below and to the side of the tip, parallel with the fuel gas stream from the tip. This prevents the ignited gas from deflecting back toward you from the cup and reduces the amount of carbon soot in the cup. Note that the flint in a striker can be replaced.

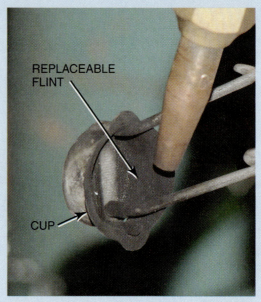

REPLACEABLE FLINT

CUP

102P0207.EPS

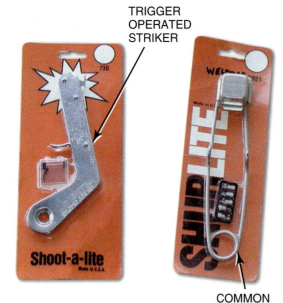

TRIGGER OPERATED STRIKER

COMMON CUP-TYPE STRIKER

102F21.EPS

Figure 21 ◆ Typical friction lighters.

cylinders. *Figure 22* shows a typical cylinder cart for a welding/cutting rig. Some carts are equipped with tool/accessory trays or boxes as well as rod holders.

3.11.0 Soapstone Markers

Because of the heat involved in welding and cutting operations, along with the tinted lenses that are required, ordinary pen or pencil marking for cutting lines or welding locations is not effective. The oldest and most common material used for marking is **soapstone** in the form of sticks or cylinders (*Figure 23*). Soapstone is soft and feels greasy and slippery. It is actually steatite, a dense, impure form of talc that is heat resistant. It also shows up well through a tinted lens under the illumination of an electric arc or gas welding/cutting flame. Some welders prefer to use silver-graphite pencils for marking dark materials and red-graphite pencils for aluminum or other bright metals. Graphite is also highly heat resistant (*Figure 23*). A few manufacturers also market heat-resistant paint/dye markers for welding.

Sharpening Soapstone Sticks

The most effective way to sharpen a soapstone stick marker is to shave it on one side with a penknife. By leaving one side flat, accurate lines can be drawn very close to a straightedge or a pattern.

102P0208.EPS

TOOL/ACCESSORY
TRAY

102F22.EPS

Figure 22 ◆ Typical cutting/welding rig cylinder cart.

SOAPSTONE
STICK AND
HOLDER

SOAPSTONE
CYLINDER
AND HOLDER

SILVER GRAPHITE
PENCILS

102F23.EPS

Figure 23 ◆ Typical soapstone and graphite markers.

3.12.0 Specialized Cutting Equipment

In addition to the common hand cutting torches, other types of equipment are used in oxyfuel cutting applications. This equipment includes mechanical guides used with a hand cutting torch, various types of motorized cutting machines, and oxygen lances. All of the motorized units use special straight body machine cutting or welding torches with a gear rack attached to the torch body to set the tip distance from the work.

3.12.1 Mechanical Guides

On long, circular, or irregular cuts, it is very difficult to control and maintain an even kerf with a hand cutting torch. Mechanical guides can help maintain an accurate and smooth kerf along the cutting line. For straight line or curved cuts, use a one- or two-wheeled accessory that clamps on the torch tip in a fixed position. The wheeled accessory maintains the proper tip distance while the tip is guided by hand along the cutting line. The fixed, two-wheeled accessory is similar to the rotating-mount wheeled unit used for a circle cutter but without the radius bar *(Figure 24)*.

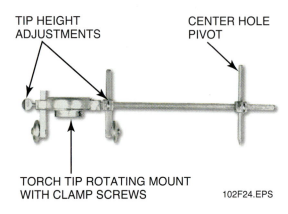

TIP HEIGHT ADJUSTMENTS

CENTER HOLE PIVOT

TORCH TIP ROTATING MOUNT WITH CLAMP SCREWS

102F24.EPS

Figure 24 ◆ Circle cutting accessory.

Perform arc or circular cuts with the circle cutting accessory shown in *Figure 24.* The torch tip fits through and is secured to a rotating mount between the two small metal wheels. The wheel heights are adjustable so that the tip distance from the work can be set. The radius of the circle is set by moving the pivot point on a radius bar. After a starting hole is cut (if needed), the torch tip is placed through and secured to the circle cutter rotating mount. Then the pivot point is placed in a drilled hole or a magnetic holder at the center of the circle. When the cut is restarted, the torch can be moved in a circle around the cut, guided by the circle cutter.

When large work with an irregular pattern must be cut, a template is often used. The torch is drawn around the edges of the template to trace the pattern as the cut is made. If multiple copies must be cut, a metal pattern, held in place and spaced for tip distance from the work by stacked magnets, is usually used. For a one- or two-time copy, a heavily weighted Masonite or aluminum template that is spaced off the workpiece could be carefully used and discarded.

3.12.2 Motor-Driven Equipment

A variety of fixed and portable motorized cutting equipment is available for straight and curved

Shop-Made Straight Line Cutting Guide

A simple solution for straight line cutting is to clamp a piece of angle iron to the work and use a band clamp around the cutting torch tip to maintain the cutting tip distance from the work. When the cut is started, the band clamp rests on the top of the vertical leg of the angle iron, and the torch is drawn along the length of the angle iron at the correct cutting speed.

102P0209.EPS

cutting/welding. The computer-controlled gantry cutting machine *(Figure 25)* and the optical pattern-tracing machine *(Figure 26)* are fixed-location machines used in industrial manufacturing applications. The computer-controlled machine can be programmed to **pierce** and then cut any pattern from flat metal stock. The optical pattern-tracing machine follows lines on a drawing using a light beam and an optical detector. They both can be rigged to cut multiple items using multiple torches operated in parallel. Both units have a motor-driven gantry that travels the length of a table and a transverse motor-driven beam or torch head that moves back and forth across the table. Both units are also equipped to use both oxyfuel cutting and plasma cutting torches. The size of the patterns can also be adjusted.

Other types of pattern-tracing machines use metal templates that are clamped in the machine. A follower wheel traces the pattern from the template. The pattern size can be increased or decreased by electrical or mechanical linkage to a moveable arm holding one or more cutting torches that cut the pattern from flat metal stock.

Portable track cutting machines (track burners) can be used in the field for straight or curved cutting and beveling. *Figures 27* and *28* show units driven by a variable-speed motor. The unit shown in *Figure 27* is available with track extensions for any length of straight cutting or beveling, along with a circle cutting attachment. The unit shown in *Figure 28* uses a somewhat flexible magnetic track for both flat straight-line or large-diameter object cutting or beveling. It is shown equipped with an optional plasma machine torch. Both units can be adapted to metal inert gas (MIG) or tungsten inert gas (TIG) welding.

102F27.EPS

Figure 27 ◆ Track burner with an oxyfuel machine torch.

102F25.EPS

Figure 25 ◆ Computer-controlled gantry cutting machine.

102F26.EPS

Figure 26 ◆ Optical pattern-tracing machine.

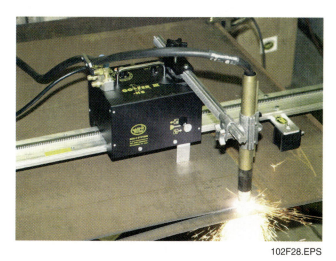

102F28.EPS

Figure 28 ◆ Track burner with a plasma machine torch.

A portable, motor-driven band track or hand-cranked ring gear cutter/beveler can be set up in the field for cutting and beveling pipe with oxyfuel or plasma machine torches (*Figures 29* and *30*). The stainless steel band track cutter uses a chain and motor sprocket drive to rotate the machine cutting torch around the pipe a full 360°. The all-aluminum ring gear type of cutter/beveler is positioned on the pipe, and then the saddle is clamped in place. In operation, the ring gear and the cutting torch rotate at different rates around the saddle for a full 360° cut.

A hand-guided oxyfuel cutting torch with an integral-precision, variable-speed motor drive (*Figure 31*), which can be used for straight line and curved cutting to achieve machine-quality cuts, is available. A circle cutting accessory is also available for the unit.

3.12.3 Exothermic Oxygen Lances

Exothermic (combustible) oxygen lances are a special oxyfuel cutting tool usually used in heavy industrial applications and demolition work. The lance is a steel pipe that contains magnesium- and aluminum-cored powder or rods (fuel). In operation, the lance is clamped into a holder (*Figure 32*) that seals the lance to a fitting that supplies oxygen to the lance through a hose at pressures of 75 to 80 psi. With the oxygen turned on, the end of the lance is ignited with an acetylene torch or flare. As long as the oxygen is applied, the lance will burn and consume itself. The oxygen-fed flame of the burning magnesium, aluminum, and steel pipe creates temperatures approaching 10,000°F. At this temperature, the lance will rapidly cut or

102F29.EPS

Figure 29 ◆ Band track pipe cutter/beveler.

102F31.EPS

Figure 31 ◆ Hand-guided motorized oxyfuel cutting torch.

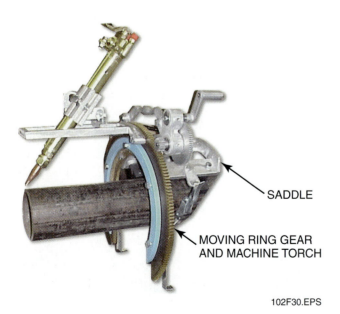

SADDLE

MOVING RING GEAR AND MACHINE TORCH

102F30.EPS

Figure 30 ◆ Ring gear pipe cutter/beveler.

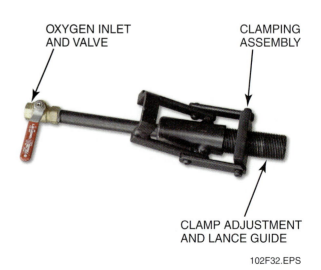

OXYGEN INLET AND VALVE

CLAMPING ASSEMBLY

CLAMP ADJUSTMENT AND LANCE GUIDE

102F32.EPS

Figure 32 ◆ Oxygen lance holder.

Cutting Through Difficult Materials

Fire brick, reinforced concrete, and large, thick iron castings or steel objects are some of the many materials that can quickly be cut or pierced with an exothermic oxygen lance. With the larger lances, use specially insulated heat-reflective suits and full head shields because of the amount of heat generated.

PIERCING CAST IRON

102P0210.EPS

pierce any material, including steel, metal alloys, and cast iron, even under water. The lances for the holder shown in *Figure 32* are 10' long and range in size from ⅜" to 1" in diameter. The larger sizes can be coupled to obtain a longer lance. A small pistol-grip heat-shielded unit that can, if desired, be used with an electric welder is also available. The arc, combined with an oxygen lance, can create temperatures exceeding 10,000°F. This small unit uses lances from ¼" to ⅜" in diameter that are 22" to 36" long and that cut very rapidly at a maximum burning time of 60 to 70 seconds. The small unit is primarily used to burn out large frozen pins and frozen headless bolts or rivets. Like a large lance, it can be used to cut any material, including concrete-lined pipe. It also can be used to remove hard-surfacing material that has been applied to wear surfaces. Both units are relatively inexpensive and can be set up in the field with only an oxygen cylinder, hose, and ignition device.

4.0.0 ◆ SETTING UP OXYFUEL EQUIPMENT

When setting up oxyfuel equipment, you must follow certain procedures to ensure that the equipment operates properly and safely. The following sections explain the procedures for setting up oxyfuel equipment.

4.1.0 Transporting and Securing Cylinders

Follow these steps to transport and secure cylinders:

 WARNING!

Always handle cylinders with care. They are under high pressure and should never be dropped, knocked over, or exposed to excessive heat. When moving cylinders, always be certain that the valve caps are in place. Use a cylinder cage to lift cylinders. Never use a sling or electromagnet.

Step 1   Transport cylinders to the workstation in the upright position on a hand truck or bottle cart (*Figure 33*).

Step 2   Secure the cylinders at the workstation.

Step 3   Remove the protective cap from each cylinder and inspect the outlet nozzles to make sure that the seat and threads are not damaged. Place the protective caps where they will not be lost and where they will be available when the cylinders are empty.

Hoisting Cylinders

Never attempt to lift a cylinder using the holes in a safety cap. Always use a lifting cage. Make sure that the cylinder is secured in the cage. Various size cages are available for high-pressure cylinders and cylinders containing liquids.

102P0211.EPS

Fuel and Oxygen Cylinder Separation for Fixed Installations

For fixed installations involving one or more cylinders coupled to a manifold, fuel and oxygen cylinders must be separated by at least 20' or be divided by a wall 5' or more high (*American National Standards Institute Z49.1*).

CAUTION

Do not transport or immediately use an acetylene cylinder found resting on its side. Stand it upright and wait at least 30 minutes to allow the acetone to settle before using it.

4.2.0 Cracking Cylinder Valves

Follow these steps to crack cylinder valves:

Step 1　Crack open the cylinder valve momentarily to remove any dirt from the valves (*Figure 34*).

SINGLE-CYLINDER HAND CART

HEAVY-DUTY TWIN-CYLINDER
HAND CART

102F33.EPS

Figure 33 ◆ Carts for transporting cylinders.

OUTLET
FACING
AWAY

102F34.EPS

Figure 34 ◆ Cracking a cylinder valve.

WARNING!

Always stand to one side of the valves when opening them to avoid injury from dirt that may be lodged in the valve.

Step 2 Wipe out the connection seat of the valves with a clean cloth. Dirt frequently collects in the outlet nozzle of a cylinder valve and must be cleaned out to keep it from entering the regulator when pressure is turned on.

WARNING!

Be sure the cloth used does not have any oil or grease on it. Oil or grease mixed with compressed oxygen will cause an explosion.

4.3.0 Attaching Regulators

Follow these steps to attach the regulators:

Step 1 Check that the regulator is closed (adjustment screw loose).

Step 2 Check the regulator fittings to ensure that they are free of oil and grease *(Figure 35)*.

Step 3 Connect and tighten the oxygen regulator to the oxygen cylinder using a torch wrench *(Figure 36)*.

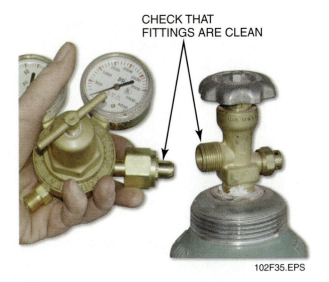

CHECK THAT
FITTINGS ARE CLEAN

102F35.EPS

Figure 35 ◆ Checking connection fittings.

TORCH
WRENCH

102F36.EPS

Figure 36 ◆ Tightening regulator connection.

Step 4 Connect and tighten the fuel gas regulator to the fuel gas cylinder. Remember that all fuel gas fittings have left-hand threads.

OPEN
REGULATOR
TO CLEAN
OUTLET

102F37.EPS

Figure 37 ◆ Cleaning the regulator.

Step 5 Crack the cylinder valve slightly and open the regulator to expel any debris from the outlet. Shut the cylinder valve and close the regulator *(Figure 37)*.

4.4.0 Installing Flashback Arrestors or Check Valves

Follow these steps to install flashback arrestors or check valves:

Step 1 Attach a flashback arrestor or check valve to the hose connection on the oxygen regulator *(Figure 38)* and tighten with a torch wrench.

Step 2 Attach and tighten a flashback arrestor or check valve to the hose connection on the fuel gas regulator. Keep in mind that all fuel gas fittings have left-hand threads.

102F38.EPS

Figure 38 ◆ Attaching a flashback arrestor.

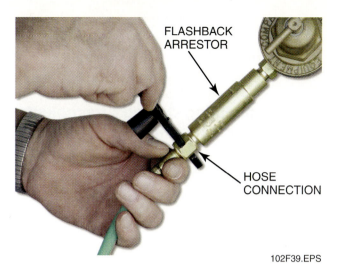

FLASHBACK
ARRESTOR

HOSE
CONNECTION

102F39.EPS

Figure 39 ◆ Connecting hose to regulator flashback arrestor.

4.5.0 Connecting Hoses to Regulators

New hoses contain talc and loose bits of rubber. These materials must be blown out of the hoses using an inert gas such as nitrogen or argon before the torch is connected. If they are not blown out, they will clog the torch needle valves.

 WARNING!

Never blow out hoses with compressed air, fuel gas, or oxygen. Compressed air often contains some oil that could explode or cause a fire when compressed in the hose with oxygen. Using fuel gas or oxygen creates a fire and explosion hazard.

Check that used hoses are not cracked, cut, damaged, or contaminated with oil or grease. Replace the hoses if these conditions exist.

Follow these steps to connect the hoses to the regulators:

Step 1 Inspect both the oxygen and fuel gas hoses for any damage, burns, cuts, or fraying.

Step 2 Replace any damaged hoses.

Step 3 Connect the green oxygen hose to the oxygen regulator flashback arrestor or check valve *(Figure 39)*.

Step 4 Connect the red or black fuel gas hose to the fuel gas regulator flashback arrestor or check valve. Keep in mind that all fuel gas fittings have left-hand threads.

4.6.0 Attaching Hoses to the Torch

Follow these steps to attach the hoses to the torch:

Step 1 Attach flashback arrestors to the oxygen and fuel gas hose connections on the torch body unless the torch has built-in flashback arrestors and check valves. Keep in mind that all fuel gas fittings have left-hand threads.

Step 2 Attach and tighten the green oxygen hose to the oxygen fitting on the flashback arrestor or torch *(Figure 40)*.

HOSE
CONNECTION

NOTE THAT THIS TORCH HAS
BUILT-IN FLASHBACK ARRESTORS
AND CHECK VALVES

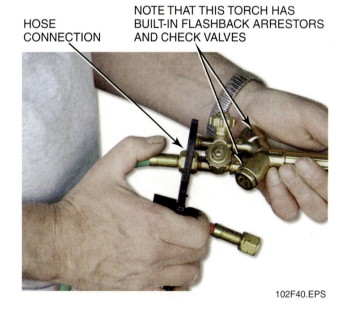

102F40.EPS

Figure 40 ◆ Connecting hoses to torch body.

Step 3 Attach and tighten the red or black hose to the fuel gas fitting on the flashback arrestor or torch. Remember that all fuel gas fittings have left-hand threads.

4.7.0 Connecting Cutting Attachments (Combination Torch Only)

If a combination torch is being used, connect cutting attachments as follows:

Step 1 Check the torch manufacturer's instructions for the correct method of installing the attachment.

Step 2 Connect the attachment and tighten with a torch wrench or by hand as required *(Figure 41)*.

4.8.0 Installing Cutting Tips

Follow these steps to install a cutting tip in the cutting torch:

Step 1 Identify the thickness of the material to be cut.

WARNING!

If acetylene fuel is being used, make sure that the maximum fuel flow rate per hour of the tip does not exceed one-tenth of the fuel cylinder capacity. If a purplish flame is observed when the torch is operating, the fuel rate is too high and acetone is being withdrawn from the acetylene cylinder along with the acetylene gas.

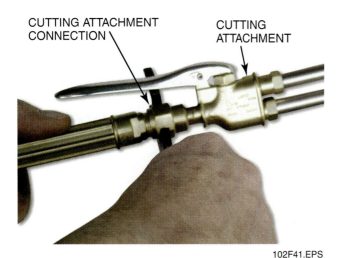

CUTTING ATTACHMENT CONNECTION

CUTTING ATTACHMENT

102F41.EPS

Figure 41 ◆ Connecting a cutting attachment.

Step 2 Identify the proper size cutting tip from the manufacturer's recommended tip size chart for the fuel being used.

Step 3 Inspect the cutting tip sealing surfaces and orifices for damage or plugged holes. If the sealing surfaces are damaged, discard the tip. If the orifices are plugged, clean them with a tip cleaner or drill.

Step 4 Check the torch manufacturer's instructions for the correct method of installing cutting tips.

Step 5 Install the cutting tip, securing it with a torch wrench or by hand as required *(Figure 42)*.

4.9.0 Closing Torch Valves and Loosening Regulator Adjusting Screws

Follow these steps to close the torch valves and loosen the regulator adjusting screws *(Figure 43)*:

Step 1 Check the fuel and oxygen valves on the torch to be sure they are closed. Closing the torch gas valves prevents gases from backing up inside the torch.

CAUTION

Loosening regulator adjusting screws closes the regulators and prevents damage to the regulator diaphragms when the cylinder valves are opened.

Step 2 Check the oxygen regulator adjusting screw to be sure it is loose (backed out).

Step 3 Check the fuel gas regulator adjusting screw to be sure it is loose.

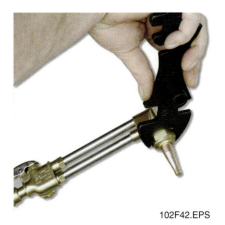

102F42.EPS

Figure 42 ◆ Installing a cutting tip.

AWS ENTRY LEVEL WELDER – TRAINEE MODULE 29102-03

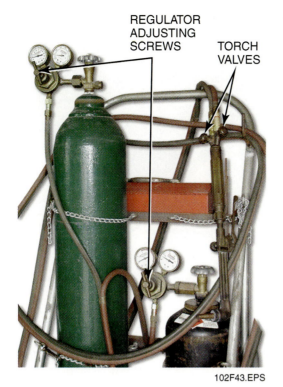

Figure 43 ◆ Torch valves and regulator adjusting screws.

4.10.0 Opening Cylinder Valves

Follow these steps to open cylinder valves *(Figure 44)*:

Figure 44 ◆ Cylinder valves and gauges.

> **WARNING!**
>
> Never stand directly in front of or behind a regulator. The regulator adjusting screw can blow out, causing serious injury. Always open the cylinder valve gradually. Quick openings can damage a regulator or gauge or even cause a gauge to explode.
>
> Oxygen cylinder valves must be opened all the way until the valve seats at the top. Seating the valve at the fully open position prevents high-pressure leaks at the valve stem.

Step 1 Standing on the cylinder valve side of the oxygen regulator, slowly open the oxygen cylinder valve all the way, allowing the pressure in the cylinder pressure gauge to rise gradually until the gauge indicates the oxygen cylinder pressure.

Step 2 Standing on the cylinder valve side of the fuel gas regulator, slowly open the fuel gas cylinder valve a quarter turn or until the cylinder pressure gauge indicates the cylinder pressure. Opening the cylinder valve a quarter turn allows it to be quickly closed in case of a fire.

4.11.0 Purging the Torch and Setting the Working Pressures

Follow these steps to purge the torch and set the working pressures *(Figure 45)*:

> **WARNING!**
>
> The working pressure gauge readings on single-stage regulators will rise after the torch valves are turned off. This is normal. However, if acetylene is being used as the fuel gas, make sure that the static pressure does not rise to 15 psig. Make sure that equipment is purged and leak tested in a well-ventilated area to avoid creating an explosive concentration of gases.

Step 1 Fully open the oxygen valve on the torch. Then depress and hold or lock open the cutting oxygen lever.

Hot Tip

Fuel Cylinder Wrench

If the fuel cylinder is equipped with a valve requiring a T-wrench, always leave the wrench in place on the valve so that the fuel can be quickly turned off. This type of valve is obsolete but still in use.

102P0212.EPS

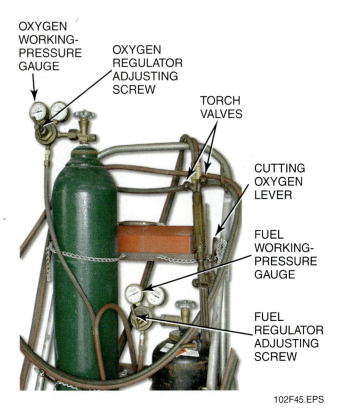

OXYGEN WORKING-PRESSURE GAUGE

OXYGEN REGULATOR ADJUSTING SCREW

TORCH VALVES

CUTTING OXYGEN LEVER

FUEL WORKING-PRESSURE GAUGE

FUEL REGULATOR ADJUSTING SCREW

102F45.EPS

Figure 45 ◆ Typical points for purging the equipment and setting the working pressures.

Step 2   Tighten the oxygen regulator adjusting screw until the working pressure gauge shows the correct oxygen gas working pressure with the gas flowing. Allow the gas to flow for five to ten seconds to purge the torch and hoses of air or fuel gas.

Step 3   At the torch, release the cutting lever and close the oxygen valve.

Step 4   Fully open the fuel valve on the torch.

Step 5   Tighten the fuel regulator adjusting screw until the working pressure gauge shows the correct fuel gas working pressure with the gas flowing. Allow the gas to flow for five to ten seconds to purge the hoses and torch of air.

Step 6   At the torch, close the fuel valve. If acetylene is used, check that the acetylene static pressure does not rise to 15 psig. If it does, immediately open the torch fuel valve and reduce the regulator output pressure as required.

4.12.0 Testing for Leaks

Equipment must be tested for leaks immediately after it is set up and periodically thereafter. The torch should be checked for leaks each time before use. Leaks could cause a fire or explosion if undetected. To test for leaks, brush a commercially prepared leak-testing formula or a solution of detergent and water on the following points. If bubbles form, a leak is present.

Always Purge the Equipment

To reduce the chances of a flashback, the hoses and torch should always be purged and the working pressures checked each time the torch is to be ignited.

Leak points include:

- Oxygen cylinder valve
- Fuel gas cylinder valve
- Oxygen regulator and regulator inlet and outlet connections
- Fuel gas regulator and regulator inlet and outlet connections
- Hose connections at the regulators, check valves/flashback arrestors, and torch
- Torch valves and cutting oxygen lever valve
- Cutting attachment connection (if used)
- Cutting tip

If there is a leak at the fuel gas cylinder valve stem, attempt to stop it by tightening the packing gland. If this does not stop the leak, mark and remove the cylinder and notify the supplier. For other leaks, tighten the connections slightly with a wrench. If this does not stop the leak, turn off the gas pressure, open all connections, and inspect the screw threads.

4.12.1 Initial and Periodic Leak Testing

Perform the following steps at initial equipment setup and periodically thereafter:

Step 1 Set the equipment to the correct working pressures with the torch valves turned off.

Step 2 Using a leak-test solution, check for leaks at the cylinder valves, regulator relief ports, and regulator gauge connections *(Figure 46)*. Also, check for leaks at hose connections, regulator connections, and check valve/flame arrestor connections up to the torch.

4.12.2 Leak-Down Testing of Regulators, Hoses, and Torch

Perform the following steps to quickly leak test the regulators, hoses, and torch before the torch is ignited.

Step 1 Set the equipment to the correct working pressures with the torch valves turned off. Then loosen both regulator adjusting screws. Check the working pressure gauges after a minute or two to see if the pressure drops. If the pressure drops, check the hose connection and regulators for leaks; otherwise, proceed to Step 2.

Step 2 Place a thumb over the cutting tip orifices and press to block the orifices.

Step 3 Turn on the torch oxygen valve and then depress and hold the cutting oxygen lever down.

Step 4 After the gauge pressure drops slightly, observe the oxygen working pressure

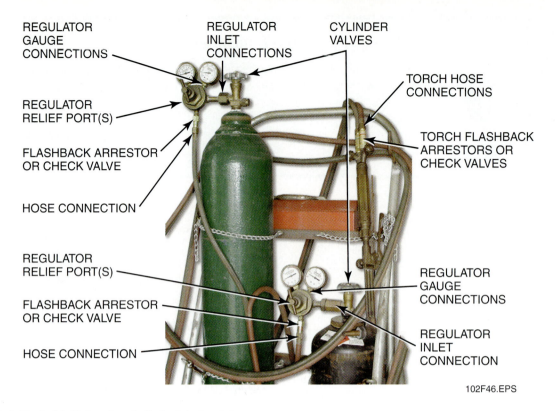

REGULATOR GAUGE CONNECTIONS

REGULATOR INLET CONNECTIONS

CYLINDER VALVES

TORCH HOSE CONNECTIONS

REGULATOR RELIEF PORT(S)

TORCH FLASHBACK ARRESTORS OR CHECK VALVES

FLASHBACK ARRESTOR OR CHECK VALVE

HOSE CONNECTION

REGULATOR RELIEF PORT(S)

REGULATOR GAUGE CONNECTIONS

FLASHBACK ARRESTOR OR CHECK VALVE

HOSE CONNECTION

REGULATOR INLET CONNECTION

102F46.EPS

Figure 46 ◆ Typical initial and periodic leak-test points.

gauge for a minute to see if the pressure continues to drop. If the pressure keeps dropping, perform the leak test described in the following section to determine the source of the leak. If the pressure does not change, close the torch oxygen valve and release the pressure at the cutting tip.

Step 5 With the tip blocked, turn on the torch fuel valve. After the gauge pressure drops slightly, carefully observe the fuel working pressure gauge for a minute. If the pressure continues to drop, perform the leak test described in the following section to determine the source of the leak. If the pressure does not change, close the torch fuel valve and release the pressure at the cutting tip.

Step 6 If no leaks are apparent, set the equipment to the correct working pressures.

4.12.3 Full Leak Testing of a Torch

Perform the following steps to test for and isolate torch leaks *(Figure 47)*.

Step 1 Set the equipment to the correct working pressures with the torch valves turned off.

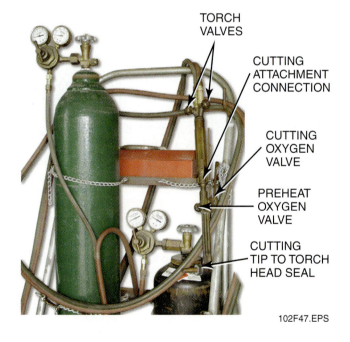

TORCH VALVES

CUTTING ATTACHMENT CONNECTION

CUTTING OXYGEN VALVE

PREHEAT OXYGEN VALVE

CUTTING TIP TO TORCH HEAD SEAL

102F47.EPS

Figure 47 ◆ Torch leak-test points.

Step 2 Place a thumb over the cutting tip orifices and press to block the orifices.

Step 3 Turn on the torch oxygen valve and then depress and lock the cutting oxygen lever down.

Step 4 With the cutting tip blocked, check for leaks using a leak-test solution at the torch oxygen valve, cutting oxygen lever valve, cutting attachment connection (if used), preheat oxygen valve (if used), and cutting tip seal at the torch head.

Step 5 Release the cutting oxygen lever and close the torch oxygen valve. Release the pressure at the cutting tip.

Step 6 With the cutting tip blocked, open the torch fuel valve.

Step 7 Using a leak-test solution, check for leaks at the torch fuel valve, cutting attachment (if used), and cutting tip seal at the torch head.

Step 8 Close the torch fuel valve and remove thumb from the cutting tip.

5.0.0 ◆ CONTROLLING THE OXYFUEL TORCH FLAME

To be able to safely use a cutting torch, the operator must understand the flame and be able to adjust it and react to unsafe conditions. The following sections will explain the oxyfuel flame and how to control it safely.

5.1.0 Oxyfuel Flames

There are three types of oxyfuel flames: **neutral flame, carburizing flame**, and **oxidizing flame**.

- *Neutral flame* – A neutral flame burns proper proportions of oxygen and fuel gas. The inner cones will be light blue in color, surrounded by a darker blue outer flame envelope that results when the oxygen in the air combines with the super-heated gases from the inner cone. A neutral flame is used for all but special cutting applications.
- *Carburizing flame* – A carburizing flame has a white feather created by excess fuel. The length

of the feather depends on the amount of excess fuel present in the flame. The outer flame envelope is longer than that of the neutral flame, and it is much brighter in color. The excess fuel in the carburizing flame (especially acetylene) produces large amounts of carbon. The carbon will combine with red-hot or molten metal, making the metal hard and brittle. The carburizing flame is cooler than a neutral flame and is never used for cutting. It is used for some special heating applications.

- *Oxidizing flame* – An oxidizing flame has an excess of oxygen. The inner cones are shorter, much bluer in color, and more pointed than a neutral flame. The outer flame envelope is very short and often fans out at the ends. An oxidizing flame is the hottest flame. A slightly oxidizing flame is recommended with some special fuel gases, but in most cases it is not used. The excess oxygen in the flame can combine with many metals, forming a hard, brittle, low-strength oxide. However, the preheat flames of a properly adjusted cutting torch will be slightly oxidizing when the cutting oxygen is shut off.

Figure 48 shows the various flames that occur at a cutting tip for both acetylene and LP gas.

5.2.0 Backfires and Flashbacks

When the torch flame goes out with a loud pop or snap, a backfire has occurred. Backfires are usually caused when the tip or nozzle touches the work surface or when a bit of hot slag briefly interrupts the flame. When a backfire occurs, you can relight the torch immediately. Sometimes the torch even relights itself. If a backfire recurs without the tip making contact with the base metal, shut off the torch and find the cause. Possible causes are:

- Improper operating pressures
- A loose torch tip
- Dirt in the torch tip seat or a bad seat

Acetylene Burning in Atmosphere
Open fuel gas valve until smoke clears from flame.

Carburizing Flame
(Excess acetylene with oxygen)
Preheat flames require more oxygen.

Neutral Flame
(Acetylene with oxygen) Temperature 5589°F (3087°C).
Proper preheat adjustment when cutting.

Neutral Flame with Cutting Jet Open
Cutting jet must be straight and clean.
If it flares, the pressure is too high for the tip size.

Oxidizing Flame
(Acetylene with excess oxygen) Not recommended for average cutting. However, if the preheat flame is adjusted for neutral with the cutting oxygen on, then this flame is normal after the cutting oxygen is off.

102F48A.EPS

LP Gas Burning in Atmosphere
Open fuel gas valve until flame begins to leave tip end.

Reducing Flame
(Excess LP-gas with oxygen) Not hot enough for cutting.

Neutral Flame
(LP-gas with oxygen) For preheating prior to cutting.

Oxidizing Flame with Cutting Jet Open
Cutting jet stream must be straight and clean.

Oxidizing Flame without Cutting Jet Open
(LP-gas with excess oxygen) The highest temperature flame for fast starts and high cutting speeds.

102F48B.EPS

Figure 48 ◆ Acetylene and LP gas flames.

When the flame goes out and burns back inside the torch with a hissing or whistling sound, a flashback is occurring. Immediately shut off the oxygen valve on the torch; the flame is burning inside the torch. If the flame is not extinguished quickly, the end of the torch will melt off. The flashback will stop as soon as the oxygen valve is closed. Therefore, quick action is crucial. Flashbacks can cause fires and explosions within the cutting rig and, therefore, are very dangerous. Flashbacks can be caused by:

- Equipment failure
- Overheated torch tip
- Slag or spatter hitting and sticking to the torch tip
- Oversized tip (tip is too large for the equipment being used)

After a flashback has occurred, wait until the torch has cooled. Then, blow oxygen (not fuel gas) through the torch for several seconds to remove soot that may have built up in the torch during the flashback before relighting it. If you hear the hissing or whistling after the torch is reignited or if the flame does not appear normal, shut off the torch immediately and have the torch serviced by a qualified technician. If the torch or tip has been damaged, replace the flashback arrestors.

5.3.0 Igniting the Torch and Adjusting the Flame

After the cutting equipment has been properly set up and purged, the torch can be ignited and the flame adjusted for cutting. Follow these steps to ignite the torch:

Step 1 Choose the appropriate cutting torch tip according to the base metal thickness you will be cutting and fuel gas you are using.

Note

Refer to the manufacturer's charts. You may have to readjust the oxygen and fuel gas pressure depending on the tip selected.

Step 2 Inspect the tip sealing surfaces and orifices. Attach the tip to the cutting torch or cutting attachment by placing it on the end of the torch and tightening the nut.

Note

Some manufacturers recommend tightening the nut with a torch wrench. Others recommend tightening the nut by hand. Check the tip manual to see if the manufacturer recommends that the nut be tightened manually or with a torch wrench.

Step 3 Put on proper protective clothing, gloves, and eye/face protection.

Step 4 Raise tinted eye protection and/or face shield.

Step 5 Release the oxygen cutting lever. If present, close the preheat oxygen valve and open the torch oxygen valve fully.

Step 6 Open the fuel gas valve on the torch handle about one-quarter turn.

Step 7 Holding the friction lighter near the side and to the front of the torch tip, ignite the torch.

WARNING!

Hold the friction lighter near the side of the tip when igniting the torch to prevent deflecting the ignited gas back toward you. Always use a friction lighter. Never use matches or cigarette lighters to light the torch because this could result in severe burns and/or could cause the lighter to explode. Always point the torch away from yourself, other people, equipment, and flammable material.

Step 8 Once the torch is lit, adjust the torch fuel gas flame by adjusting the flow of fuel gas with the fuel gas valve. Increase the flow of fuel gas until the flame stops smoking or pulls slightly away from the tip. Decrease the flow until the flame returns to the tip.

Step 9 Open the preheat oxygen valve (if present) or the oxygen torch valve very slowly and adjust the torch flame to a neutral flame.

Step 10 Press the cutting oxygen lever all the way down and observe the flame. It should have a long, thin, high-pressure oxygen cutting jet up to 8" long extending from the cutting oxygen hole in the center of the tip. If it does not:
• Check that the working pressures are set as recommended on the manufacturer's chart.
• Clean the cutting tip. If this does not clear up the problem, change the cutting tip.

Step 11 With the cutting oxygen on, observe the preheat flame. If it has changed slightly to a carburizing flame, increase the preheat oxygen until the flame is neutral. After this adjustment, the preheat flame will change slightly to an acceptable oxidizing flame when the cutting oxygen is shut off.

Obtaining Maximum Fuel Flow

Increasing the fuel flow until the flame pulls away from the tip and then decreasing the flow until the flame returns to the tip sets the maximum fuel flow for the tip size in use.

5.4.0 Shutting Off the Torch

Follow these steps to shut off the torch after a cutting operation is completed:

WARNING!
Always turn off the oxygen flow first to prevent a possible flashback into the torch.

Step 1 Release the cutting oxygen lever. Then close the torch or preheat oxygen valves.

Step 2 Close the torch fuel gas valve quickly to extinguish the flame.

6.0.0 ◆ SHUTTING DOWN OXYFUEL CUTTING EQUIPMENT

When a cutting job is completed and the oxyfuel equipment is no longer needed, it must be shut down. Follow these steps to shut down oxyfuel cutting equipment *(Figure 49)*:

Step 1 Close the fuel gas and oxygen cylinder valves.

Step 2 Open the fuel gas and oxygen torch valves to allow gas to escape. Do not proceed to Step 3 until all pressure is released and all regulator gauges read zero.

Step 3 Back out the fuel gas and oxygen regulator adjusting screws until they are loose.

Step 4 Close the fuel gas and oxygen torch valves.

Step 5 Coil up the hose and secure the torch to prevent damage.

7.0.0 ◆ DISASSEMBLING OXYFUEL EQUIPMENT

Follow these steps if the oxyfuel equipment must be disassembled after use:

Step 1 Check to be sure the equipment has been properly shut down. This includes checking that:
- The cylinder valves are closed.
- All pressure gauges read zero.

Step 2 Remove both hoses from the torch.

Step 3 Remove both hoses from the regulators.

Step 4 Remove both regulators from the cylinder valves.

Step 5 Replace the protective caps on the cylinders.

Step 6 Return the oxygen cylinder to its proper storage place.

WARNING!
Always transport and store gas cylinders in the upright position. Be sure they are properly secured (chained) and capped.

Step 7 Return the fuel gas cylinder to its proper storage place.

WARNING!
Regardless if the cylinders are empty or full, never store fuel gas cylinders and oxygen cylinders together. Storing cylinders together is a violation of OSHA and local fire regulations and could result in a fire and explosion.

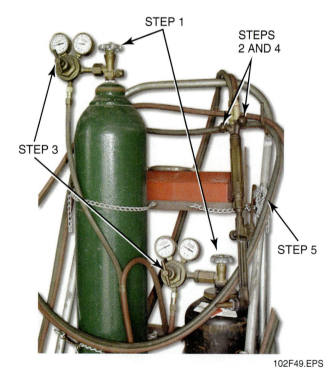

STEP 1

STEPS 2 AND 4

STEP 3

STEP 5

102F49.EPS

Figure 49 ◆ Shutting down oxyfuel cutting equipment.

Marking and Tagging Cylinders

Do not use permanent markers on cylinders; use soapstone or another temporary marker. If a cylinder is defective, place a warning tag on it.

8.0.0 ◆ CHANGING EMPTY CYLINDERS

Follow these procedures to change a cylinder when it is empty:

> **WARNING!**
>
> When moving cylinders, always be certain that they are in the upright position and the valve caps are secured in place. Never use a sling or electromagnet to lift cylinders. To lift cylinders, use a cylinder cage.

Step 1   Check to be sure equipment has been properly shut down. This includes checking that:
  • The cylinder valves are closed.
  • All pressure gauges read zero.

Step 2   Remove the regulator from the empty cylinder.

Step 3   Replace the protective cap on the empty cylinder.

Step 4   Transport the empty cylinder from the workstation to the storage area.

Step 5   Mark *MT* (empty) and the date (or the accepted site notation for indicating an empty cylinder) near the top of the cylinder using soapstone *(Figure 50)*.

Step 6   Place the empty cylinder in the empty cylinder section of the cylinder storage area for the type of gas in the cylinder.

9.0.0 ◆ PERFORMING CUTTING PROCEDURES

The following sections explain how to recognize good and bad cuts, how to prepare for cutting operations, and how to perform straight-line cutting, piercing, bevel cutting, washing, and gouging.

102F50.EPS

Figure 50 ◆ Typical empty cylinder marking.

9.1.0 Inspecting the Cut

Before attempting to make a cut, you must be able to recognize good and bad cuts and know what causes bad cuts. This is explained in the following list and illustrated in *Figure 51:*

• A good cut features a square top edge that is sharp and straight, not ragged. The bottom edge can have some slag adhering to it but not an excessive amount. What slag there is should be easily removable with a chipping hammer. The **drag lines** should be near vertical and not very pronounced.

• When preheat is insufficient, bad gouging results at the bottom of the cut because of slow travel speed.

• Too much preheat will result in the top surface melting over the cut, an irregular cut edge, and an excessive amount of slag.

DIRECTION OF TRAVEL →

GOOD CUT

PREHEAT
INSUFFICIENT

TOO MUCH
PREHEAT

CUTTING PRESSURE
TOO LOW

OXYGEN PRESSURE
TOO HIGH AND
UNDERSIZE TIP

TRAVEL SPEED
TOO SLOW

TRAVEL SPEED
TOO FAST

TORCH HELD OR
MOVED UNSTEADILY

CUT NOT RESTARTED
CAREFULLY, CAUSING
GOUGES AT RESTARTING
POINTS (CIRCLED)

102F51.EPS

Figure 51 ◆ Examples of good and bad cuts.

- When the cutting oxygen pressure is too low, the top edge will melt over because of the resulting slow cutting speed.
- Using cutting oxygen pressure that is too high will cause the operator to lose control of the cut, resulting in an uneven kerf.
- A travel speed that is too slow results in bad gouging at the bottom of the cut and irregular drag lines.
- When the travel speed is too fast, there will be gouging at the bottom of the cut, a pronounced break in the drag line, and an irregular kerf.
- A torch that is held or moved unsteadily across the metal being cut can result in a wavy and irregular kerf.
- When a cut is lost and then not restarted carefully, bad gouges will result at the point where the cut is restarted.

Note

The following tasks are designed to develop your skills with a cutting torch. Practice each task until you are thoroughly familiar with the procedure. As you complete each task, take it to your instructor for evaluation. Do not proceed to the next task until your instructor tells you to continue.

9.2.0 Preparing for Oxyfuel Cutting with a Hand Cutting Torch

Before metal can be cut, the equipment must be set up and the metal prepared. One important step is to properly lay out the cut by marking it with soapstone or punch marks. The few minutes this takes will result in a quality job, reflecting craftsmanship and pride in your work. Follow these steps to prepare to make a cut:

Step 1   Prepare the metal to be cut by cleaning any rust, scale, or other foreign matter from the surface.

Step 2   If possible, position the work so you will be comfortable when cutting.

Step 3   Mark the lines to be cut with soapstone or a punch.

Step 4   Select the correct cutting torch tip according to the thickness of the metal to be cut, the type of cut to be made, the amount of preheat needed, and the type of fuel gas to be used.

Step 5   Ignite the torch.

Step 6   Use the procedures outlined in the following sections for performing particular types of cutting operations.

9.3.0 Cutting Thin Steel

Thin steel is ³⁄₁₆" thick or less. When cutting by hand, use a tip one size larger than is recommended. A major concern when cutting thin steel is distortion caused by the heat of the torch and the cutting process. To minimize distortion, move as quickly as you can without losing the cut. Follow these steps to cut thin steel:

Step 1   Prepare the metal surface.

Step 2   Light the torch.

Step 3   Hold the torch so that the tip is pointing in the direction the torch is traveling at a 15° to 20° angle. Make sure that a preheat orifice and the cutting orifice are centered on the line of travel next to the metal (*Figure 52*).

CAUTION
Holding the tip upright when cutting thin steel will overheat the metal, causing distortion.

Step 4   Preheat the metal to a dull red. Use care not to overheat thin steel because this will cause distortion.

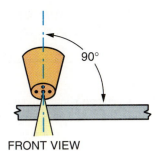

FRONT VIEW

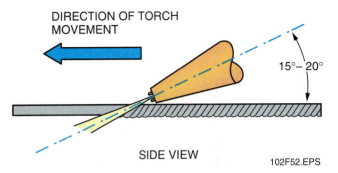

DIRECTION OF TORCH MOVEMENT

15°–20°

SIDE VIEW

102F52.EPS

Figure 52 ◆ Cutting thin steel.

Note
The edge of the tip can be rested on the surface of the metal being cut and then slid along the surface when making the cut.

Step 5   Press the cutting oxygen lever to start the cut, and then move quickly along the line to be cut. To minimize distortion, move as quickly as you can without losing the cut.

9.4.0 Cutting Thick Steel

Most oxyfuel cutting will be on steel that is more than ³⁄₁₆" thick. Whenever heat is applied to metal, distortion is a problem, but as the steel gets thicker, it becomes less of a problem. Follow these steps to cut thick steel with a hand cutting torch:

Step 1   Prepare the metal surface and torch. Light the torch.

Step 2   Ignite and adjust the torch flame.

Note
The torch can be moved from either right to left or left to right. Choose the direction that is the most comfortable for you.

Step 3 Follow the number sequence shown in *Figure 53* to perform the cut.

Note
When cutting begins, the tips of the preheat flame should be held ¹⁄₁₆" to ⅛" above the workpiece. For steel up to ⅜" thick, the first and third procedures can usually be omitted.

9.5.0 Piercing a Plate

Before holes or slots can be cut in a plate, the plate must be pierced. Piercing puts a small hole through the metal where the cut can be started. Because more preheat is necessary on the surface of a plate than at the edge, choose the next-larger cutting tip than is recommended for the thickness to be pierced. When piercing steel that is more than 3" thick, it may help to preheat the bottom side of the plate directly under the spot to be pierced. Follow these steps to pierce a plate for cutting:

Step 1 Prepare the metal surface and torch.

Step 2 Ignite the torch and adjust the flame.

Step 3 Hold the torch tip ¼" to ⁵⁄₁₆" above the spot to be pierced until the surface is a bright cherry red *(Figure 54)*.

Step 4 Slowly press the cutting oxygen lever. As the cut starts, raise the tip about ½" above the metal surface and tilt the torch slightly so that molten metal does not blow back onto the tip. The tip should be raised and tipped before the cutting oxygen lever is fully depressed.

Step 5 Maintain the tipped position until a hole burns through the plate. Then rotate the tip vertically.

Step 6 Lower the torch tip to about ³⁄₁₆" above the metal surface and continue to cut outward from the original hole to the edge of the line to be cut.

9.6.0 Cutting Bevels

Bevel cutting is often performed to prepare the edge of steel plate for welding. Follow these steps to perform bevel cutting *(Figure 55)*:

Step 1 Prepare the metal surface and the torch.

Step 2 Ignite the torch and adjust the flame.

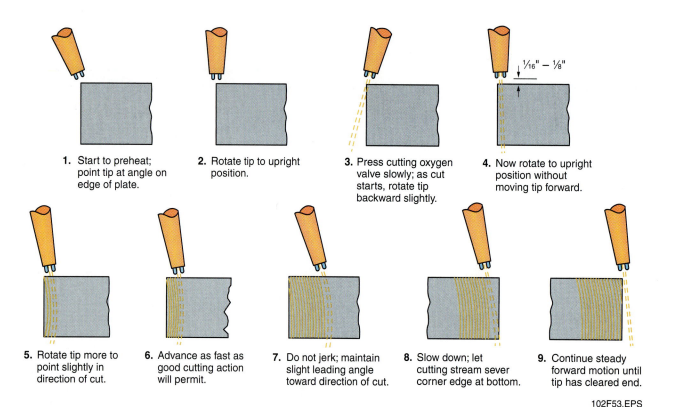

1. Start to preheat; point tip at angle on edge of plate.

2. Rotate tip to upright position.

3. Press cutting oxygen valve slowly; as cut starts, rotate tip backward slightly.

4. Now rotate to upright position without moving tip forward.

5. Rotate tip more to point slightly in direction of cut.

6. Advance as fast as good cutting action will permit.

7. Do not jerk; maintain slight leading angle toward direction of cut.

8. Slow down; let cutting stream sever corner edge at bottom.

9. Continue steady forward motion until tip has cleared end.

102F53.EPS

Figure 53 ◆ Procedures for flame cutting with a hand torch.

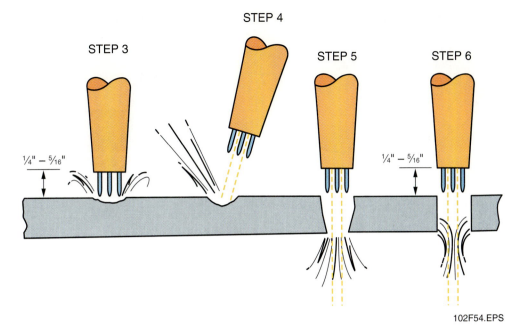

STEP 3 STEP 4 STEP 5 STEP 6

1/4" – 5/16"

1/4" – 5/16"

102F54.EPS

Figure 54 ◆ Steps for piercing steel.

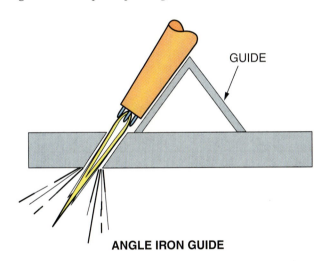

GUIDE

ANGLE IRON GUIDE

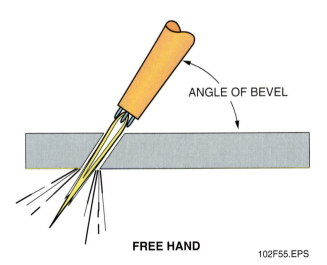

ANGLE OF BEVEL

FREE HAND

102F55.EPS

Figure 55 ◆ Cutting a bevel.

Step 3 Hold the torch so that the tip faces the metal at the desired bevel angle.

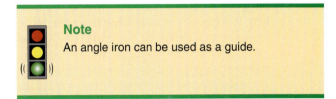

Note
An angle iron can be used as a guide.

Step 4 Preheat the edge to a bright cherry red.

Step 5 Press the cutting oxygen lever to start the cut.

Step 6 As cutting begins, move the torch tip at a steady rate along the line to be cut. Pay particular attention to the torch angle to ensure it is uniform along the entire length of the cut.

9.7.0 Washing

Washing is a term used to describe the process of cutting out bolts or rivets. Washing operations use a special tip with a large cutting hole that produces a low-velocity stream of oxygen. The low-velocity oxygen stream helps prevent cutting into the surrounding base metal. Washing tips can also be used to remove items such as blocks, angles, or channels that are welded onto a surface. Follow these steps to perform washing *(Figure 56):*

Step 1 Prepare the metal surface and torch.

Step 2 Ignite the torch and adjust the flame.

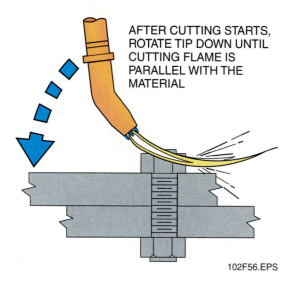

AFTER CUTTING STARTS, ROTATE TIP DOWN UNTIL CUTTING FLAME IS PARALLEL WITH THE MATERIAL

102F56.EPS

Figure 56 ◆ Washing.

102F57.EPS

Figure 57 ◆ Gouging.

Step 3 Preheat the metal to be cut until it is a bright cherry red.

Step 4 Move the cutting torch at a 55° angle to the metal surface.

Step 5 At the top of the material, press the cutting oxygen lever to cut the material to be removed. Continue moving back and forth across the material while rotating the tip to a position parallel with the material. Move the tip back and forth and down to the surrounding metal. Use care not to cut into the surrounding metal.

CAUTION

As the surrounding metal heats up, there is a greater danger of cutting into it. Try to complete the washing operation as quickly as possible. If the surrounding metal gets too hot, stop and let it cool down.

9.8.0 Gouging

Gouging is the process of cutting a groove into a surface. Gouging operations use a special curved tip that produces a low-velocity stream of oxygen that curves up, allowing the operator to control the depth and width of the groove. It is an effective means to gouge out cracks or weld defects for welding. Gouging tips can also be used to remove steel backing from welds or to wash off bolt or rivet heads. Gouging tips are not as effective as washing tips for removing the shank of the bolt or rivet. Follow these steps to gouge *(Figure 57)*:

Step 1 Prepare the metal surface and torch.

Step 2 Ignite the torch and adjust the flame.

Step 3 Holding the torch so that the preheat holes are pointed directly at the metal, preheat the surface until it becomes a bright cherry red.

Step 4 When the steel has been heated to a bright cherry red, slowly roll the torch away from the metal so that the holes are at an angle that will enable you to cut the gouge to the correct depth. As you roll the torch away, press the cutting oxygen lever gradually.

Step 5 Move the cutting torch along the line of the gouge. As you move the torch, rock it back and forth as necessary to create a gouge of the required depth and width.

Note

The travel speed and torch angle are very important when gouging. If the travel speed or torch angle is incorrect, the gouge will be irregular and there will be a buildup of slag inside the gouge. Practice until the gouge is clean and even.

10.0.0 ◆ PORTABLE OXYFUEL CUTTING MACHINE OPERATION

As explained previously, machine oxyfuel gas cutters or track burners are basic guidance systems

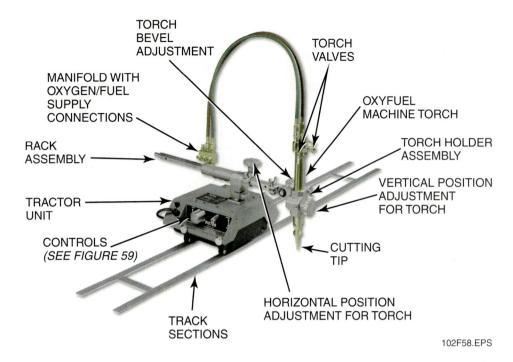

Figure 58 ♦ Victor track burner.

driven by a variable speed electric motor to enable the operator to cut or bevel straight lines at any desired speed. The device *(Figure 58)* is usually mounted on a track or used with a circle-cutting attachment to enable the operator to cut various diameters from 4" to 96". It consists of a heavy-duty tractor unit fitted with an adjustable torch mount and gas hose attachments. It is also equipped with an ON/OFF switch, a reversing switch, a clutch, and a speed-adjusting dial calibrated in feet/meters per minute.

The device shown in *Figure 58* offers the following operational features:

- Makes straight-line cuts of any length
- Makes circle cuts up to 96" in diameter
- Makes bevel or chamfer cuts
- Has an infinitely variable cutting speed from 1 to 60 seconds per minute
- Has dual speed and clutch controls to enable operation of the machine from either end

10.1.0 Machine Controls

Figure 59 shows the location of the following controls:

- *Directional control* – Set the machine direction by toggling the FWD-OFF-REV toggle switch located next to the power cord.
- *Speed control* – Turn the large knob on either end of the machine to position the speed indicator at the desired cutting speed.

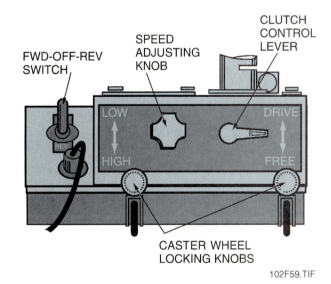

Figure 59 ♦ Victor track burner controls.

- *Clutch operation* – Engage the clutch by rotating one of the two clutch levers, located on either end of the machine, to the DRIVE position. Place the clutch lever in the FREE position to permit easy manual positioning of the machine prior to or after the actual cutting operation.

10.2.0 Torch Adjustment

The rack assembly permits the torch holder assembly to move toward or away from the tractor unit. The torch holder allows vertical positioning of the torch. The torch bevel adjustment allows torch positioning at any angle from +90° to −90° in a plane perpendicular to the track. After adjusting the torch to the desired position, tighten all clamping screws to prevent the torch from making any unexpected movements.

10.3.0 Straight Line Cutting

Cut straight lines using the following procedure:

WARNING!
Most cutting machines are not designed to detect the end of their track or workpiece. Take care that an unattended machine does not fall from an elevated workpiece.

Step 1 Place the machine track on the workpiece and line it up before placing the machine on the track.

Step 2 Be sure the track is long enough for the cut to be made. If not, install additional track. Connect track sections carefully. Extend the track on both sides of the cut and support the track. When properly connected, the machine should travel smoothly from one track section to the next. If the cut is long, the track may have to be clamped at both ends beyond the cut to keep the track from moving during the cut.

Step 3 Place the machine on the track. Place the clutch lever in the FREE position. Be sure that the supply gas hoses and the power lines are long enough and free to move with the machine so that it can complete the cut properly.

Step 4 Move the machine to the approximate point where the cut will start. Set the drive speed control to the desired cutting speed. Set the FWD-OFF-REV switch to the OFF position. Plug the power cord into a 115 alternating current (AC), 60 Hertz (Hz) power outlet.

Step 5 Ensure that all clamping screws are properly tightened. Ignite and properly adjust the torch, then preheat the start of the cut. Set the FWD-OFF-REV switch to the desired direction of travel. Simultaneously turn on the cutting oxygen and set the clutch lever to the DRIVE position.

Step 6 When the cut is completed, stop the machine and shut off the torch.

10.4.0 Bevel Cutting

Perform bevel cutting operations using the following procedure:

WARNING!
Most cutting machines are not designed to detect the end of their track or workpiece. Take care that an unattended machine does not fall from an elevated workpiece.

Step 1 Place the machine track on the workpiece and line it up before placing the machine on the track.

Step 2 Be sure the track is long enough for the cut to be made. If not, install additional track. Connect track sections carefully. Extend the track on both sides of the cut and support the track. When properly connected, the machine should travel smoothly from one track section to the next. If the cut is long, the track may have to be clamped at both ends beyond the cut to keep the track from moving during the cut.

Step 3 Place the machine on the track. Place the clutch lever in the FREE position. Be sure that the supply gas hoses and the power lines are long enough and free to move with the machine so that it can complete the cut properly.

Step 4 Loosen the bevel adjusting knob, set the torch angle to the desired bevel angle, and then tighten the bevel adjusting knob.

Step 5 Move the machine to the approximate point where the cut will start. Set the drive speed control to the desired cutting speed. Set the FWD-OFF-REV switch to the OFF position. Plug the power cord into a 115AC, 60Hz power outlet.

Step 6 Ensure that all clamping screws are properly tightened. Ignite and properly adjust the torch, then preheat the start of the cut. Set the FWD-OFF-REV switch to the desired direction of travel. Simultaneously turn on the cutting oxygen and set the clutch lever to the DRIVE position.

Step 7 When the cut is completed, stop the machine and shut off the torch.

Summary

Oxyfuel cutting has many uses on job sites. It can be used to cut plate and shapes to size, prepare joints for welding, clean metals or welds, and disassemble structures. Because of the high pressures and flammable gases involved, there is a danger of fire and explosion when using oxyfuel equipment. However, these risks can be minimized when the oxyfuel cutting operator is well trained and knowledgeable. Be sure you know and understand the safety precautions and equipment presented in this module before using oxyfuel equipment.

Review Questions

1. For gas welding, either the face shield or the welding goggle lenses must be an approved shade _____ filter.
   a. 3 or 4
   b. 5 or 6
   c. 7 or 8
   d. 9 or 10

2. Most work site fires caused by welding or cutting are started by _____.
   a. improper disposal of slag
   b. escape of acetylene
   c. welding torches
   d. cutting torches

3. When mixed with air or oxygen, _____ will explode at much lower concentrations than any other fuel.
   a. propylene
   b. propane
   c. acetylene
   d. MAPP®

4. When pure oxygen is combined with fuel gases, the pure oxygen produces _____.
   a. argon
   b. hydrogen
   c. a colorless, odorless, and tasteless gas
   d. a high-temperature flame needed for flame cutting

5. The smallest standard oxygen cylinder holds about _____ cubic feet of oxygen.
   a. 40
   b. 65
   c. 80
   d. 85

6. Acetylene gas must be withdrawn from a cylinder at an hourly rate that does not exceed _____ of the cylinder capacity.
   a. ½
   b. ⅓
   c. ⅕
   d. ⅒

7. Safety fuse plugs in the top and bottom of an acetylene cylinder are designed to _____.
   a. release acetylene gas in the event of a fire
   b. prevent the release of acetylene gas
   c. prevent the withdrawal of acetone
   d. release acetone from the cylinder

8. Methylacetylene propadiene (MAPP®) gas, a liquefied fuel used in oxyfuel cutting, burns at temperatures almost as high as acetylene and has the stability of _____.
   a. natural gas
   b. propylene
   c. propane
   d. oxygen

9. The amount of liquefied gas remaining in a cylinder is determined by the _____ of the cylinder.
   a. color
   b. heat
   c. weight
   d. pressure

10. The regulators used on fuel gas cylinders are often painted red and always have _____ on all connections.
    a. right-hand threads
    b. left-hand threads
    c. metric threads
    d. safety latches

11. The attachment on the top of a fuel gas cylinder that allows the gas to flow only in one direction is called a _____.
    a. single-stage regulator
    b. two-stage regulator
    c. flashback arrestor
    d. check valve

12. The cutting tips used with liquefied fuel gases must have at least _____ preheat holes.
    a. four
    b. five
    c. six
    d. seven

13. When lifting oxyfuel cutting cylinders, always use a(n) _____.
    a. sling cable
    b. sling strap
    c. cylinder cage
    d. electromagnet

14. To avoid injury from dirt that may be lodged in the valve and regulator seat of a gas cylinder, always stand _____ the valve when opening the valve to clear the regulator seat.
    a. to the side of
    b. in front of
    c. behind
    d. above

15. When clearing debris from new oxyfuel cutting equipment hoses, blow the hoses out with _____.
    a. nitrogen
    b. propane
    c. propylene
    d. compressed air

16. The first step in installing a cutting tip is to _____.
    a. inspect the cutting tip sealing surfaces and orifices for damage
    b. determine the size of cutting tip to use
    c. determine the kind of gas being used
    d. identify the thickness of the material to be cut

17. Before opening cylinder valves, verify that the adjusting screws on the oxygen and fuel gas regulators are _____.
    a. tight
    b. fully clockwise
    c. loose
    d. fully counterclockwise

18. When a cutting flame has an excess of fuel, the flame is called a(n) _____ flame.
    a. cold
    b. neutral
    c. oxidizing
    d. carburizing

19. When disassembling oxyfuel equipment, verify that all pressure gauges read _____ before starting to take the equipment apart.
    a. −1.0
    b. 0
    c. within 0.3 of 0
    d. within 0.2 of 0

20. When inspecting a completed cut made with oxyfuel cutting equipment, the drag lines of the cut should be near _____ and not very pronounced.
    a. 30°
    b. 45°
    c. horizontal
    d. vertical

Ray French

Execution Coordinator, Special Projects
Flint Hills Resource

As a young man, Ray French got off on the wrong foot, but eventually he found his niche in the welding trade.

How did you get your start in the welding trade?
I received a college football scholarship but flunked out of college in my freshman year. I then spent a year loading ships and barges until my father suggested that I learn a trade. My uncle had a welding shop, so I decided to try welding. I really liked it! I went to welding school for two years while working part time at my uncle's welding shop.

How did your career progress after welding school?
I was hired by an oil refinery and worked in the welding shop until I received my first-class welder classification. Then I was assigned to a hot-shot team made up of welders, pipefitters, boilermakers, pump mechanics, and electricians. During my time on that team, I had the opportunity to learn a lot about each of these trades. I supervised several shutdowns—that's where a unit is taken out of service and completely overhauled—and eventually was promoted to maintenance foreman. Today, my job title is Execution Coordinator, and I supervise large projects at a major oil refinery.

What do you like most about your job?
I like the pride that comes from doing a good job and being able to see the results of my work every day. I also like the respect I get because of my skills and my knowledge of the trade.

What do you think it takes to be a success?
You have to take pride in your work. Do a quality job every time and don't take shortcuts. Your reputation is everything. It will follow you from job to job. No matter where you go, someone will know you or know someone who knows you.

Trade Terms Introduced in This Module

Backfire: A loud snap or pop as a torch flame is extinguished.

Carburizing flame: A flame burning with an excess amount of fuel; also called a reducing flame.

Drag lines: The lines on the kerf that result from the travel of the cutting oxygen stream into, through, and out of the metal.

Ferrous metals: Metals containing iron.

Flashback: The flame burning back into the tip, torch, hose, or regulator, causing a high-pitched whistling or hissing sound.

Kerf: The edge of the cut.

Neutral flame: A flame burning with correct proportions of fuel gas and oxygen.

Oxidizing flame: A flame burning with an excess amount of oxygen.

Pierce: To penetrate through metal plate with an oxyfuel cutting torch.

Slag: The material that is expelled from the kerf when oxyfuel cutting.

Soapstone: Soft, white stone used to mark metal.

Performance Accreditation Tasks

The Performance Accreditation Tasks (PATCs) correspond to and support the learning objectives in AWS EG2.0-95, Guide for the Training and Qualification of Welding Personnel: Entry-Level Welder.

Note that in order to satisfy all learning objectives in AWS EG2.0-95, the instructor must also use the PATCs contained in the second level in the Welding Contren™ Learning Series.

PATCs provide specific acceptable critera for performance and help to ensure a true competency-based welding program for students.

The following tasks are designed to test your competency with a cutting torch. Do not perform a cutting task until directed to do so by your instructor.

SETTING UP, IGNITING, ADJUSTING, AND SHUTTING DOWN OXYFUEL EQUIPMENT

Using oxyfuel equipment that has been completely disassembled, demonstrate how to:
• Set up oxyfuel equipment
• Ignite and adjust the flame
  – Carburizing
  – Neutral
  – Oxidizing
• Shut off the torch
• Shut down the oxyfuel equipment

Criteria for Acceptance:

• Set up the oxyfuel equipment in the correct sequence　　　　　_____
• Demonstrate that there are no leaks　　　　　_____
• Properly adjust all three flames　　　　　_____
• Shut off the torch in the correct sequence　　　　　_____
• Shut down the oxyfuel equipment　　　　　_____

CUTTING A SHAPE FROM THIN STEEL

Using a carbon steel plate, lay out and cut the shape and holes shown in the figure. If available, use a machine track cutter to straight cut the 6" dimensions.

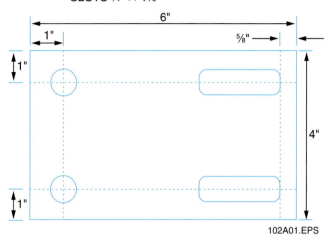

NOTE: MATERIAL – CARBON STEEL ⅛" TO ¼" THICK
HOLES ¾" DIAMETER
SLOTS ¾" × 1½"

102A01.EPS

Criteria for Acceptance:

- Outside dimensions ±¹⁄₁₆"
- Inside dimensions (holes and slots) ±⅛"
- Square ±5º
- Minimal amount of slag sticking to plate which can be easily removed
- Square kerf face with minimal notching not exceeding ¹⁄₁₆" deep

CUTTING A SHAPE FROM THICK STEEL

Using a carbon steel plate, lay out and cut the shape and holes shown in the figure.
If available, use a machine track cutter to bevel and straight cut the 6" dimensions.

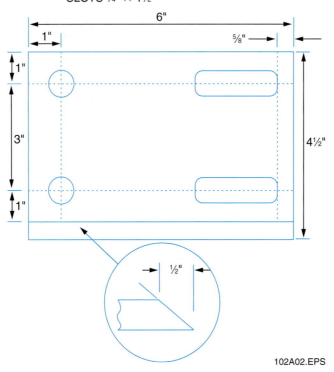

NOTE: MATERIAL – CARBON STEEL OVER ¼" THICK
　　　　　HOLES ¾" DIAMETER
　　　　　SLOTS ¾" × 1½"

102A02.EPS

Criteria for Acceptance:

- Outside dimensions ±1⁄16"
- Inside dimensions (holes and slots) ±1⁄8"
- Square ±5º
- Bevel ±2º
- Minimal amount of slag sticking to plate which can be easily removed
- Square kerf face with minimal notching not exceeding 1⁄16" deep

_____
_____
_____
_____
_____
_____

Additional Resources

This module is intended to present thorough resources for task training. The following reference work is suggested for further study. This is optional material for continued education rather than for task training.

Modern Welding, 2000. A. D. Althouse, C. H. Turnquist, W. A. Bowditch, and K. E. Bowditch. Tinley Park, IL: The Goodheart Willcox Company, Inc.

Base Metal Preparation

Course Map

This course map shows all of the modules in the AWS Entry Level Welder – Phase 1 curriculum. The suggested training order begins at the bottom and proceeds up. Skill levels increase as you advance on the course map. The local Training Program Sponsor may adjust the training order.

AWS ENTRY LEVEL WELDER—PHASE 1

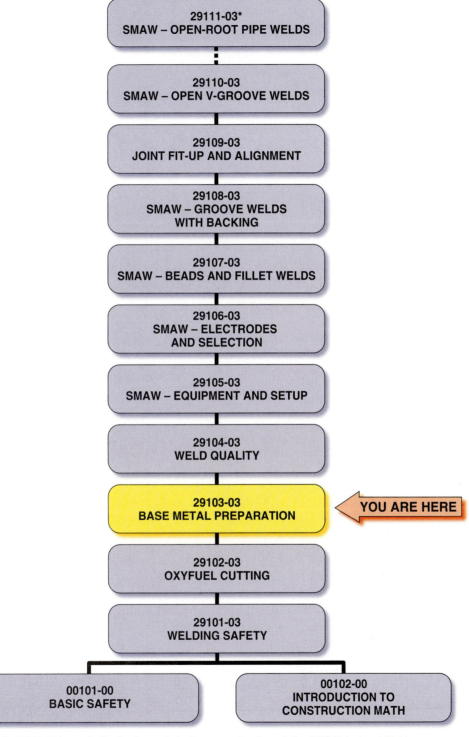

29111-03*
SMAW – OPEN-ROOT PIPE WELDS

29110-03
SMAW – OPEN V-GROOVE WELDS

29109-03
JOINT FIT-UP AND ALIGNMENT

29108-03
SMAW – GROOVE WELDS
WITH BACKING

29107-03
SMAW – BEADS AND FILLET WELDS

29106-03
SMAW – ELECTRODES
AND SELECTION

29105-03
SMAW – EQUIPMENT AND SETUP

29104-03
WELD QUALITY

29103-03
BASE METAL PREPARATION ← YOU ARE HERE

29102-03
OXYFUEL CUTTING

29101-03
WELDING SAFETY

00101-00
BASIC SAFETY

00102-00
INTRODUCTION TO
CONSTRUCTION MATH

*Module 29111-03 is an elective for those students progressing through the AWS Entry Level Welder program.
29103CMAP.EPS

Figures

Base Metal Preparation

Objectives

When you have completed this module, you will be able to do the following:

1. Clean base metal for welding or cutting.
2. Identify and explain joint design.
3. Explain joint design considerations.
4. Using a nibbler, cutter, or grinder, mechanically prepare the edge of a mild steel plate ¼" to ¾" thick at 22½° (or 30° depending on equipment available).
5. Using a nibbler, cutter, or grinder, mechanically prepare the end of a pipe with a 30° or 37½° bevel (depending on equipment available) and a ³⁄₃₂" land. Use 6" , 8", or 10" Schedule 40 or Schedule 80 mild steel pipe.
6. Select the proper joint design based on a welding procedure specification (WPS) or instructor direction.

Prerequisites

Before you begin this module, it is recommended that you successfully complete Modules 00101 through 29102.

Required Trainee Materials

1. Pencil and paper
2. Appropriate personal protective equipment

1.0.0 ◆ INTRODUCTION

To ensure that the highest quality welds are produced and to comply with welding codes, base metals must be properly prepared prior to welding. The type of preparation required depends on the governing code requirements, the base metal type, the condition of the base metal, the welding process to be used, and the equipment available. This module will explain how to properly prepare various base metals to conform to the appropriate welding codes.

2.0.0 ◆ BASE METAL CLEANING

All base metals should be cleaned before welding. Even new materials that may look clean often are not. They pick up contaminants and surface corrosion during shipping and handling. When performing maintenance welding, a welder often comes across components that have been exposed to surface contamination such as corrosion, paint, oil, and grease. To ensure quality welds and to conform to code requirements, surface contaminants and oxides must be removed prior to welding. In addition, the heat generated by the welding process could cause the contaminants to give off toxic fumes that could endanger the person performing the welding.

2.1.0 Surface Corrosion

All metal has surface corrosion. Corrosion occurs when metal is exposed to air. Some corrosion exists as a thin, hard film that merely stains the metal and acts as a retardant to further corrosion. Examples of this type of corrosion are found on stainless steel, copper, and aluminum. The thin, hard layer of corrosion bonds tightly to the surface and protects the metal. A more recognizable type of corrosion is found on mild steel and is called rust. It is a very coarse type of corrosion that tends to flake easily, exposing more of the base metal to the corrosion. Alloys such as chromium and copper are often added to mild steel to protect it from this coarse type of corrosion. The alloys cause the corrosion (rust) to be very fine and to bond to the surface, protecting the base metal under the film. These types of steels are called alloy steels. **Weathering steel** is an example of a copper-alloyed steel that protects itself from corrosion by forming a hard, tough layer of brown corrosion (rust) on its surface. Weathering steels are used outdoors and require no painting.

Regardless of the type of corrosion, it must be removed prior to welding. Any type of corrosion can cause weld defects. *Figure 1* shows examples of rust and corrosion on metals that must be removed prior to welding.

2.2.0 Defects Caused by Surface Contamination

The most common defect caused by surface contamination is **porosity.** Porosity occurs when gas pockets or voids appear in the weld metal. When the porosity is elongated or extends into more than one layer of weld metal, it is called piping porosity. Piping porosity is formed as the gas

pocket floats toward the surface of the weld, leaving a void behind. The gas pocket is trapped as the weld metal solidifies, but as the next layer of weld is deposited, it continues to float up. Porosity may not be visible on the surface of the weld. An example of uniformly distributed porosity is shown in *Figure 2(A)*. *Figure 2(B)* shows surface-breaking porosity.

Because of the defects caused by surface contaminants and the possibility of toxic fumes, all welding codes require surface cleaning. In general, the codes state that the surface to be thermally cut or welded must be clean and free from paint, oil, rust, scale, and other material that

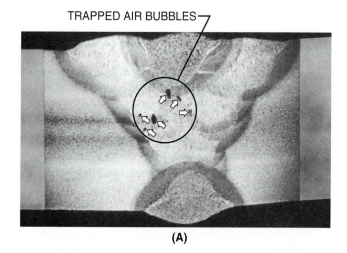

TRAPPED AIR BUBBLES

(A)

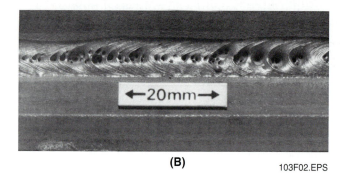

←20mm→

(B)

103F02.EPS

103F01.EPS

Figure 1 ◆ Examples of rust and corrosion.

Figure 2 ◆ Porosity.

Appearances Can Be Deceiving

Do not assume a metal's surface is corrosion-free just because it is shiny. Corrosion on stainless steel and aluminum is silver. The metal may appear shiny, but to obtain high-quality welds, the corrosion must be removed.

would be harmful either to the weld or to the base metal when heat is applied. Cleaning is typically performed by mechanical and/or chemical means.

2.3.0 Mechanical Cleaning

Mechanical cleaning is the most common method of removing surface contamination. Tools used for mechanical cleaning include hand tools, power tools, and special sandblasting equipment. When performing mechanical cleaning, be sure to wear safety glasses and a face shield for protection from the flying particles produced during the cleaning operation. In addition, special clothing is required for sandblasting.

2.3.1 Hand Tools

Flexible scrapers are used to remove dirt and grease from **weldments** or weldment components. Rigid scrapers can be used to remove hardened paint or dirt.

Wire brushes are used to remove paint or surface corrosion. When cleaning stainless steel or aluminum, wire brushes with stainless steel bristles must be used.

Wire brushing will remove light to medium corrosion but will not remove tight corrosion. Tight corrosion must be removed by filing, grinding, or sandblasting.

Files are also an excellent way to remove surface corrosion. When using a file, be sure it has a handle on the file tang.

WARNING!
Using a file without a handle can result in serious injury from the pointed file tang.

Use a fine file and keep it clean to prevent scratching the base metal surface. Take care not to damage or scratch the base metal outside the weld zone.

Clean files by using stainless steel brushes that are either new or have only been used to brush the same type of metal to which the file has been exposed.

Figure 3 shows a brush and file used for cleaning metal.

2.3.2 Power Tools

Hand tools work well for small jobs, but for large jobs or jobs where speed is important, power tools are best. Power tools may be electrically or pneumatically powered. Angle grinders and end grinders are very effective in removing large areas of surface contamination. Die grinders and small angle grinders work very well for small areas such as weld grooves and bevel edges. Grinders also have attachments that can be used for special applications (*Figure 4*). These include grinding disks and wheels, wire brushes, rotary files, flapper wheels, and cutoff wheels. Grinding disks and wheels are made for specific types of metals. Be sure you have the correct disk or wheel for the

103F03.EPS

Figure 3 ◆ Typical brush and file used for cleaning metal.

Cleaning Stainless Steel and Aluminum

When brushing stainless steel, use only stainless steel brushes that have been used on stainless steel. When brushing aluminum, use only stainless steel brushes that have been used on aluminum. If a standard wire brush with steel bristles is used, it will contaminate the surface with mild steel, causing weld defects. When filing aluminum or stainless steel, do not use files that have been used on other metals. Filings often stick to the file and could contaminate the metal being filed, resulting in weld defects.

Hammers in Welding

Welders often use versions of the hammers shown in the photo. What do you think they are used for in welding operations?

103P0301.EPS

Using a File

Should the file shown in *Figure 3* be used in the condition shown? Why or why not?

type of metal being ground. Always use aluminum oxide disks for grinding aluminum or stainless steel. Do not use wheels that have been used on other metals for stainless steel or aluminum. This will contaminate the surface of the aluminum or the stainless steel.

When using wire brush attachments on stainless steel or aluminum, brushes with stainless steel bristles must be used. Use only stainless steel brushes that have been used on stainless steel for brushing stainless steel. Use only stainless steel brushes that have been used on aluminum for brushing aluminum. Mild steel bristles will contaminate stainless steel or aluminum, causing weld defects.

When grinding, care must be taken to prevent grinding the base metal below the minimum allowable base metal thickness. If the base metal is ground below the minimum allowable base metal thickness, the base metal will have to be replaced or, if allowable, built up with weld. Both of these are expensive alternatives.

Cutoff wheels are used to cut metals and other materials. They are made for specific types of metals or other materials such as stone. Be sure you have the correct cutoff wheel for the type of material being cut.

Weld flux chippers, needle scalers, and chipping hammers are air (pneumatically) powered. They are used by welders to clean surfaces and to remove slag from cuts and welds. Weld flux chippers, needle scalers, and chipping hammers are also excellent for removing paint, heavy scale, or hardened dirt but are not very effective for removing surface corrosion. Weld flux chippers have a single chisel; needle scalers have about 18 to 20 blunt steel needles about 10" long. Most weld flux chippers can be converted to needle scalers with the needle scaler attachment. Pneumatic chipping hammers, which hit much harder than weld flux chippers, are used for **peening** or gouging weld joints. *Figure 5* shows a weld flux chipper and a needle scaler.

RASP   ROTARY FILE   DIE GRINDER STONE CONE   SNAGGING WHEELS   FLAPPER WHEEL   WIRE CAP BRUSH

END GRINDER   ANGLE GRINDER   DIE GRINDER

WIRE WHEEL BRUSH   ABRASIVE GRINDING DISC   CUTOFF WHEEL   RAISED HUB WHEEL

103F04.EPS

Figure 4 ◆ Grinders and grinder attachments.

Facts about Grinding Wheels and Disks

If a grinding wheel or disk shatters while in use, the pieces of the wheel or disk become deadly projectiles as the weights of the pieces are multiplied by the centrifugal force and the revolutions per minute (rpm) of the grinding motor. The force with which a piece of wheel or disk is released is equivalent to that of an armor-piercing bullet. Never grind without full face protection. Avoid standing in the orbital path of the wheel or disk in case it disintegrates.

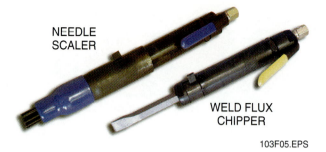

Figure 5 ◆ Weld flux chipper and needle scaler.

3.0.0 ◆ JOINT DESIGN

Welded joints are selected primarily for the safety and strength required for the conditions to be encountered. When selecting the joint for a particular application, many factors must be taken into account, including **load** considerations, environments, materials, processes, and cost.

3.1.0 Load Considerations

The loads in a welded steel structure or component are transferred through the welded joint. The welded joint must be designed to withstand the stresses that these loads cause. The forces that apply the stress are tensile forces (*Figure 6*).

3.2.0 Types of Joints

There are five basic types of joints. They are the butt joint, lap joint, corner joint (inside and outside), T-joint, and edge joint. *Figure 7* shows illustrations of each of these joints. Each of these joints may be accomplished using various types of welds, which will now be discussed.

3.3.0 Types of Welds

Keeping in mind the five basic types of joints that are used to join base metals, we will shift our focus to the types of welds applied to these joints and the base metal preparation that may be required. Most types of welds require some degree of base metal edge preparation depending on the type of joint being welded. Common types of welds include surfacing, plug or slot, fillet, square groove, bevel groove, V-groove, J-groove, and U-groove. If the type of weld requires a specified root opening, it is typically shown on the welding drawing if one is available. Base metals may require some type of groove to be cut on one or both edges. Edge preparations will be discussed for each type of weld as the type of weld is introduced in this section.

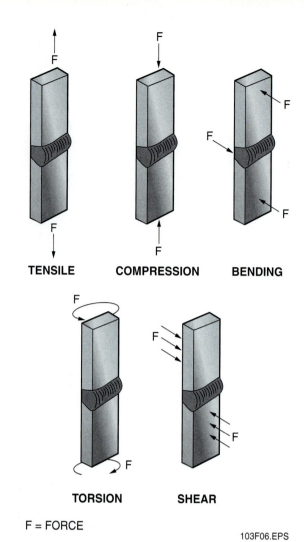

F = FORCE

Figure 6 ◆ Forces that act on welded joints.

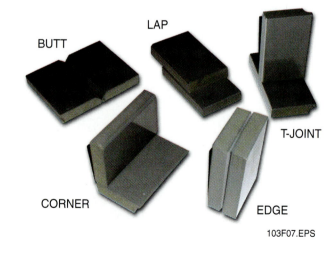

Figure 7 ◆ Five basic joint types.

3.3.1 Surfacing Welds

Surfacing is the only type of weld that is applied to only one base metal surface. It is used to build up a base surface that has either become worn below the desired thickness or dimension or to add strength and stability to a base by adding material to it. An example of surfacing is shown in *Figure 8*.

Before applying the first layer, the base metal surface should be cleaned to remove contaminants—such as oxidation and dirt—using manual or power cleaning tools as previously described. When applying the surfacing weld using an oxyfuel process, the surface should be preheated with the torch to eliminate warping the base material.

Each layer of surfacing must be properly scraped and cleaned before applying additional layers of surfacing in order for proper penetration and layering to take place.

Surfacing is often applied in the tool and die industry to rebuild relatively expensive tools and dies that have worn down. It is also used in the repair of heavy equipment to build up areas that are exposed to various excessive stress factors such as shear, tensile stress, and torsion.

3.3.2 Plug and Slot Welds

Plug and slot welds are used to join metal pieces when the edges cannot be welded. The plug or slot weld may be applied on lap, corner, and

103F08.EPS

Figure 8 ◆ Surfacing welding on an inclined vertical surface.

T-joints. The hole or slot can be either completely or partially filled with weld material in joining it to the other base metal. If the hole is round, the weld is referred to as a plug weld. If the weld is slotted or elongated, it is referred to as a slot weld. *Figure 9* illustrates how a plug or slot weld may be used on a lap joint. A typical application is to overfill the hole or slot, then grind the excess molten material to a level finished surface.

Preparation of the base metals for this type of weld requires proper cleaning and removal of any oxides or dirt on both pieces, then drilling the hole

Surfacing

Surfacing, as defined by the American Welding Society, is the application by welding, brazing, or thermal spraying of a layer of material to a surface to obtain desired properties or dimensions, as opposed to making a joint.

Preheating Surface Welds

For the shielded metal arc welding (SMAW) process, obtain a higher current setting than would normally be used for joint welding with the designated electrode diameter. This preheats the metal ahead of the arc and allows the surface to remain in a molten state long enough for any impurities to surface.

Hot Tip

One Application of Surface Welding

These shafts are located within gearboxes used in the marine industry. They turn at relatively high rpms and will corrode and wear rapidly if subjected to salt water. It is not always cost-effective to discard and replace each shaft with a new one. Surface welding may be used to bring the shafts back to the desired dimensions by building them up and re-machining them. They would then be classified as rebuilt shafts.

REBUILT SHAFT

103P0302.EPS

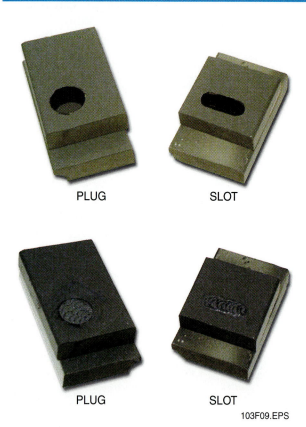

PLUG   SLOT

PLUG   SLOT

103F09.EPS

Figure 9 ◆ Plug and slot welds.

or slot to receive the weld. The walls of the hole or slot must also be cleaned of any oxides, dirt, and oils prior to welding.

Plug and slot welds are commonly used where a finished surface is required or in areas that do not permit overall dimensions to exceed those of the plate thickness.

3.3.3 Fillet Welds

Fillet welds may be applied to lap joints, T-joints, or corner joints. The fillet weld does not require any base metal edge preparation other than removing contaminants by cleaning with a brush or other appropriate cleaning tools. The two base metals are positioned together with or without a root opening, and one or more passes (welding beads) are applied at the intersection of the two base metal edges. Any time more than one pass is made on a fillet weld, or any weld, the previous pass must be cleaned of all contaminants such as slag, dirt, oxides, and porosity before the next pass is applied.

When using a fillet weld on outside corner welds, a corner-to-corner or a half-lap fit can be used, as illustrated in *Figure 10.* The corner-to-corner joint is difficult to assemble because neither plate can support the other. Care must also be used when welding to prevent burning through the corner. The half-lap joint is easier to assemble, requires less welding material, and has less chance of burning through the corner. The half-lap does require a second weld on the inside of the corner. If a half-lap fit is used, allowances in the plate dimensions must be made for the lap.

3.3.4 Square-Groove Welds

Square-groove welds can be used with butt joints, corner joints (inside and outside), T-joints, and edge joints. The difference between the fillet and the square groove is that a square groove requires a root opening to be set or prepared between the two base pieces prior to welding, as illustrated in *Figure 11*. This allows penetration into a greater amount of the surface area of the two base metals. A partial joint penetration weld will have much less strength than a complete joint penetration weld. Welding codes impose restrictions for complete joint penetration welds. With shielded metal arc welding (SMAW), the maximum base metal thickness is ¼", and welding from both sides is required. In addition, a root opening of half the thickness of the base metal is required, and the root of the first weld must be gouged before the second weld is made. For gas metal arc welding (GMAW) and flux-cored arc welding (FCAW), the maximum base metal thickness is ⅜", with the same requirements for welding from both sides and back gouging. Welding codes and specifications will be discussed later in this module.

3.3.5 Bevel-Groove Welds

Bevel-groove welds are commonplace in the welding industry because the base metal preparation is relatively simple and provides greater surface penetration than the square groove. It can be applied to all five of the basic weld joints. Typically, a single bevel is cut on the edge of one of the base metal pieces. Preparations for a bevel groove are shown in *Figure 12*. A bevel-groove weld preparation may include a specified root opening along with the beveled cut, as shown in the butt weld arrangement in *Figure 12*. Cutting the bevel may be accomplished by various cutting methods, which will be discussed later in this module.

3.3.6 V-Groove Welds

V-groove welds are generally applied to inside or outside corner joints, butt joints, or edge joints. *Figure 13* shows an illustration of a prepared

CORNER-TO-CORNER

HALF LAP   103F10.EPS

Figure 10 ◆ Fit-up of corner-to-corner and half-lap fillet welds.

103F11.EPS

Figure 11 ◆ Preparation for a square groove on a butt weld.

103F12.EPS

Figure 12 ◆ Preparations for corner and butt bevel-groove welds.

V-groove on a butt joint to be welded. V-grooves and bevel grooves are more economical to prepare than other types of welds. Typically, a 45° angle must be cut on both edges to prepare the base metals to receive the weld. In preparing inside or outside corner joints or butt joints for V-groove welds, the angle may not extend from the top of the bevel edge to the bottom of the bevel edge on each piece, but may be cut short of the bottom corner if a root face is required. In this edge preparation, the flat surface from where the angle stops to the bottom of the piece is referred to as the root face. This is illustrated in *Figure 13*.

3.3.7 Single versus Double V-Groove Welds

When possible, a double V-groove weld should be used in place of a single V-groove. The double V-groove requires half the weld metal compared to the single V-groove. Welding from both sides also reduces **distortion** because the forces of distortion work against each other. *Figure 14* shows a comparison between a double V-groove and a single V-groove preparation.

3.3.8 J- and U-Groove Welds

J- and U-groove welds are similar. The main difference is that a J-groove weld requires only one base metal to have its edge grooved in the form of a J (*Figure 15*). A U-groove is formed by preparing two matching J-grooves, one on each base metal edge. Even though they require much less weld material than bevel or V-groove welds, they require much more preparation time because of the shape of the grooves. A J-groove weld can be applied to all five types of weld joints. The U-groove can only be used on butt, inside and outside corner, and edge joints.

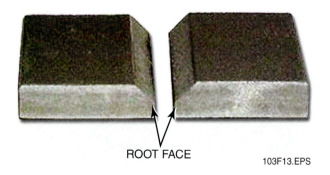

Figure 13 ◆ V-groove weld.

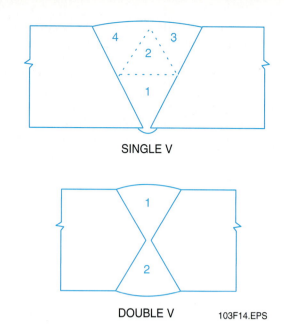

SINGLE V

DOUBLE V   103F14.EPS

Figure 14 ◆ Comparison of double and single V-grooves.

J-GROOVE

U-GROOVE
103F15.EPS

Figure 15 ◆ J- and U-groove welds.

3.3.9 Combination Fillet and Groove Welds

A combination groove and fillet weld is sometimes used in place of a fillet weld alone. A groove weld requires more preparation time than a fillet weld. A fillet weld requires more time and material than a groove weld. The combination weld

Distortion

Distortion results from the shrinkage that occurs as the weld cools, as shown in the photo. The example on the left is an outside weld that has pulled the workpiece out as the weld cooled. The example on the right is an inside weld that has pulled the workpiece in as the weld cooled. The small angle gap on the opposite edge of the joint is an indication of the degree of distortion. A tack weld on the opposite side of the joint would have prevented this problem.

DISTORTION CAUSED
BY OUTSIDE WELD

DISTORTION CAUSED
BY INSIDE WELD

103P0303.EPS

requires more preparation time than a fillet weld alone, but saves time and material in the welding process itself. The combination weld also helps to reduce distortion. *Figure 16* shows combination groove and fillet welds.

3.3.10 Groove Angles and Root Openings

The purpose of the groove angle is to allow access to the root of the weld. The root preparation is sized to control **melt-through.** Increasing or decreasing the root preparation (**root opening**

NON-BEVELED
T-JOINT

DOUBLE-BEVELED
T-JOINT

SINGLE-BEVELED
T-JOINT

103F16.EPS

Figure 16 ◆ Combination fillet and groove welds.

and **land**) will result in excess melt-through or insufficient root penetration. As a general rule, as the groove angle decreases, the root opening increases to compensate. Lands are used with open root joints but not when metal **backing** strips are used. *Figure 17* shows groove angles and root openings.

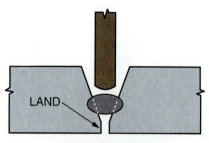

GROOVE ANGLE AND ROOT OPENING TOO SMALL

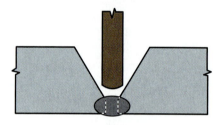

CORRECT GROOVE ANGLE AND ROOT OPENING

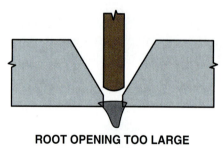

ROOT OPENING TOO LARGE

103F17.EPS

Figure 17 ◆ Groove angles and root openings.

3.3.11   Open Root Welds

For open root welds on plate, the groove angle should be 60°, and the maximum size of the root opening and land is ⅛". For open root welds on pipe, the groove angle is 60° or 75°, depending on the code or specifications used. The maximum size of the root opening and land is ⅛". *Figure 18* shows open root joint preparation.

3.3.12   Welds with Backing on Plate

Backing for plate can be strips made from the same material as the base metal, flux-coated tape, fiberglass-coated tape, ceramic tape, or gas. When gas is used as a backing, the same groove angles and root preparation used for open root welds are

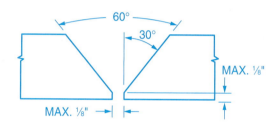

OPEN-ROOT GROOVE ANGLE ON PLATE

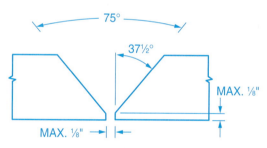

OPEN-ROOT GROOVE ANGLE ON PIPE

103F18.EPS

Figure 18 ◆ Open root joint preparation.

Groove Angles

If the groove angle is larger than necessary, it will require additional weld metal to fill. This will increase the time and cost to complete the weld, as well as increase distortion. If the groove angle is too small, it may result in weld defects.

AWS ENTRY LEVEL WELDER – TRAINEE MODULE 29103-03

Inert Gas Hazards

Nitrogen and argon are used as backing and purge gases for some welding applications. These gases will displace all the oxygen in the space where they are used.

At a refinery in Corpus Christi, Texas, while the welder was at lunch, a helper entered a pipe in which argon had been used as a backing gas. When the welder returned, he found the helper dead.

Nitrogen is lighter than air and, if allowed, will float upward. Argon, on the other hand, is heavier than air and will settle in low-lying areas. When removing nitrogen from a confined space, open the top to allow it to float out. When removing argon from a confined space, provide some way for it to drain out the bottom.

used. When flux-coated tape, fiberglass tape, or ceramic tape is used as a backing, the groove angle should be 60°, the maximum root opening should be ³⁄₁₆", and the maximum land should be ³⁄₁₆". When applying tape, be sure to clean the surface to ensure the tape adheres tightly. Also make sure that the tape is centered on the weld joint.

> **Note**
> Check the **welding procedure specification (WPS)** or manufacturer's recommendations for joint preparation before using flux-coated, fiberglass, or ceramic tape.

When strips made from the same material as the base metal are used, a groove angle of 45°, a root opening of ¼", and a feather edge (no land) is used. The backing strip must be thick enough and wide enough to prevent burn-through at the root and to absorb and dissipate the heat of the root pass. The recommendation for a backing strip for welding mild steel up to ¾" thick is ⅜" thick by 3" wide.

> **Note**
> Check the WPS or site procedures for the backing size requirements for your site.

The backing strip must be tack welded so that it will be held securely in place during welding. It must also be centered on the groove and be in close contact (no gaps) with the back of the plate being welded. *Figure 19* shows plate joint preparation for backing.

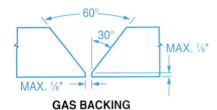

GAS BACKING

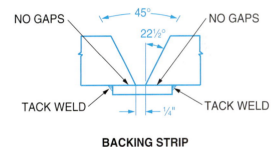

BACKING STRIP

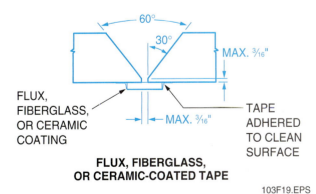

FLUX, FIBERGLASS, OR CERAMIC-COATED TAPE

103F19.EPS

Figure 19 ◆ Plate joint preparation for backing.

Out-of-Position Welding

Out-of-position welding generally requires smaller electrodes with more passes to fill the joint. Because out-of-position welding is more difficult, there is also a greater likelihood of weld defects.

3.4.0 Welding Position

It is easier and faster to weld in the flat position than it is to weld out of position. Always try to position the weldment so that welding is performed in the flat position. If a weldment is an assembly of parts, look for welds that can be made on the bench before tacking the assembly together. When welding pipe, weld as many fittings as possible before tacking the pipe in position, and try to leave welds that will be the most accessible and the easiest to perform. When preparing weld joints, try to prepare the joint so that welding takes place from the easiest position. *Figure 20* shows the positioning of weld joints for welding.

3.5.0 Codes and Welding Procedure Specifications

A welding code is a detailed listing of the rules and principles that apply to specific welded products. Codes ensure that safe and reliable welded products will be produced and that persons asso-

ciated with the welding operation will be safe. Clients may specify which welding codes should be followed when they place their orders or award contracts.

If codes are mandated, a welding procedure specification (WPS) must be written for each weld. A WPS is a written set of instructions for producing reliable welds. It includes the type of joint to be used, as well as the type of weld and any groove preparation that may be required for that particular weld. Each WPS is written and tested in accordance with a particular welding code or specification. All welding requires that acceptable industry standards be followed, but not all welds are mandated by codes. The requirement for use of a WPS is often listed on blueprints as a note or in the tail of the welding symbol. Reading welding symbols will be covered later in this level. If you are unsure whether or not the welding being performed requires a WPS, you should not proceed until you verify this information. *Figure 21* illustrates part of a WPS that shows blanks that would be used to specify the type of joint, groove design, backing material, position of the groove, and acceptable range of base material thickness.

4.0.0 ◆ WELDING JOINT PREPARATION

There are many different ways to prepare a joint for welding. The joint can be prepared mechanically using nibblers, grinders, or cutters. It can also be prepared thermally using an oxyfuel, plasma arc, or carbon arc cutting torch. The method used to prepare the joint will generally be based on the base metal type, ease of use, and code or procedure specifications.

4.1.0 Identify Joint Specification

Before starting a joint preparation, determine if the joint is covered by a WPS or site quality stan-

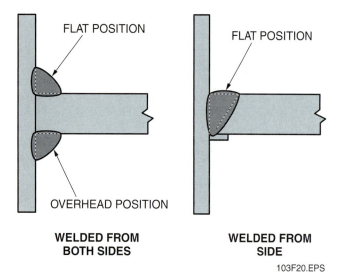

Figure 20 ◆ Positioning welds.

Welding Procedure Specification

TITLE: _____

PROCESS	APPROXIMATE NO. OF PASSES	ROD OR ELECT SIZE	CURRENT	VOLTAGE	FILLER METALS SPA SPEC CLASS	TYPE	
						F NO.	A NO.

JOINTS (QW-402)

Groove Design _____

Backing   Yes _____   No _____

Backing Material (Type) _____

Other _____

_____

_____

BASE METALS (QW-403)

P. No. _____ Group _____ to P. No. _____

Thickness Range _____

POSTWELD HEAT TREATMENT (QW-407)

Temperature _____

Time Range _____

Other _____

GAS (QW-408)

Shielding Gas(es) _____

% Composition (mixtures) _____

_____

Flow Rate _____

103F21.EPS

Figure 21 ◆ Partial WPS.

dard. If the joint is covered by a WPS or site quality standard, the method of preparing the joint, as well as the joint type and parameters, will be specified. These specifications must be followed. If you are unsure whether or not a joint is covered by a WPS or site quality standard, check with your supervisor before proceeding.

4.2.0 Mechanical Joint Preparation

Mechanical joint preparation is used most often on alloy steel, stainless steel, and **nonferrous metal** piping or plate. It is slower than thermal methods such as oxyfuel, carbon arc, and plasma arc, but it has the advantages of high precision with low heat input and the absence of oxides commonly left by thermal methods.

Thermally Sensitive Material Joint Preparation

Mechanical preparation is often required for materials that could be affected by the heat of thermal processes.

4.2.1 Grinders

Hand-held electric or air-operated grinders are used in welding shops and even more often in the field to cut and bevel pipe and plate to prepare them for welding. *Figure 22* shows an example of a hand-held electric grinder being used to prepare a root bead on a test pipe.

WARNING!

Cutoff and grinding wheels must be selected based on the material to be cut or beveled. Using an improper wheel may damage the wheel or workpiece and can sometimes create hazardous conditions for the operator should the wheel shatter during operation. Refer to the manufacturer's recommendations and warnings when selecting grinding wheels.

103F22.EPS

Figure 22 ◆ Hand-held grinder in use.

4.2.2 Pipe Beveling Machines

Welded piping is found extensively in the utilities industry and even more in the petrochemical industry. Nearly every piece of pipe that is welded to a fitting, another piece of pipe, or any other connection requires that the edge be cut square and beveled according to specifications before welding. If a mechanical cutting and beveling process is used to accomplish this, it may be done using an electrically or pneumatically powered beveling machine. Many of these are portable machines that operate much like a lathe. They have mandrels that hold various cutting tools, as well as numerous adjusting mechanisms that make it easy to set the bevel angle and depth. Various models are available to cut and bevel 2" to 60" pipe. When specifications call for tubing to be welded, machines similar to pipe beveling machines are available to face, square, and chamfer the ends of tubes in preparation for welding. Boiler tube ends are typically prepared using these smaller machines before welding the tubes in place.

Special cutoff machines are also made to be mounted on the outside of the pipe. They can be mounted with a ring or a special chain with rollers. An electrically or pneumatically operated grinder with a cutoff blade is mounted and manually or electrically powered around the pipe to cut it off.

WARNING!

Always follow all manufacturers' safety procedures when using grinders. Failure to follow the manufacturer's safety recommendations could result in serious personal injury.

Figure 23 illustrates a pipe-beveling machine being properly applied to the end of a pipe to prepare it for welding.

Portable Pipe-Beveling Machine Floor Stands

Floor stands are available to convert portable pipe-beveling machines to more permanent stationary devices.

Cutters use round cutting tools similar to mill cutting tools. The bevel angle is set by adjusting the cutter or changing the cutter blade. Cutters leave the surface much smoother than nibblers and can be used for cutoff operations. Another advantage of cutters is that they can easily be used to prepare J- or compound bevels by changing the shape of the cutter. Cutters made for pipe are sometimes called pipe-end-prep laths.

WARNING!
Always follow all manufacturers' safety procedures when using nibblers and cutters. Failure to follow the manufacturer's safety recommendations could result in serious personal injury.

4.3.0 Thermal Joint Preparation

Thermal joint preparation includes preparing a joint with the oxyfuel, plasma arc, or carbon arc cutting process. Although the carbon arc cutting process can be used for beveling plate and pipe, the results are not as satisfactory as beveling performed with oxyfuel or plasma arc cutting. The carbon arc cutting process is best for gouging seams, cracks, or weld repairs.

Note
Other modules in the Welding curriculum cover the specifics of oxyfuel cutting, plasma arc cutting, and carbon arc cutting in more detail.

The torch used for oxyfuel or plasma arc cutting can be hand-held or mounted on a motorized carriage. The motorized carriage for plate cutting runs on flat tracks positioned on the surface of the plate to be cut. The carriage has on/off and forward/reverse switches and a speed adjustment. A handwheel on top of the carriage adjusts the torch holder **transversely,** and a handwheel at the torch holder adjusts the torch vertically. The bevel angle is set by pivoting the torch holder. The carriage can carry a special oxyfuel or plasma arc cutting torch designed to fit into the torch holder. *Figure 25* shows a motorized carriage for cutting and beveling plate.

103F23.EPS

Figure 23 ◆ Operating a beveling machine.

4.2.3 Nibblers and Cutters

Nibblers prepare the edge of a plate or pipe with a reciprocal punch that cuts off a chip with each stroke. The bevel angle is set by adjusting the nibbler. Nibblers must have access to an edge to be used. *Figure 24* shows a nibbler.

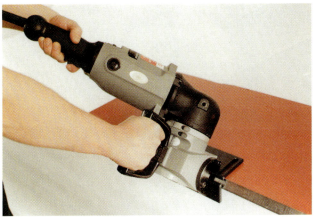

103F24.EPS

Figure 24 ◆ A nibbler.

Thermal Joint Preparation Precautions

When preparing a joint with the oxyfuel, plasma arc, or carbon arc cutting process, all slag must be removed prior to welding. Any slag remaining on the joint during welding will cause porosity in the weld. Joints prepared with the carbon arc cutting torch must be carefully inspected for carbon deposits. It is common for small carbon deposits to be left behind as the carbon electrode is consumed during the cutting or gouging operation. These carbon deposits will cause defects such as hard spots, loss of ductility, and cracking in the weld. Before welding, use a grinder to clean surfaces prepared with the carbon arc torch.

Special equipment is used for cutting pipe. A steel ring or special chain with rollers is attached to the outside of the pipe. The torch is carried around the pipe on a torch holder that is powered by electricity, air, or a hand crank. The bevel angle is set by pivoting the torch holder, and the torch is adjusted vertically with a handwheel on the torch holder. An out-of-round attachment can be used to compensate for pipe that is out-of-round. For large-diameter pipe (54" or larger), special internal equipment is available. With this special equipment, the torch mechanism is mounted on the inside of the pipe. Special plasma or oxyfuel torches can be mounted in the torch holders of the external or internal pipe cutting equipment. *Figure 26* shows pipe cutting and beveling equipment.

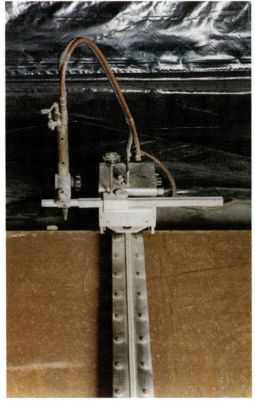

103F25.EPS

Figure 25 ◆ Motorized carriage for cutting and beveling plate.

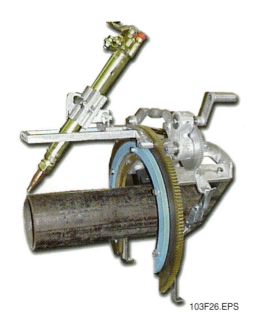

103F26.EPS

Figure 26 ◆ Pipe cutting and beveling equipment.

Automatic Pipe Bevelers

Automatic versions of pipe bevelers and cutters are an improvement over pattern cutters and other similar equipment. This is because there is no need to reset the preheat flame before each cut. Once an automatic pipe beveler is initially set up, all subsequent cuts can be made with the same settings. These systems also save time and cutting gases because all gases are extinguished immediately when the system is switched off.

Summary

The importance of proper joint preparation cannot be overemphasized. If a joint is not properly prepared, the resulting weld will not perform as designed. This causes an unsafe weldment. Chemical or mechanical cleaning can be performed to remove base metal surface contamination prior to welding. It is important to select the proper type of joint and use proper joint preparation methods to ensure acceptable welds. In addition, welding codes set guidelines that must be followed during joint preparation. Following the procedures presented in this module will ensure that the joints you prepare will follow established industry practices and meet code requirements.

Review Questions

1. Corrosion of metals occurs when they are exposed to _____.
   a. shielding gas
   b. welding
   c. air
   d. polishing

2. The most common method for cleaning contaminated surfaces is _____ cleaning.
   a. chemical
   b. ultrasonic
   c. thermal
   d. mechanical

3. A pneumatic tool that uses a number of blunt steel needles to remove surface corrosion is called a _____.
   a. weld flux chipper
   b. needle scaler
   c. needle nose chipper
   d. corrosion needler

4. The five basic types of joints are butt, lap, corner, T-joint, and _____.
   a. chamfer
   b. edge
   c. bevel
   d. square

5. The written set of specifications for producing sound welds is a(n) _____.
   a. WPS
   b. PQR
   c. WRS
   d. SWS

6. Thermal preparation of joints includes the use of oxyfuel, plasma arc, and _____ cutting.
   a. electric
   b. low heat
   c. continuous-arc
   d. carbon arc

7. Grinding wheels that are used improperly or are poorly selected can cause injury due to _____.
   a. radiation
   b. reversing
   c. shattering
   d. air-locking

8. Before a pipe can be welded to a fitting or other connection, its edge usually must be _____.
   a. burnished
   b. hardened
   c. ground to fit
   d. cut square and beveled

9. Nibblers must have access to an _____ to be used.
   a. inverter
   b. excess pipe flange
   c. edge
   d. axis

10. The best cutting process for gouging seams, cracks, or making repair welds is the _____ cutting process.
  a. plasma arc
  b. carbon steel
  c. grinding
  d. carbon arc

Identify each type of weld shown in the figure.

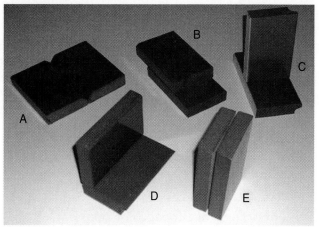

103E01.EPS

11. Edge _____

12. Lap _____

13. Butt _____

14. T-joint _____

15. Corner _____

Trade Terms Introduced in This Module

Backing: A weldable or nonweldable material used behind a root opening to allow defect-free welding at the open root of a joint.

Base metal: Metal to be welded, cut, or brazed.

Code: A set of regulations covering permissible materials, service limitations, fabrication, inspection, testing procedures, and qualifications of welders.

Distortion: The expansion and contraction of welded parts caused by the heating and subsequent cooling of the weld joint.

Land: A small facing on the end of a chamfer. Another term for root face.

Load: The amount of force applied to a material or a structure.

Melt-through: Complete joint penetration.

Nonferrous metal: A metal, such as aluminum, copper, or brass, lacking sufficient quantities of iron to have any effect on its properties.

Oxide: The scale that forms on metal surfaces when they are exposed to air.

Peen: The rounded end of a ball peen hammer used to strike a weld as it cools in order to reduce stress on the weld.

Porosity: Gas pockets, inclusions, or voids in metal.

Root opening: The space between the base metal pieces at the bottom or root of the joint.

Transverse: Placed or running crosswise.

Weathering steel: Steel alloy that, under specific conditions, is designed to form a very dense oxide layer on its outer surfaces, which retards further corrosion.

Welding procedure specification (WPS): The document containing all the detailed methods and practices required to produce a sound weld.

Weldment: An assembly that is fastened together by welded joints.

Performance Accreditation Tasks

The Performance Accreditation Tasks (PATCs) correspond to and support the learning objectives in AWS EG2.0-95, Guide for the Training and Qualification of Welding Personnel: Entry-Level Welder.

Note that in order to satisfy all learning objectives in AWS EG2.0-95, the instructor must also use the PATCs contained in the second level in the Welding Contren™ Learning Series.

PATCs provide specific acceptable critera for performance and help to ensure a true competency-based welding program for students.

The following tasks are designed to develop your competency in preparing base metal. Practice each task until you are thoroughly familiar with the procedure.

As you complete each task, take it to your instructor for evaluation. Do not proceed to the next task until instructed to do so by your instructor.

PREPARE PLATE JOINTS MECHANICALLY

Using a nibbler or cutter, mechanically prepare the edge of a ¼" to ¾" carbon steel plate with a 22½° or 30° bevel (depending on equipment available) and a ³⁄₃₂" root face as shown.

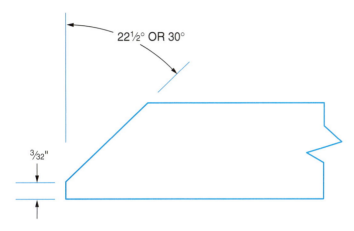

22½° OR 30°

³⁄₃₂"

NOTE: BASE METAL = CARBON STEEL PLATE

103A01.EPS

Criteria for Acceptance:

- Bevel angle ±2½°   _____
- Bevel face smooth and uniform to ¹⁄₁₆"   _____
- Root face ±¹⁄₃₂"   _____

PREPARE PIPE JOINTS MECHANICALLY

Using a nibbler or cutter, mechanically prepare the end of a pipe with a 30° or 37½° bevel (depending on the equipment available) and a ³⁄₃₂" root face. Use 6", 8", or 10" Schedule 40 or Schedule 80 carbon steel pipe. The figure shows how to prepare the pipe in this manner.

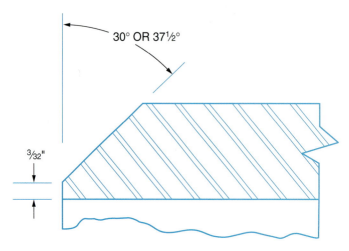

30° OR 37½°

³⁄₃₂"

NOTE: BASE METAL = CARBON STEEL PIPE

103A02.EPS

Criteria for Acceptance:

- Bevel angle ±2½°
- Bevel face smooth and uniform to ¹⁄₁₆"
- Root face ±¹⁄₃₂"

_____

_____

_____

PREPARE PLATE JOINTS THERMALLY

Using oxyfuel cutting equipment and a motorized carriage, cut ¼" to ¾" thick carbon steel plate into 3" strips at least 8" long with a bevel of 22½° as shown. Grind a ³⁄₃₂" root face after cutting.

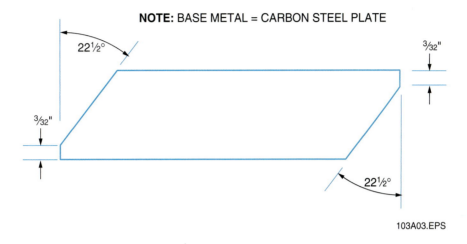

NOTE: BASE METAL = CARBON STEEL PLATE

103A03.EPS

Criteria for Acceptance:

- Bevel angle ±2½°
- No slag
- Minimal notching not exceeding ¹⁄₁₆" deep on the kerf face
- Minimum of ¹⁄₁₆" radius at the top edge and bottom edge of the kerf
- Root face ±¹⁄₃₂"

_____
_____
_____
_____
_____

PREPARE PIPE JOINTS THERMALLY

Using oxyfuel cutting equipment and a motorized carriage, bevel cut 6", 8", or 10" Schedule 40 or Schedule 80 carbon steel pipe into 4" wide rings. The bevel should be 30° with a ³⁄₃₂" root face prepared with a grinder. Prepare the pipe joint as shown.

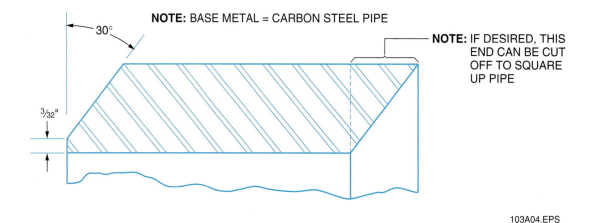

NOTE: BASE METAL = CARBON STEEL PIPE

30°

³⁄₃₂"

NOTE: IF DESIRED, THIS END CAN BE CUT OFF TO SQUARE UP PIPE

103A04.EPS

Criteria for Acceptance:

- Bevel angle ±2½°
- Root face ±¹⁄₃₂"
- No slag
- Minimal notching not exceeding ¹⁄₁₆" deep on the kerf face
- Minimum of ¹⁄₁₆" radius at the top edge of the kerf

Additional Resources

This module is intended to present thorough resources for task training. The following reference works are suggested for further study. These are optional materials for continued education rather than for task training.

Welding Handbook, Volume 5, 2001. Miami, FL: The American Welding Society.

The Procedure Handbook of Arc Welding, 1994. Cleveland, OH: The Lincoln Electric Company.

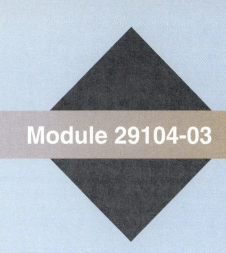

Module 29104-03

Weld Quality

Course Map

This course map shows all of the modules in the AWS Entry Level Welder – Phase 1 curriculum. The suggested training order begins at the bottom and proceeds up. Skill levels increase as you advance on the course map. The local Training Program Sponsor may adjust the training order.

AWS ENTRY LEVEL WELDER—PHASE 1

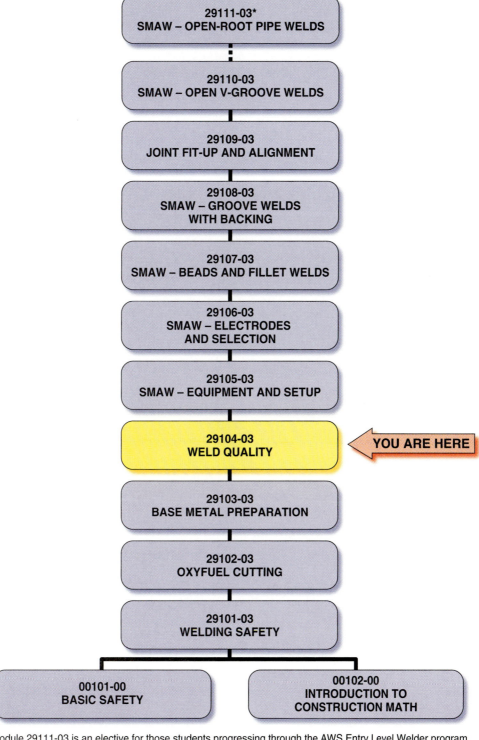

29111-03*
SMAW – OPEN-ROOT PIPE WELDS

29110-03
SMAW – OPEN V-GROOVE WELDS

29109-03
JOINT FIT-UP AND ALIGNMENT

29108-03
SMAW – GROOVE WELDS
WITH BACKING

29107-03
SMAW – BEADS AND FILLET WELDS

29106-03
SMAW – ELECTRODES
AND SELECTION

29105-03
SMAW – EQUIPMENT AND SETUP

29104-03
WELD QUALITY ◄ YOU ARE HERE

29103-03
BASE METAL PREPARATION

29102-03
OXYFUEL CUTTING

29101-03
WELDING SAFETY

00101-00
BASIC SAFETY

00102-00
INTRODUCTION TO
CONSTRUCTION MATH

*Module 29111-03 is an elective for those students progressing through the AWS Entry Level Welder program.

29104CMAP.EPS

MODULE 29104-03 CONTENTS

Figures

Table

Weld Quality

Objectives

When you have completed this module, you will be able to do the following:

1. Identify and explain codes governing welding.
2. Identify and explain weld imperfections and their causes.
3. Identify and explain nondestructive examination practices.
4. Identify and explain welder qualification tests.
5. Explain the importance of quality workmanship.
6. Identify common destructive testing methods.

Prerequisites

Before you begin this module, it is recommended that you successfully complete the following: Modules 00100-00 through 29104-03.

Required Trainee Materials

1. Pencil and paper
2. Appropriate personal protective equipment

1.0.0 ◆ INTRODUCTION

Quality is an important aspect of the welder's job. A weld that can pass or be acceptable for one welding application may not be acceptable for another. Acceptable welding criteria have been established in codes and standards. Several codes govern welding activities, qualification requirements, and tests that can be performed on weldments to determine and identify various weld imperfections.

The following sections will identify and explain applicable codes, various weld imperfections and their causes, types of weldment testing, and welder qualification test requirements. The importance of quality workmanship will also be discussed, along with typical site quality organizational structures for ensuring that quality welds are achieved.

2.0.0 ◆ CODES GOVERNING WELDING

A code is a set of requirements covering permissible materials, service limitations, fabrication, inspection, testing procedures, and qualifications of welders. Welding codes ensure that safe and

The Importance of Welding Inspection

It is important that you learn to recognize the various types of weld discontinuities and identify their causes so that you can be the first line of defense against problem welds.

Welding is used to join metals in many critical applications, including buildings, bridges, pipelines, motor vehicles, and heavy machinery. Failure of a weld in any of these applications could mean loss of life or major property damage. For this reason, welding inspectors will frequently check your work to make sure your welds meet the standards established for the project. The greater the potential hazard or cost risk, the more frequent and intense the inspections will be.

reliable welded products will be produced and that persons associated with the welding operation will be safe. Clients specify in the contract which codes will be used on the project. All welding must then be performed following the guidelines and specifications outlined in that code. Since there are a number of codes and each is updated periodically, you should know which welding codes and code year apply to your job.

Codes and specifications that apply to welding safety and quality have been developed and are published by a number of nationally recognized agencies. To eliminate the necessity of writing a code for each new job, sections of these existing codes are referenced by the project contract. Agencies and societies that have established codes include:

- American Society of Mechanical Engineers (ASME)
- American Welding Society (AWS)
- American Petroleum Institute (API)
- American National Standards Institute (ANSI)

This module will not cover all the detailed requirements of each individual code or code section. Instead, general considerations that apply to weld quality will be covered.

2.1.0 American Society of Mechanical Engineers (ASME)

The American Society of Mechanical Engineers (ASME) has two codes: the *ASME Boiler and Pressure Vessel Code* and *ASME B31, Code for Pressure Piping*. Both of these codes are endorsed by the American National Standards Institute.

The *ASME Boiler and Pressure Vessel Code (BPVC)* contains eleven sections. The sections most frequently referenced by welders are:

- *Section II, Material Specifications* – This section contains the specifications for acceptable ferrous (Part A) and nonferrous (Part B) base metals and for acceptable welding and brazing filler metals and fluxes (Part C). Many of these specifications are identical to and have the same number designation as AWS specifications for welding consumables. This section is used to match base metals and filler metals.
- *Section V, Nondestructive Examination* – This section covers the methods and standards for nondestructive examination of boilers and pressure vessels.
- *Section IX, Welding and Brazing Qualifications* – This section covers the qualification of welders, welding operators, brazers, and brazing operators. It also covers the welding and brazing procedures that must be used for welding or brazing boilers or pressure vessels. This section of the code is often cited in other codes and standards as the welding qualification standard.

ASME B31, Code for Pressure Piping, consists of eight sections. Each section gives the minimum requirements for the design, materials, fabrication, erection, testing, and inspection of a particular type of piping system. In particular, *B31.1, Power Piping*, covers power and auxiliary service systems for electric generation stations. *B31.3* covers chemical plant and petroleum refining piping.

All sections of *ASME B31, Code for Pressure Piping*, require qualification of the welding procedures and testing of welders and welding operators. Some sections require these qualifications to be performed in accordance with Section IX of the *ASME Boiler and Pressure Vessel Code*, while in others it is optional.

2.2.0 American Welding Society (AWS)

The American Welding Society (AWS) publishes numerous documents covering welding. These documents include codes, standards, specifications, recommended practices, and guides. *AWS D1.1, Structural Welding Code – Steel*, is the code most frequently referenced. It covers welding and qualification requirements for welded structures of carbon and low-alloy steels. It is not intended to apply to pressure vessels, pressure piping, or base metals less than ⅛" thick.

Inspection Standards

Among the many documents and standards published by the American Welding Society are documents covering weld inspection. These include:

- *AWS B1.10, Guide for the Nondestructive Inspection of Welds*
- *AWS B1.11, Guide for the Visual Inspection of Welds*
- *Certification Manual for Welding Inspectors*

2.3.0 American Petroleum Institute (API)

The American Petroleum Institute (API) publishes documents in all areas related to petroleum production. *API 1104, Standard for Welding of Pipelines and Related Facilities,* applies to arc and oxyfuel gas welding of piping, pumping, transmission, and distribution systems for petroleum. It presents methods for making acceptable welds by qualified welders using approved welding procedures, materials, and equipment. It also presents suitable methods to ensure proper analysis of weld quality.

2.4.0 American National Standards Institute (ANSI)

The American National Standards Institute (ANSI) is a private organization that does not actually prepare standards. Instead, it adopts standards that it feels are of value to the public interest. ANSI standards deal with dimensions, ratings, terminology and symbols, test methods, performance, and safety specifications for materials, equipment, components, and products in many fields, including construction. Many codes used today have been adopted as ANSI standards.

3.0.0 ◆ BASIC ELEMENTS OF WELDING CODES

All welding codes provide detailed information about qualification in three general areas. These are:

- Welding **procedure qualification**
- Welder performance qualification
- Welding operator qualification

Machine welding is covered in some codes but is not common to all codes. Each of the above types of qualification is different and subject to different requirements.

Note
The information in this module is provided as a general guideline only. Check with your supervisor if you are unsure of the codes and specification requirements for your project.

3.1.0 Welding Procedure Qualification

A welding procedure is a written document that contains materials, methods, processes, electrode types, techniques, and all other necessary and relevant information about the weldment. Welding procedures must be qualified before they can be used. Procedure qualification has nothing to do with the skills of the individual welder.

Welding procedure qualifications are limiting instructions written to explain how welding will be done. These limiting instructions are listed in a document known as a welding procedure specification (WPS). The purpose of the WPS is to define and document in detail the variables involved in welding a certain base metal. The WPS lists the following in detail:

- The various base metals to be joined by welding
- The filler metal to be used
- The range of preheat and postheat treatment
- The thickness and other variables described for each welding process

WPS variables are identified either as essential or nonessential variables. Essential variables are items in the welding procedure specification that cannot be changed without requalifying the welding procedure. Nonessential variables are items in the WPS that may be changed within a range identified by the code but that do not affect the qualification status. The following are considered essential variables in a welding procedure:

- Filler metal classification
- Material thickness
- Joint design
- Type of base metal
- Welding process

Examples of nonessential variables that may be changed without having to requalify the welding procedure include:

- Amperage
- Travel speed
- Shielding gas flow (if applicable)
- Electrode and filler wire size

Note
Do not change any variable (essential or nonessential) without discussing it with your supervisor.

The WPS is qualified for use by welding test coupons (samples of the welded material) and by testing the coupons in accordance with the applicable code. A test weld is made and test coupons are cut from it. The test coupons are used to make tensile tests, root bends, and face bends as required by the code. The test results are then recorded on a document known as a **procedure qualification record** (PQR). If the weldment produced by the particular procedure meets the standards of the code, the procedure becomes

The WPS

The WPS tells the welder what materials and methods are to be used.

ANNEX B—SAMPLE FORMS

WELDING PROCEDURE SPECIFICATION (WPS)   Yes ☐
PREQUALIFIED _____   QUALIFIED BY TESTING _____
or PROCEDURE QUALIFICATION RECORDS (PQR)   Yes ☐

Identification # _____

Revision _____ Date _____ By _____

Authorized by _____ Date _____

Company Name _____

Welding Process(es) _____

Supporting PQR No.(s) _____

Type—Manual ☐   Semi-Automatic ☐
Machine ☐   Automatic ☐

JOINT DESIGN USED

Type:

Single ☐   Double Weld ☐

Backing: Yes ☐ No ☐

Backing Material:

Root Opening _____ Root Face Dimension _____

Groove Angle: _____ Radius (J–U) _____

Back Gouging: Yes ☐ No ☐   Method _____

BASE METALS

Material Spec. _____

Type or Grade _____

Thickness: Groove _____ Fillet _____

Diameter (Pipe) _____

FILLER METALS

AWS Specification _____

AWS Classification _____

SHIELDING

Flux _____   Gas _____

Composition _____

Electrode-Flux (Class) _____ Flow Rate _____

_____ Gas Cup Size _____

PREHEAT

Preheat Temp., Min _____

Interpass Temp., Min _____   Max _____

POSITION

Position of Groove: _____ Fillet: _____

Vertical Progression: Up ☐   Down ☐

ELECTRICAL CHARACTERISTICS

_____

Transfer Mode (GMAW)   Short-Circuiting ☐
Globular ☐   Spray ☐

Current: AC ☐   DCEP ☐   DCEN ☐   Pulsed ☐

Other _____

Tungsten Electrode (GTAW)

Size: _____

Type: _____

TECHNIQUE

Stringer or Weave Bead: _____

Multi-pass or Single Pass (per side) _____

Number of Electrodes _____

Electrode Spacing   Longitudinal _____

Lateral _____

Angle _____

Contact Tube to Work Distance _____

Peening _____

Interpass Cleaning: _____

POSTWELD HEAT TREATMENT

Temp. _____

Time _____

WELDING PROCEDURE

| Pass or Weld Layer(s) | Process | Filler Metals | | Current | | Volts | Travel Speed | Joint Details |
		Class	Diam.	Type & Polarity	Amps or Wire Feed Speed			

Form E-1 (Front)

104UA0401.EPS

---

qualified. Each WPS must have a PQR to document the quality of the weld produced. An example PQR is shown in the *Appendix* in the back of this module.

The methods used to qualify procedures are more detailed and thorough than those used to qualify either welders or welding operators. This is because procedures must qualify physical and metallurgical properties

3.2.0 Welder Performance Qualification

Once a procedure is qualified, the welder using it must be qualified to use that procedure by passing a welding performance qualification test. Because no single performance test can qualify welders for all the different types of welding that must be done, a welder may be required to pass additional performance qualification tests. Performance tests used to qualify welders are covered later in this module.

3.3.0 Welder Operator Qualification

When fully automatic welding equipment is used, the operators of the equipment must demonstrate their ability to set up and monitor the equipment so that it will produce acceptable welds. The codes also contain qualification tests for these operators.

4.0.0 ◆ WELD DISCONTINUITIES AND THEIR CAUSES

Codes and standards define the quality requirements necessary to achieve the integrity and reliability of the weldment. These quality requirements help ensure that welded joints are capable of serving their intended function for the intended life of the weldment. Weld **discontinuities** can prevent a weld from meeting the minimum quality requirements.

AWS defines a discontinuity as an interruption of the typical structure of a weldment, such as a lack of **homogeneity** in the mechanical, metallurgical, or physical characteristics of the material or weldment. A discontinuity is not necessarily a **defect**. A defect found during inspection will require the weld to be rejected. A single excessive discontinuity or a combination of discontinuities can make the weldment defective (unable to meet minimum quality requirements). However, a weld can have one or more discontinuities and be acceptable.

The welder should be able to identify discontinuities and understand the effect they have on weld integrity. Some can be seen by visual inspection. Those that are internal to the weldment can only be detected through other testing methods. The most common weld discontinuities are:

- Porosity
- **Inclusions**
- Cracks
- Incomplete joint penetration
- Incomplete fusion
- Undercuts
- Arc strikes
- Spatter
- Unacceptable weld profiles

4.1.0 Porosity

Porosity is the presence of voids or empty spots in the weld metal. It is the result of gas pockets being trapped in the weld as it is being made. As the molten metal hardens, the gas pockets form voids. Unless the gas pockets work up to the surface of the weld before it hardens, porosity cannot be detected through visual inspection.

Porosity can be grouped into four major types:

- *Uniformly scattered porosity* – May be located throughout single-pass welds or throughout several passes in multiple-pass welds (*Figure 1B*).
- *Clustered porosity* – A localized grouping of pores that results from improperly starting or stopping the welding.
- *Linear porosity* – May be aligned along a weld interface, the root of a weld, or a boundary between weld beads (*Figure 1A*).
- *Piping porosity* – Normally extends from the root of the weld toward the face. These elongated gas pores are also called wormholes. They do not often extend to the surface, and the porosity cannot be visually detected (*Figure 1C*).

Discontinuities

Ideally, a weld should not have any discontinuities, but most will have one or more. When evaluating a weld, it is important to note the type, size, and location of the discontinuity. Any one of these factors, or all three, can change a discontinuity to a defect, requiring the weld to be rejected during the inspection process. For example, discontinuities located where stresses exist tend to expand and thus are more detrimental than those in other locations. Surface or near-surface discontinuities may be more detrimental than similarly shaped, internal discontinuities.

(A) LINEAR POROSITY

(B) SCATTERED SURFACE POROSITY

(C) PIPING POROSITY

104F01.EPS

Figure 1 ◆ Examples of porosity.

Most porosity is caused by improper welding techniques or contamination. Improper welding techniques may cause inadequate shielding gas to be formed. As a result, parts of the weld are left unprotected. Oxygen in the air or moisture in the flux or on the base metal that dissolves in the weld pool can become trapped and produce porosity.

The intense heat of the weld can decompose paint, dirt, oil, or other contaminants, producing hydrogen. This gas can become trapped in the solidifying weld pool and produce porosity.

4.2.0 Inclusions

Inclusions are foreign matter trapped in the weld metal (*Figure 2*), between weld beads, or between the weld metal and the base metal. Inclusions are sometimes jagged and irregularly shaped. Sometimes they form in a continuous line. This concentrates stresses in one area and reduces the structural integrity (strength) of the weld.

Inclusions generally result from faulty welding techniques, improper access to the joint for welding, or both. A typical example of an inclusion is slag, which normally forms over a deposited weld. If the electrode is not manipulated correctly, the force of the arc will cause some of the slag particles to be blown into the molten pool. If the pool solidifies before the inclusions can float to the top, they become lodged in the metal, producing a

SURFACE SLAG INCLUSIONS

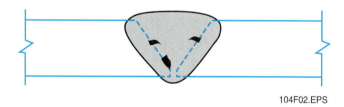

104F02.EPS

Figure 2 ◆ Examples of nonmetallic inclusions.

Avoiding Porosity

Excessive porosity has a serious effect on the mechanical properties of the joint. Although some codes permit a certain amount of porosity in welds, it is best to have as little as possible. This can be accomplished by properly cleaning the base metal, avoiding excessive moisture in the electrode covering, and using proper welding techniques.

AWS ENTRY LEVEL WELDER – TRAINEE MODULE 29104-03

defective weld. Sharp notches in joint boundaries or between weld passes also can result in slag entrapment.

Inclusions are more likely to occur in out-of-position welding because the tendency is to keep the molten pool small and allow it to solidify rapidly to prevent it from sagging.

With proper welding techniques and the correct electrode used with the proper setting, inclusions can be avoided or kept to a minimum. Other remedies include:

- Positioning the work to maintain slag control
- Changing the electrode to improve control of molten slag
- Thoroughly removing slag between weld passes
- Grinding the weld surface if it is rough and likely to entrap slag
- Removing heavy mill scale or rust on weld preparations

- Avoiding the use of electrodes with damaged coverings

4.3.0 Cracks

Cracks are narrow breaks that occur in the weld metal, in the base metal, or in the crater formed at the end of a weld bead (*Figure 3*). They are caused when localized stresses exceed the ultimate strength of the metal. Cracks are generally located near other weld or base metal discontinuities.

4.3.1 Weld Metal Cracks

Three basic types of cracks can occur in weld metal: transverse, longitudinal, and crater. As seen in *Figure 3*, weld metal cracks are named to correspond with their location and direction.

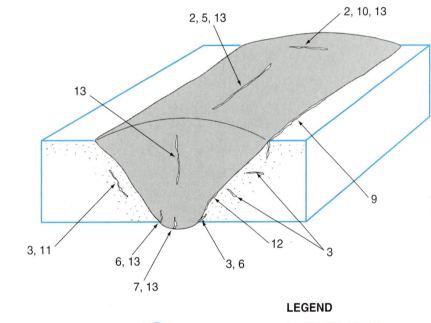

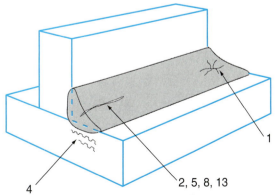

LEGEND

1. CRATER CRACK
2. FACE CRACK
3. HEAT-AFFECTED ZONE CRACK
4. LAMELLAR TEAR
5. LONGITUDINAL CRACK
6. ROOT CRACK
7. ROOT SURFACE CRACK
8. THROAT CRACK
9. TOE CRACK
10. TRANSVERSE CRACK
11. UNDERBEAD CRACK
12. WELD INTERFACE CRACK
13. WELD METAL CRACK

104F03.EPS

Figure 3 ◆ Types of cracks.

Transverse cracks run across the face of the weld and may extend into the base metal. They are more common in joints that have a high degree of restraint.

Longitudinal cracks are usually located in the center of the weld deposit. They may be the continuation of crater cracks or cracks in the first layer of welding. Cracking of the first pass is likely to occur if the bead is thin. If this cracking is not eliminated before the other layers are deposited, the crack will progress through the entire weld deposit.

Crater cracks have a tendency to form in the crater whenever the welding operation is interrupted. These cracks usually proceed to the edge of the crater and may be the starting point for longitudinal weld cracks. Crater cracks can be minimized or prevented by filling craters to a slightly convex shape prior to breaking the welding arc.

Figure 4 shows examples of various types of weld metal cracks.

Weld metal cracking can usually be reduced by taking one or more of the following actions:

- Improving the contour or composition of the weld deposit by changing the electrode manipulation or electrical conditions
- Increasing the thickness of the deposit and providing more weld metal to resist the stresses by decreasing the travel speed
- Reducing thermal stress by preheating
- Using low-hydrogen electrodes

- Balancing shrinkage stress by sequencing welds
- Avoiding rapid cooling conditions

4.3.2 Base Metal Cracks

Base metal cracking usually occurs within the heat-affected zone of the metal being welded. The possibility of cracking increases when working with **hardenable materials**. These cracks usually occur along the edges of the weld and through the heat-affected zone into the base metal. Types of base metal cracking include **underbead cracking** and toe cracking.

Underbead cracks are limited mainly to steel. They are usually found at regular intervals under the weld metal and usually do not extend to the surface. Because of this, they cannot be detected by visual methods of inspection.

Toe cracks are generally the result of strains caused by thermal shrinkage acting on a heat-affected zone that has been **embrittled**. They sometimes occur when the base metal cannot accommodate the shrinkage strains that are imposed by welding.

Base metal cracking can usually be reduced or eliminated by one of the following means:

- Controlling the cooling rate by preheating
- Controlling heat input
- Using the correct electrode
- Controlling welding materials

Hot and Cold Cracks

Hot cracks occur while the weld is solidifying. They can be caused by insufficient ductility at high temperature. Cold cracks occur after the weld has solidified. They are often caused by improper welding technique.

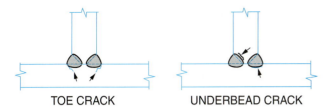

TOE CRACK

UNDERBEAD CRACK

TOE CRACK

LONGITUDINAL CRACK AND LINEAR POROSITY

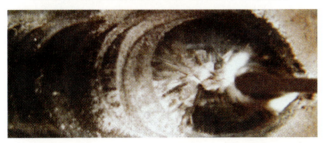

CRATER CRACK

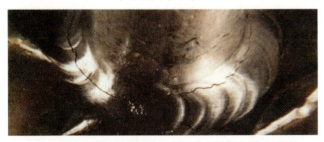

LONGITUDINAL CRACK OUT OF CRATER CRACK

FILLET WELD THROAT CRACK

104F04.EPS

Figure 4 ◆ Types of weld metal cracks.

4.4.0 Incomplete Joint Penetration

Incomplete joint penetration (*Figure 5*) occurs when the filler metal fails to penetrate and fuse with an area of the weld joint. This incomplete penetration will cause weld failure if the weld is subjected to tension or bending stresses.

Insufficient heat at the root of the joint is a frequent cause of incomplete joint penetration. If the metal being joined first reaches the melting point at the surfaces above the root of the joint, molten metal may bridge the gap between these surfaces and screen off the heat source before the metal at the root melts.

Improper joint design is another leading cause of incomplete joint penetration. If the joint is not prepared or fitted accurately, an excessively thick root face or an insufficient root gap may cause incomplete penetration. Incomplete joint penetration is likely to occur under the following conditions:

- If the root face dimension is too big even though the root opening is adequate
- If the root opening is too small
- If the included angle of a V-groove is too small

Figure 6 shows correct and incorrect joint designs.

Even if the welding heat is correct and the joint design is adequate, incomplete penetration can result from poor control of the welding arc. Examples of poor control include:

- Using an electrode that is too large
- Traveling too fast
- Using a welding current that is too low

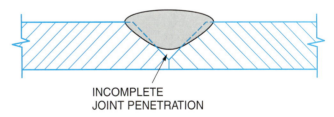

INCOMPLETE
JOINT PENETRATION

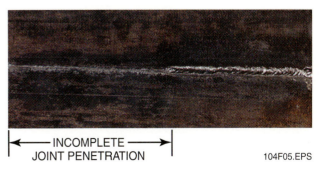

INCOMPLETE
JOINT PENETRATION

104F05.EPS

Figure 5 ◆ Incomplete joint penetration.

Causes of Incomplete Joint Penetration

Incomplete joint penetration is generally associated with groove welds. It may result from insufficient welding heat, improper joint design (too much metal for the welding arc to penetrate), or poor control of the welding arc.

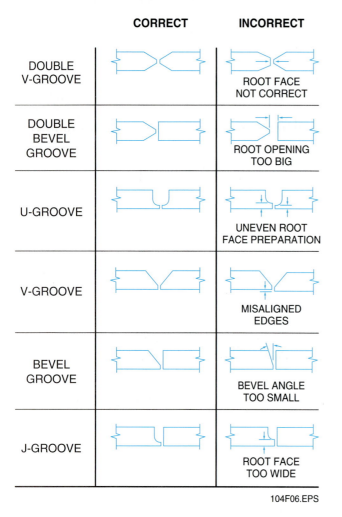

	CORRECT	INCORRECT
DOUBLE V-GROOVE		ROOT FACE NOT CORRECT
DOUBLE BEVEL GROOVE		ROOT OPENING TOO BIG
U-GROOVE		UNEVEN ROOT FACE PREPARATION
V-GROOVE		MISALIGNED EDGES
BEVEL GROOVE		BEVEL ANGLE TOO SMALL
J-GROOVE		ROOT FACE TOO WIDE

104F06.EPS

Figure 6 ◆ Correct and incorrect joint designs.

Incomplete penetration is always undesirable in welds, especially in single-groove welds if the root of the weld is subject either to tension or bending stresses. It can lead directly to weld failure or can cause a crack to start at the unfused area.

4.5.0 Incomplete Fusion

Many welders confuse incomplete joint penetration with incomplete fusion. It is possible to have good penetration without complete root fusion. Incomplete fusion is the failure of a welding process to fuse, or join together, layers of weld metal or weld metal and base metal.

Incomplete fusion may occur at any point in a groove or fillet weld, including the root of the weld. Often the weld metal simply rolls over onto the plate surface. This is generally referred to as overlap. In many cases, the weld has good fusion at the root and at the plate surface, but because of poor technique and insufficient heat, the toe of the weld does not fuse. *Figure 7* shows incomplete fusion and overlap. Causes for incomplete fusion include:

- Insufficient heat as a result of low welding current, high travel speeds, or an arc gap that is too close
- Wrong size or type of electrode
- Failure to remove oxides or slag from groove faces or previously deposited beads
- Improper joint design
- Inadequate gas shielding

Incomplete fusion discontinuities affect weld joint integrity in much the same way as porosity and slag inclusion.

4.6.0 Undercut

Undercut is a groove melted into the base metal beside the weld. It is the result of the arc removing more metal from the joint face than is replaced by weld metal. On multilayer welds, it may also occur at the point where a layer meets the wall of a groove. *Figure 8* shows undercut and overlap.

Undercutting is usually due to improper electrode manipulation. Other causes of undercutting include:

- Using a current adjustment that is too high
- Having an arc gap that is too long
- Failing to fill up the crater completely with weld metal

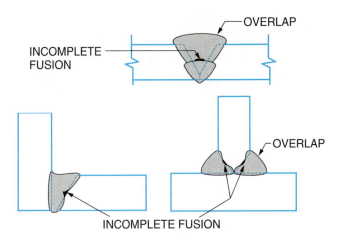

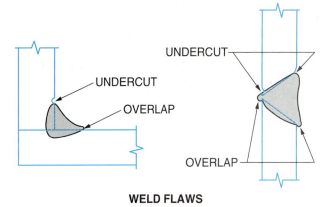

WELD FLAWS

INCOMPLETE FUSION AT WELD FACE

INCOMPLETE FUSION BETWEEN INDIVIDUAL WELD BEADS

104F07.EPS

Figure 7 ◆ Incomplete fusion and overlap.

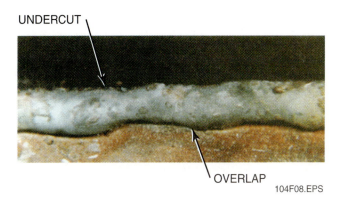

104F08.EPS

Figure 8 ◆ Undercut and overlap.

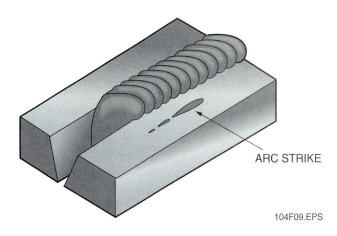

104F09.EPS

Figure 9 ◆ Arc strikes.

Most welds have some undercut that can be found upon careful examination. When it is controlled within the limits of the specifications and does not create a sharp or deep notch, undercut is usually not considered a weld defect. However, when it exceeds the limits, undercutting can be a serious defect because it reduces the strength of the joint.

4.7.0 Arc Strikes

Arc strikes are small, localized points where surface melting occurs away from the joint. These spots may be caused by accidentally striking the arc in the wrong place or by faulty ground connections.

Striking an arc on base metal that will not be fused into the weld metal should be avoided. Arc strikes can cause hardness zones in the base metal and can become the starting point for cracking.

Figure 9 shows an example of arc strikes.

4.8.0 Spatter

Spatter (*Figure 10*) is made up of very fine particles of metal on the plate surface adjoining the weld

WELD SPATTER

104F10.EPS

Figure 10 ◆ Weld spatter.

area. It is usually caused by high current, a long arc, an irregular and unstable arc, or improper shielding. Spatter makes a poor appearance on the weld and base metal and can make it difficult to inspect the weld.

4.9.0 Acceptable and Unacceptable Weld Profiles

The profile of a finished weld can affect the performance of the joint under load as much as other discontinuities affect it. This applies to the profile of a single-pass and to a layer of a multiple-pass weld. An unacceptable profile for a single-pass or multiple-pass weld could lead to the formation of discontinuities such as incomplete fusion or slag inclusions as the other layers are deposited. *Figure 11* shows acceptable and unacceptable weld profiles for both fillet and groove welds.

The ideal fillet has a uniform concave or convex face, although a slightly nonuniform face is acceptable. The convexity of a fillet weld or individual surface bead will be approximately 0.07 times the actual face width or the width of the individual surface bead, plus ¹⁄₁₆".

Butt welds should be made with slight reinforcement (not exceeding ⅛") and a gradual transition to the base metal at each toe. Butt welds should not have excess convexity, insufficient throat, excessive undercut, or overlap. If a butt weld has any of these defects, it should be repaired. The bead width should not exceed the groove width by more than ⅛".

Note

Refer to your site's WPS for specific requirements on fillet or butt welds. The information provided here is only a general guideline. The site WPS or quality specifications must be followed for all welds. Check with your supervisor if you are unsure of the specifications for your application.

5.0.0 ◆ NONDESTRUCTIVE EXAMINATION (NDE) PRACTICES

Nondestructive examination (NDE), sometimes referred to as nondestructive testing or nondestructive inspection, is a term used for those inspection methods that allow materials to be examined without changing or destroying them. NDE methods can usually detect the discontinuities and defects described earlier.

Nondestructive examination is usually performed by the site quality group as part of the site quality program. Certified welding inspectors trained in the proper test methods conduct the examinations.

The welder should be familiar with basic nondestructive examination practices. These include:

- Visual inspection (VT)
- Liquid penetrant inspection (PT)
- Magnetic particle inspection (MT)
- Radiographic inspection (RT)
- Ultrasonic inspection (UT)
- Electromagnetic (eddy current) inspection (ET)
- Leak examination

5.1.0 Visual Inspection

In visual inspection, the surface of the weld and the base metal are observed for visual imperfections. Certain tools and gauges may be used during the inspection. Visual inspection is the examination method most commonly used by welders and inspectors. It is the fastest and most inexpensive method for examining a weld. However, it is limited to what can be detected by the naked eye or through a magnifying glass.

Properly done before, during, and after welding, visual inspection can detect more than 75% of discontinuities before they are found by more expensive and time-consuming nondestructive examination methods.

Prior to welding, the base metal should be examined for conditions that may cause weld defects. Dimensions, including edge preparation, should also be confirmed by measurements. If problems or potential problems are found, corrections should be made before proceeding any further.

After the parts are assembled for welding, the weld joint should be visually checked for a proper root opening and any other aspects that might affect the quality of the weld. The following should be visually examined:

- Proper cleaning
- Joint preparation and dimensions

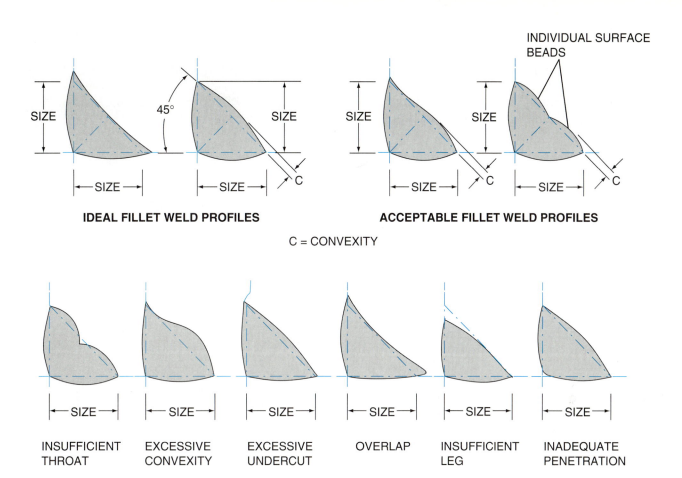

IDEAL FILLET WELD PROFILES

ACCEPTABLE FILLET WELD PROFILES

INDIVIDUAL SURFACE BEADS

SIZE

45°

SIZE

SIZE

C

C = CONVEXITY

SIZE

SIZE

C

SIZE

C

| INSUFFICIENT THROAT | EXCESSIVE CONVEXITY | EXCESSIVE UNDERCUT | OVERLAP | INSUFFICIENT LEG | INADEQUATE PENETRATION |

UNACCEPTABLE FILLET WELD PROFILES

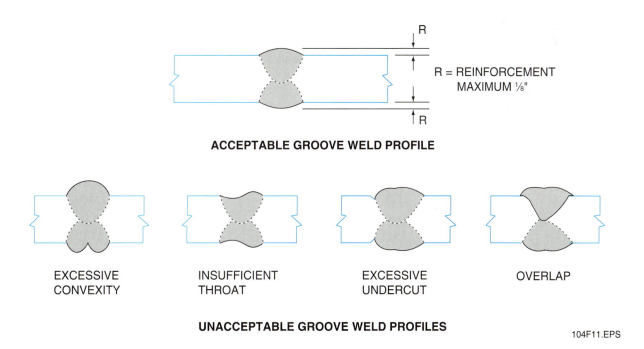

R

R = REINFORCEMENT MAXIMUM ⅛"

R

ACCEPTABLE GROOVE WELD PROFILE

EXCESSIVE CONVEXITY

INSUFFICIENT THROAT

EXCESSIVE UNDERCUT

OVERLAP

UNACCEPTABLE GROOVE WELD PROFILES

104F11.EPS

Figure 11 ◆ Acceptable and unacceptable weld profiles.

- Clearance dimensions for backing strips, rings, or consumable inserts
- Alignment and fit-up of the pieces being welded
- Welding procedures and machine settings
- Specified preheat temperature (if applicable)
- Tack weld quality

During the welding process, visual inspection is the primary method for controlling quality. Some of the aspects that should be visually examined include:

- Quality of the root pass and the succeeding weld layers
- Sequence of weld passes
- Interpass cleaning
- Root preparation prior to welding a second side
- Conformance to the applicable procedure

After the weld is completed, the weld surface should be thoroughly cleaned. A thorough visual examination may disclose weld surface defects such as cracks, shrinkage cavities, undercuts, incomplete penetration, nonfusion, overlap, and crater deficiencies before they are discovered using other nondestructive inspection methods.

An important aspect of visual examination is checking the dimensional accuracy of the weld after it is completed. Dimensional accuracy is determined by conventional measuring gauges. The purpose of using the gauges is to determine if the weld is within allowable limits as defined by the applicable codes and specifications.

Some of the more common welding gauges are:

- Undercut gauge
- Butt weld reinforcement gauge
- Fillet weld blade gauge set

5.1.1 Undercut Gauge

An undercut gauge is used to measure the amount of undercut on the base metal. Typically, codes allow for undercut to be no more than 0.010" deep when the weld is transverse to the primary stress in the part that is undercut. Several types of gauges can be used to check the amount of undercut. These gauges have a pointed end that is pushed into the undercut. The back side of the gauge indicates the measurement in either inches or millimeters. Two types of undercut gauges currently used are the bridge cam gauge and the V-WAC gauge (*Figure 12*). The V-WAC gauge is used only for measuring undercut. The bridge cam gauge is used for many other measurements.

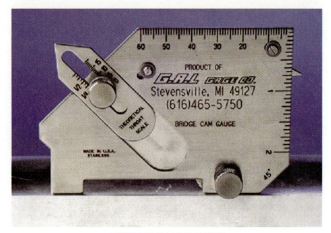

BRIDGE CAM GAUGE

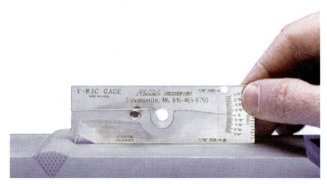

V-WAC GAUGE   104F12.EPS

Figure 12 ◆ Undercut gauges.

5.1.2 Butt Weld Reinforcement Gauge

The butt weld reinforcement gauge has a sliding pointer calibrated to several different scales that are used to measure the size of a fillet weld or the reinforcement of a butt weld. To use the gauge for a fillet weld, position it as shown in *Figure 13* and slide the pointer to contact the base metal or weld metal. Be sure to read the correct scale for the measurement being taken. The other end of the gauge is used for butt welds.

5.1.3 Fillet Weld Blade Gauge Set

The fillet weld blade gauge set has seven individual blade gauges for measuring convex and concave fillet welds. The individual gauges are held together by a screw secured with a knurled nut. The seven individual blade gauges can measure eleven concave and convex fillet weld sizes: ⅛", ³⁄₁₆", ¼", ⁵⁄₁₆", ⅜", ⁷⁄₁₆", ½", ⅝", ¾", ⅞", and 1" and their metric equivalents. The same blade size cannot be used for measuring both concave and convex fillet welds.

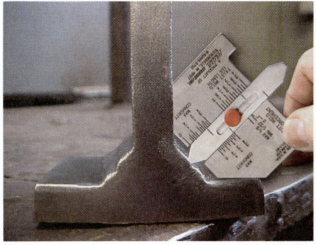

104F13.EPS

Figure 13 ◆ Butt weld reinforcement gauge.

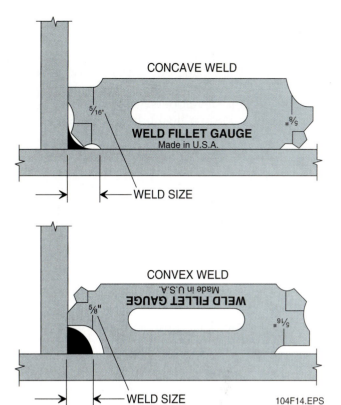

CONCAVE WELD

⁵/₁₆"

WELD FILLET GAUGE
Made in U.S.A.

⁵/₈"

WELD SIZE

CONVEX WELD

Made in U.S.A.
WELD FILLET GAUGE

⁵/₈"

⁵/₁₆"

WELD SIZE

104F14.EPS

Figure 14 ◆ Fillet weld blade gauge.

To use the fillet weld blade gauge set, identify the type of fillet weld to be measured (concave or convex) and the size. Select the appropriate blade and position it. Be sure the gauge blade is flush to the base metal with the tip touching the vertical member.

Figure 14 shows an application of a fillet weld blade gauge.

5.2.0 Liquid Penetrant Inspection

Liquid penetrant inspection (PT) is a nondestructive method for locating defects that are open to the surface. It cannot detect internal defects. The technique is based on the ability of a penetrating liquid, which is usually red in color, to wet the surface opening of a discontinuity and to be drawn into it. A liquid or dry powder developer, which is usually white in color, is then applied

Hot Tip

Fixed Weld Fillet Gauge

This fixed weld fillet gauge can be used to measure 11 fillet weld sizes.

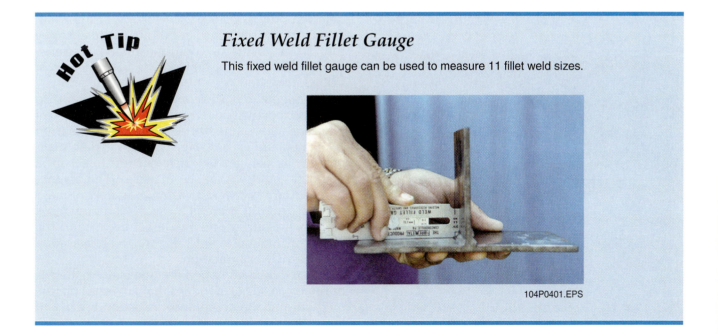

104P0401.EPS

over the metal. If the flaw is significant, red penetrant bleeds through the white developer to indicate a discontinuity or defect.

The dye, cleaner, and developer are available in spray cans for convenience. Some solvents used in the cleaners and developers contain high amounts of chlorine, a known health hazard, to make the liquids nonflammable.

WARNING!

Refer to the material safety data sheet (MSDS) for hazards associated with the liquid penetrant solvent.

The most common defects found using this process are surface cracks. Most cracks exhibit an irregular shape. The width of the bleed-out (the red dye bleeding through the white developer) is a relative measure of the depth of a crack.

Surface porosity, metallic oxides, and slag will also hold penetrant and cause bleed-out. These indications are usually more circular and have less width than a crack. *Figure 15* shows liquid penetrant materials and an example of the results of liquid penetrant inspection.

5.2.1 *Advantages of Liquid Penetrant Inspection*

The advantages of liquid penetrant inspection are that it can find small defects not visible to the naked eye, it can be used on most types of metals, it is basically inexpensive, and it is fairly easy to use and interpret. It is most useful to examine

104F15.EPS

Figure 15 ◆ Liquid penetrant materials and inspection example.

welds that are susceptible to surface cracks. Except for visual inspection, it is perhaps the most commonly used nondestructive examination method for surface inspection of nonmagnetic parts.

5.2.2 *Disadvantages of Liquid Penetrant Inspection*

The disadvantages of liquid penetrant inspection are that it takes more time to use than visual inspection, and it can only find surface defects. The presence of weld bead ripples and other irregularities can also hinder the interpretation of indications. Because chemicals are used, care must be taken when performing the inspection. When testing the rough, irregular surfaces produced by welding, the presence of nonrelevant indications may also make interpretation difficult.

5.3.0 Magnetic Particle Inspection

Magnetic particle inspection (MT) is a nondestructive examination method that uses electricity to magnetize the weld to be examined. After the metal is magnetized, metal particles are sprinkled onto the weld surface. If there are defects in the surface or just below the surface, the metal particles will be grouped into a pattern around the defect. The defect can be identified by the shape, width, and height of the particle pattern.

Magnetic particle inspection is used to test welds for such defects as surface cracks, nonfusion, porosity, undercut, and slag inclusions. It can also be used to inspect plate edges for surface imperfections prior to welding. Defects can be detected only at or near the surface of the weld. Defects much deeper than this are not likely to be found. Certain discontinuities exhibit characteristic powder patterns that can be identified by a skilled inspector.

For magnetic particle examination, the part to be inspected must be ferromagnetic (made of steel or a steel alloy), smooth, clean, dry, and free from oil, water, and excess slag. The part is then magnetized by using an electric current to set up a magnetic field within the material. The magnetized surface is covered with a thin layer of magnetic powder. If there is a defect, the powder is held to the surface at the defect because of the powerful magnetic field. *Figure 16* shows magnetic particle examination.

When this examination method is used, there is normally a code or standard that governs both the method and the acceptance/rejection criteria of indications.

Liquid Penetrant Test

These three photos show the sequence of an actual liquid penetrant test.

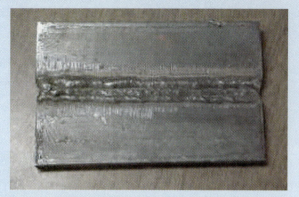

1. WELD READY FOR PENETRANT TEST

2. PENETRANT APPLIED

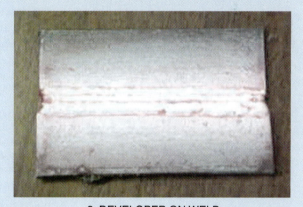

3. DEVELOPER ON WELD

104P0402.EPS

5.3.1 Advantages of Magnetic Particle Inspection

The advantages of magnetic particle inspection are that it can find small defects not visible to the naked eye, and it is faster than liquid penetrant inspection.

5.3.2 Disadvantages of Magnetic Particle Inspection

The disadvantages of magnetic particle inspection are that the materials must be capable of being magnetized, the inspector must be skilled in interpreting indications, rough surfaces can

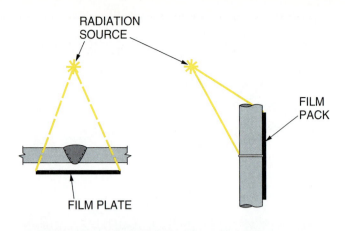

Figure 16 ◆ Magnetic particle examination.

interfere with the results, the method requires an electrical power source, and it cannot find internal discontinuities located deep in the weld.

5.4.0 Radiographic Inspection

Radiography (RT) is a nondestructive examination method that uses radiation (X-rays or gamma rays) to penetrate the weld and produce an image on film. When a joint is radiographed, the radiation source is placed on one side of the weld and the film on the other. The joint is then exposed to the radiation source. The radiation penetrates the metal and produces an image on the film. The film is called a radiograph and provides a permanent record of the weld quality. Radiography should only be used and interpreted by trained, qualified personnel. *Figure 17* shows a radiography examination.

Radiographic inspection can produce a visible image of weld discontinuities, both surface and subsurface, when they are different in density from the base metal and different in thickness parallel to the radiation. Surface discontinuities are better identified by visual, penetrant, or magnetic particle examination.

5.4.1 Advantages of Radiographic Inspection

The advantages of radiographic inspection are that the film gives a permanent record of the weld quality, the entire thickness can be examined, and it can be used on all types of metals.

5.4.2 Disadvantages of Radiographic Inspection

The disadvantages of radiographic inspection are that it is a slow and expensive method for inspect-

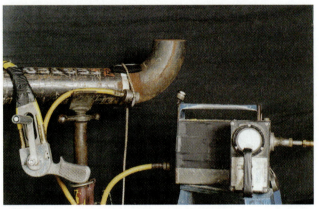

Figure 17 ◆ Radiography examination.

ing welds, some joints are inaccessible to radiography, and radiation of any type is very hazardous to humans. Cracks can frequently be missed if they are very small or are not aligned with the radiation beam.

5.5.0 Ultrasonic Inspection

Ultrasonic inspection (UT) is a relatively low-cost nondestructive examination method that uses soundwave vibrations to find surface and subsurface defects in the weld material. Ultrasonic waves are passed through the material being tested and are reflected back by any density change caused by a defect. The reflected signal is shown on the screen display of an oscilloscope.

The term ultrasonic indicates that these frequencies are above those heard by the human ear. Ultrasonic devices operate very much like depth sounders or fish finders.

Ultrasonic examination can be used to detect and locate cracks, **laminations**, shrinkage cavities, pores, slag inclusions, incomplete fusion, and incomplete joint penetration as well as other discontinuities in the weld. A qualified inspector can

AWS ENTRY LEVEL WELDER – TRAINEE MODULE 29104-03

interpret the signal on a screen to determine the approximate position, depth, and size of the discontinuity. *Figure 18* shows a portable ultrasonic device.

5.5.1 Advantages of Ultrasonic Inspection

The advantages of ultrasonic inspection are that it can find defects throughout the material being examined, it can be used to check materials that cannot be radiographed, it is nonhazardous to personnel and equipment, and it can detect even small defects.

5.5.2 Disadvantages of Ultrasonic Inspection

The disadvantages of ultrasonic inspection are that it requires a high degree of skill to properly interpret the patterns, and very small or thin weldments are difficult to inspect. Also, a permanent record is not readily obtained.

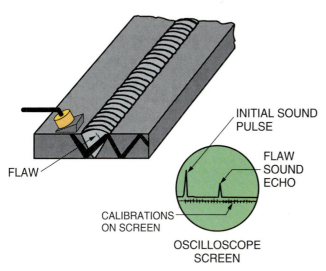

INITIAL SOUND PULSE

FLAW SOUND ECHO

FLAW

CALIBRATIONS ON SCREEN

OSCILLOSCOPE SCREEN

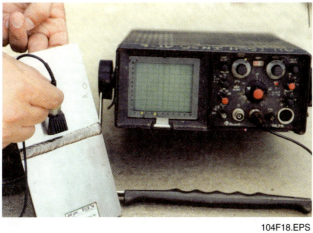

104F18.EPS

Figure 18 ◆ Portable ultrasonic device.

5.6.0 Electromagnetic (Eddy Current) Inspection

Like magnetic particle testing, electromagnetic, or eddy current, inspection (ET) uses electromagnetic energy to detect defects in the joint. An alternating current (AC) coil, which produces a magnetic field, is placed on or around the part being tested. After being calibrated, the coil is moved over the part to be inspected. The coil produces a current in the metal through induction. The induced current is called an eddy current. If a discontinuity is present in the test part, it will interrupt the flow of the eddy currents. This change can be observed on the oscilloscope.

Eddy currents only detect discontinuities near the surface of the part. This method is suitable for both ferrous and nonferrous materials and is used in testing welded tubing and pipe. It can determine the physical characteristics of a material and the wall thickness in tubing. It can check for porosity, pinholes, slag inclusions, internal and external cracks, and nonfusion.

5.6.1 Advantage of Eddy Current Inspection

The advantage of eddy current inspection is that it can detect surface and near-surface weld defects. It is particularly useful in inspecting circular parts like pipes and tubing.

5.6.2 Disadvantage of Eddy Current Inspection

The disadvantage of eddy current inspection is that eddy currents decrease with depth, so defects farther from the surface may go undetected. The accuracy of the examination depends in large part on the calibration of the instrument and the qualification of the inspector.

5.7.0 Leak Testing

Leak testing is used to determine the ability of a pipe or vessel to contain a gas or liquid under pressure. Testing methods vary depending on the application of the weldment. In some cases, the vessel is pressurized and tested by immersing it in water or applying a soap bubble solution to the weld (*Figure 19*). An open tank can be tested using water that contains fluorescein, which can be detected by ultraviolet light.

A method called the vacuum box test is used to test a vessel where only one side of the weld is accessible. The base of a storage tank is an example.

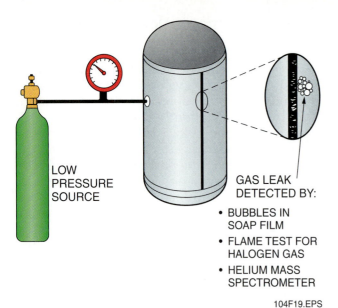

LOW
PRESSURE
SOURCE

GAS LEAK
DETECTED BY:

- BUBBLES IN
  SOAP FILM
- FLAME TEST FOR
  HALOGEN GAS
- HELIUM MASS
  SPECTROMETER

104F19.EPS

Figure 19 ◆ Leak testing.

The vacuum box is a transparent box with a soft rubber seal. A vacuum pump is used to extract all the air from the box. A leak is indicated by the presence of bubbles.

In the helium spectrometer leak test, helium is used as a tracer gas. Because of the small size of helium atoms, they can pass through an opening so small that it might not be detectable by other test methods. Sensitive instruments are used to detect the presence of helium.

6.0.0 ◆ DESTRUCTIVE TESTING

There is considerable overlap in the application of nondestructive and destructive tests. Destructive tests, which destroy the weld, are frequently used on several sample weldments to supplement, confirm, or establish the limits of nondestructive tests. Destructive testing is also used to provide supporting information. Once this information has been established, nondestructive examinations can be made on similar welds to locate all discontinuities above the critical defect size that was determined by the destructive tests.

Destructive tests are so called because the test sample is destroyed or damaged in the testing process and is no longer suitable for use. Destructive testing is often done in a laboratory setting using machines designed to apply different stresses to the weld or the base metal, depending upon which is being tested. Some of the tests can be accomplished by striking the sample with a hammer. Examples of destructive tests commonly used to examine welds and base metals include:

- *Tensile test* – In this test, a sample is placed in a tensile testing machine and is pulled until it breaks. The tensile test is sometimes used to determine if the weld performs as well as the base metal. In most cases, however, the test is performed to determine specific characteristics of the weldment, such as strength and ductility (the amount the sample will stretch before breaking).

- *Hardness test* – Whereas strength testing examines the ability of the sample to transmit a load, hardness testing examines its ability to resist penetration. Hardness testing is usually done using a penetrating device that leaves an indentation in the sample. The depth or diameter of the indentation, depending on the type of hardness test being conducted, is then measured to determine the hardness.

- *Impact test* – The ability of a weld to withstand an impact, such as a hammer strike, is measured by this test. The characteristic being tested is referred to as toughness. In a laboratory environment, a pendulum-type machine that simulates a heavy hammer blow is used. A notch of a specified size is made in the sample. Then the sample is placed in the jaws of the testing machine and struck with a pendulum blow.

Jig Plunger and Die

The jig has a plunger and die that are dimensioned for the thickness of the specimen being bent. Refer to the welding code being used for testing for the required dimensions of the bending jig's plunger and die.

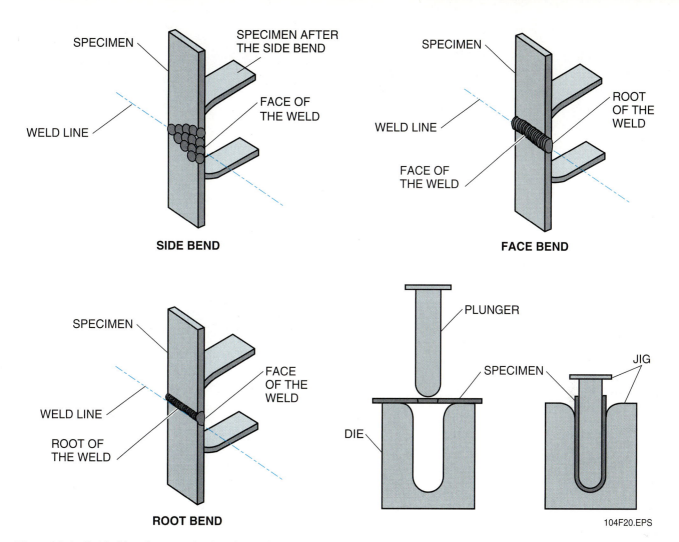

SIDE BEND

SPECIMEN

SPECIMEN AFTER THE SIDE BEND

FACE OF THE WELD

WELD LINE

FACE BEND

SPECIMEN

ROOT OF THE WELD

WELD LINE

FACE OF THE WELD

ROOT BEND

SPECIMEN

FACE OF THE WELD

WELD LINE

ROOT OF THE WELD

PLUNGER

SPECIMEN

JIG

DIE

104F20.EPS

Figure 20 ◆ Guided bend test method and samples.

• *Soundness test* – The three types of soundness tests include bend, nick-break, and fillet weld break tests.
  – Bend testing is the most commonly used soundness test used in determining the qualifications of a welder or welding procedures. Bend tests are performed by placing the sample in a special fixture and applying stress in a way that causes the sample to bend 180°. The bend is then inspected for weld defects. Guided bend tests are used to evaluate groove test welds on both plates and pipes. In this method, specimens are bent into a U-shape with a device called a jig. The bending action places stress on the weld metal and reveals any discontinuities in the weld.
  The three types of tests performed on the jig are:
  • *Root bends* – test the penetration and fusion throughout the root of the joint.

• *Face bends* – test for porosity, slag inclusion, gas pockets, or other defects as well as the quality of the fusion to the side walls and the face of the weld joint.
• *Side bends* – test for soundness and fusion.

Face and root bend tests are used for materials up to ⅜" thick. For materials thicker than ⅜", side bend testing may be used. *Figure 20* shows examples of guided bend testing.

  – Nick-break testing is used primarily in the pipeline industry. The specimen is saw-cut so it will break in a specific place. Then it is broken, and the weld zone is examined for defects.
  – In a fillet weld break test, a fillet weld is made on one side of a T-joint (*Figure 21*). Stress is then applied to the T-joint until the weld fractures. The weld is then examined for defects. This is not a weld strength test; it is intended to break the weld so that it can be examined for discontinuities.

Machines Used in Destructive Testing

A wide variety of specialized machines are used in destructive testing.

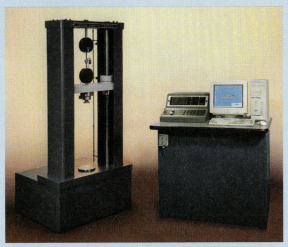

TENSILE TEST MACHINE

IMPACT TEST MACHINE   ROCKWELL HARDNESS
TESTER

104P0403.EPS

7.0.0 ◆ WELDER PERFORMANCE QUALIFICATION TESTS

The purpose of welder performance qualification is to measure the proficiency of individual welders. As previously discussed, codes require that welders take a test to qualify to perform a welding procedure.

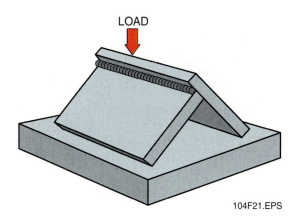

Figure 21 ◆ Fillet weld break test.

Note

Various codes and specifications often require similar methods for qualifying welders. The applicable code or specification should be consulted for specific details and requirements. Ask your supervisor if you are unsure about which codes or specifications apply to your project.

7.1.0 Welding Positions Qualification

The welder or welding operator is qualified by welding position. Welders may be qualified to perform a welding procedure in only one, or possibly all, position(s) by taking a welding test. The qualification tests are designed to measure the welder's ability to make groove and fillet welds in different positions on plate, pipe, or both in accordance with the applicable code. Each welding position is designated by a number and a letter, such as 1G. These designations are standard for all codes.

The letter G designates a groove weld. The letter F designates a fillet weld. For plate welding, the positions are designated by the following numbers:

- *1* – Flat position welding
- *2* – Horizontal position welding
- *3* – Vertical position welding
- *4* – Overhead position welding

Figure 22 shows the plate welding positions for both fillet welds and groove welds.

For pipe welding, there are additional positions: 5F, 5G, and 6G. The numbers 5 and 6 indicate that multiple position welds are required. Also, in the 1G (flat groove) and 1F (flat fillet) positions, the pipe is rotated during welding. *Figure 23* shows the pipe welding positions for both fillet welds and groove welds.

A welder who qualifies in one position does not automatically qualify to weld in all positions. However, in most cases qualification for groove welds will qualify the welder for fillet welds; qualification for pipe will qualify the welder for plate. Qualification in one code may or may not qualify the welder in other codes. The qualification requirements between codes may not match. Refer to your site requirements/code for qualifying requirements (*Figure 24*).

7.2.0 AWS Structural Steel Code

The AWS structural steel code provides information concerning the qualification of welding procedures, welders, and welding operators for the types of welding done by contractors and fabricators in building and bridge construction. Qualification for plate welding also qualifies the welder for rectangular tubing.

The mild steel electrodes used with shielded metal arc welding (SMAW) are classified by F numbers: F1, F2, F3, and F4. Qualification with an electrode in a particular F-number classification will qualify the welder with all electrodes identified in that classification and in lower F-number classifications. *Table 1* shows AWS F-number electrode classifications.

Material thickness is an essential variable in qualification tests under AWS code. Some of the tests in the code qualify the welder only up to

Table 1 AWS F-Number Electrode Classification

Group	AWS Electrode Classification				Type
F4	EXX15	EXX16	EXX18		Low-Hydrogen
F3	EXX10	EXX11			Fast-Freeze
F2	EXX12	EXX13	EXX14		Fill-Freeze
F1	EXX20	EXX24	EXX27	EXX28	Fast-Fill

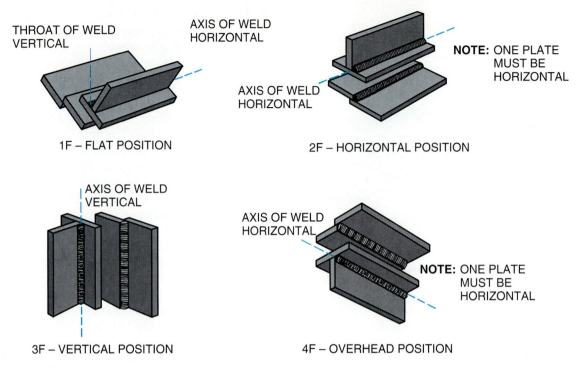

THROAT OF WELD
VERTICAL

AXIS OF WELD
HORIZONTAL

1F – FLAT POSITION

AXIS OF WELD
HORIZONTAL

NOTE: ONE PLATE
MUST BE
HORIZONTAL

2F – HORIZONTAL POSITION

AXIS OF WELD
VERTICAL

3F – VERTICAL POSITION

AXIS OF WELD
HORIZONTAL

NOTE: ONE PLATE
MUST BE
HORIZONTAL

4F – OVERHEAD POSITION

FILLET WELDS

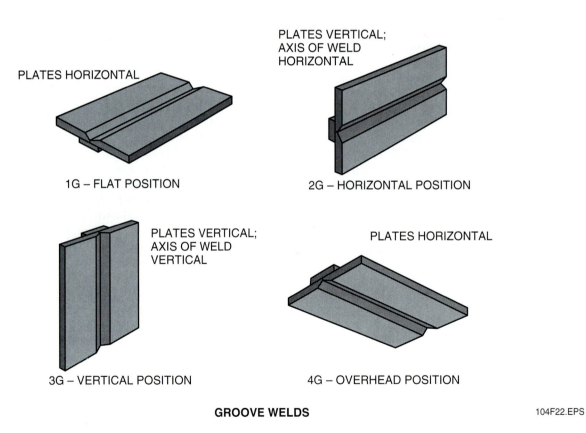

PLATES HORIZONTAL

1G – FLAT POSITION

PLATES VERTICAL;
AXIS OF WELD
HORIZONTAL

2G – HORIZONTAL POSITION

PLATES VERTICAL;
AXIS OF WELD
VERTICAL

3G – VERTICAL POSITION

PLATES HORIZONTAL

4G – OVERHEAD POSITION

GROOVE WELDS

104F22.EPS

Figure 22 ◆ Welding positions for plate.

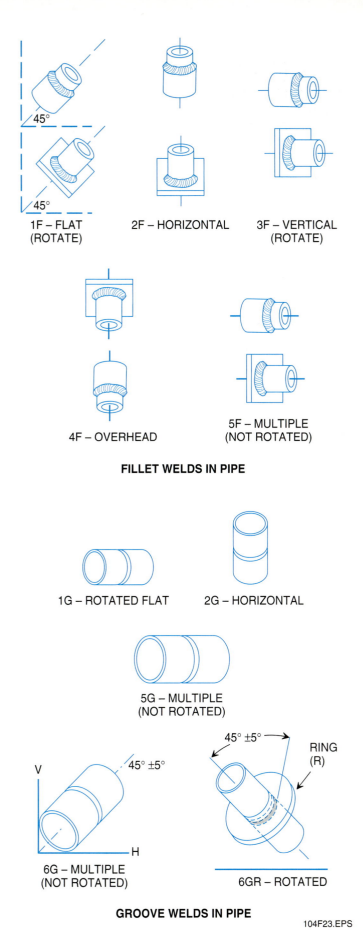

FILLET WELDS IN PIPE

1F – FLAT (ROTATE)

2F – HORIZONTAL

3F – VERTICAL (ROTATE)

4F – OVERHEAD

5F – MULTIPLE (NOT ROTATED)

1G – ROTATED FLAT

2G – HORIZONTAL

5G – MULTIPLE (NOT ROTATED)

6G – MULTIPLE (NOT ROTATED)

6GR – ROTATED

GROOVE WELDS IN PIPE

104F23.EPS

Figure 23 ◆ Welding positions for pipe.

twice the thickness of the test piece. Others qualify the welder for unlimited thicknesses.

A typical AWS welder qualification test is a V-groove weld with metal backing in the 3G and 4G positions using an F4 electrode. Passing this test qualifies the welder to weld with F4 or lower electrodes and make groove and fillet welds in all positions. *Figure 25* shows a typical AWS plate test.

7.3.0 ASME Code

Individual welders and welding operators who are required to weld to ASME code must qualify in accordance with Section IX of the *ASME Boiler and Pressure Vessel Code* on either plate or pipe. Qualification on pipe also qualifies the welder to weld plate, but not vice versa. Qualification with groove welds also qualifies the welder for fillet welds, but not vice versa. It is possible under the code to qualify for fillet welds only.

The typical ASME welder qualification test is to weld pipe in the 6G position using an open root. Passing this test qualifies the welder to weld pipe in all positions and plate in all positions (fillet and groove). If F3 electrodes are used all the way out, the welder only qualifies on F3 electrodes on both plate and pipe. If F3 electrodes are used for welding the root and F4 electrodes are used for filler, the welder qualifies to weld pipe or plate with F3 electrodes and pipe or plate with F4 electrodes as long as backing is used. *Figure 26* shows a typical ASME pipe test.

7.4.0 Welder Qualification Tests

The welder becomes qualified by successfully completing a weld made in accordance with the WPS. It is general practice for code welding to qualify welders on the groove weld tests. Passing these tests also permits the welder to weld fillet welds.

7.4.1 Making the Test Weld

Although the qualification tests are designed to determine the capability of welders, welders have failed for reasons not related to their welding ability. This is due principally to carelessness in the application of the weld and in the preparation of the test specimen. It is important to note prior to welding where the test strips will be cut from the weld coupon. By doing this, you can avoid potential problems such as restarts in the area of the test strips. The following sections will explain how to prepare a test specimen.

WELDING PROCEDURE SPECIFICATION (WPS)   Yes ☐
PREQUALIFIED _____   QUALIFIED BY TESTING _____
or PROCEDURE QUALIFICATION RECORDS (PQR)   Yes ☒

Identification # **PQR 231**
Revision **1**____ Date **12-1-02** By **W. Lye**
Authorized by **J. Jones**____ Date **1-18-03**
Type—Manual ☐   Semi-Automatic ☐
Machine ☐   Automatic ☐

Company Name **Red Inc.**
Welding Process(es) **FCAW**
Supporting PQR No.(s) **–**

JOINT DESIGN USED
Type: **Butt**
Single ☒   Double Weld ☐
Backing: Yes ☒ No ☐
        Backing Material: **ASTM A131A**
Root Opening **1/4"** Root Face Dimension _____**–**_____
Groove Angle: **52-1/2°** Radius (J–U) _____**–**
Back Gouging: Yes ☐ No ☒   Method ____**–**

BASE METALS
Material Spec. **ASTM A131**
Type or (Grade) **A**
Thickness: Groove **1"**   Fillet _____**–**
Diameter (Pipe) _____**–**

FILLER METALS
AWS Specification **A5.20**
AWS Classification **E71T-1**

SHIELDING
Flux _**–**_____   Gas **CO$_2$**
                    Composition _____
Electrode-Flux (Class)_____ Flow Rate _____
_____ Gas Cup Size _____

PREHEAT
Preheat Temp., Min **75° Ambient**
Interpass Temp., Min **75°F**_____ Max **350° F**

POSITION
Position of Groove: **O.H.**_____ Fillet: **–**
Vertical Progression: Up ☐ Down ☐

ELECTRICAL CHARACTERISTICS

Transfer Mode (GMAW)   Short-Circuiting ☐
                          Globular ☒ Spray ☐
Current: AC ☐   DCEP ☒   DCEN ☐   Pulsed ☐
Other _____

Tungsten Electrode (GTAW)
        Size: _____
        Type: _____

TECHNIQUE
Stringer or Weave Bead: **Stringer**
Multi-pass or Single Pass (per side) **Multipass**
Number of Electrodes **1**
Electrode Spacing   Longitudinal **–**
                       Lateral _____**–**
                       Angle _____**–**

Contact Tube to Work Distance **3/4-1"**
Peening **None**
Interpass Cleaning: **Wire Brush**

POSTWELD HEAT TREATMENT
Temp. **N.A.**
Time _____

WELDING PROCEDURE

Pass or Weld Layer(s)	Process	Filler Metals		Current		Volts	Travel Speed	Joint Details
		Class	Diam.	Type & Polarity	Amps or Wire Feed Speed			
1	FCAW	E71T-1	.045"	DC+	180	26	8	
2–8	"	"	"	"	200	27	10	
9–11	"	"	"	"	200	27	11	
12–15	"	"	"	"	200	27	9	
16	"	"	"	"	200	27	11	

Form E-1 (Front)

104F24.TIF

Figure 24 ◆ Example of welding code.

104F25.EPS

Figure 25 ◆ Typical AWS plate test.

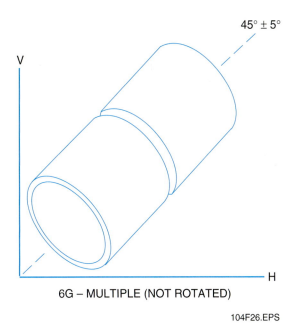

6G – MULTIPLE (NOT ROTATED)

104F26.EPS

Figure 26 ◆ Typical ASME pipe test.

7.4.2 Removing Test Specimens

After making the qualification test weld, the test specimens are cut from the test pipe or plate by any suitable means. There are specific locations where the test specimen is cut from the pipe or plate.

For pipe welded in the 1G or 2G positions, two specimens are required. For material ⅜" thick and under, a face bend and a root bend are required. For ⅜" only, side bends can be substituted. For material more than ⅜" thick, two side bends are required. *Figure 27* shows where the specimens are cut from the test weld.

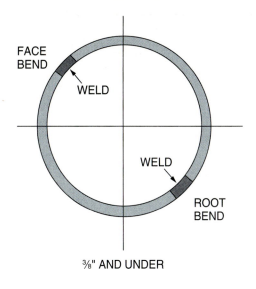

⅜" AND UNDER

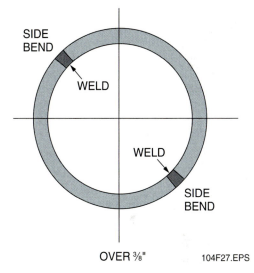

OVER ⅜" 104F27.EPS

Figure 27 ◆ Locations of pipe specimens for 1G and 2G positions.

For pipe welded in the 5G or 6G positions, four specimens are required. For material ¹⁄₁₆" thick up to ⅜" thick, two face bends and two root bends are required. For material more than ⅜" thick, four side bends are required. *Figure 28* shows where the specimens are cut from the test weld.

Typical specimen locations for plate welds are shown in *Figure 29*. For material ⅜" thick, a face bend and a root bend are required. For material more than ⅜" thick, two side bends are required.

Note

Tests are usually given on ⅜"-thick plate for limited thickness and 1"-thick plate for unlimited thickness qualifications.

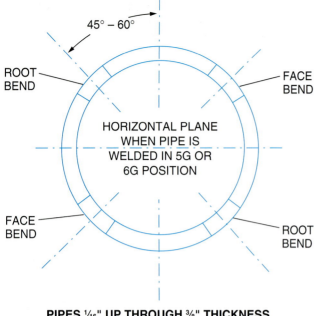

PIPES 1/16" UP THROUGH 3/8" THICKNESS

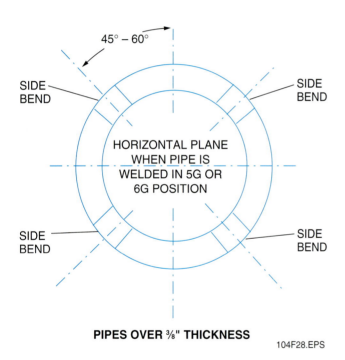

PIPES OVER 3/8" THICKNESS

104F28.EPS

Figure 28 ◆ Locations of pipe specimens for 5G and 6G positions.

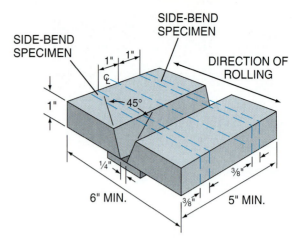

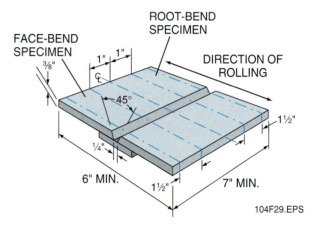

104F29.EPS

Figure 29 ◆ Specimen locations for plate welds.

7.4.3 Preparing the Specimens for Testing

After the specimen has been cut from the test piece, it must be properly prepared for testing. Poor specimen preparation can cause a sound weld metal to fail. For example, a slight nick may open up under the severe bending stress of the test, causing the specimen to fail. To properly prepare the test specimen:

- Grind or machine the surface to a smooth finish. All grinding and machining marks must be lengthwise on the sample. Otherwise, they produce a notch effect, which may cause failure.
- Remove any bead reinforcement on either the face or the root side of the weldment. This is part of the test requirement and, more important to the welder, failure to do so can cause the failure of a good weld.
- Round the edges to a smooth 1/16" radius. This can be done with a file. Rounded edges help prevent failure caused by cracks starting at a sharp corner.
- When grinding specimens, do not water-quench them when they are hot. Quenching may create small surface cracks that become larger during the bend test.

Figure 30 shows a prepared test specimen.

After bending, the specimen is evaluated by measuring the discontinuities that are exposed. The criteria for acceptance can vary by code or site quality standards. The *AWS D1.1* states that for

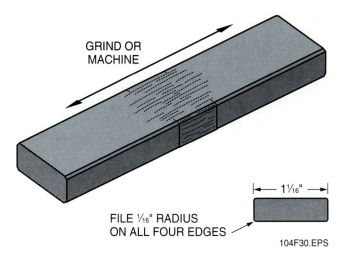

GRIND OR
MACHINE

1¹⁄₁₆"

FILE ¹⁄₁₆" RADIUS
ON ALL FOUR EDGES

104F30.EPS

Figure 30 ◆ Example of a prepared test specimen.

acceptance, the surface shall contain no discontinuities exceeding the following dimensions:

- ⅛" *(3.2 mm)* – Measured in any direction on the surface.
- ⅜" *(9.5 mm)* – Sum of the greatest dimensions of all discontinuities exceeding ¹⁄₃₂" (0.8 mm) but less than or equal to ⅛" (3.2 mm).
- ¼" *(6.4 mm)* – Maximum corner crack, except when the corner crack results from visible slag inclusion or other fusion-type discontinuities, then a ⅛" (3.2 mm) maximum shall apply. A specimen with corner cracks exceeding ¼" (6.4 mm), with no evidence of slag inclusions or other fusion-type discontinuities may be discarded, and a replacement test specimen from the original weldment shall be tested.

Note

In some cases, a radiographic inspection will be used instead of the guided bend test. This allows the entire weld to be examined and can detect small discontinuities at any location within the weld.

When the welder passes the qualification test(s), the test results and the procedure(s) that the welder may weld to are listed on a record that is kept by the company. This record becomes part of the quality documentation, and the welder becomes certified to weld to that procedure.

7.4.4 *Welder Qualification Limits*

Welders may retest if they initially fail the test. An immediate retest consists of two test welds of each

type that failed. All the test specimens must pass this retest. A complete retest may be made if the welder has had further training or practice since the last test.

Note

Retest requirements may vary depending on site quality standards. Check your site's quality standards for specific retest requirements.

After welders have qualified, they may have to requalify if they have not used the specific process for a certain time period. This time period varies from three months under ASME to six months under AWS. They may also be required to requalify if there is a reason to question their ability to make welds that meet the WPS. Also, since welder performance qualification is limited to the essential variables of a particular procedure, any change in one or more of the essential variables requires the welder to requalify with the new procedure.

8.0.0 ◆ QUALITY WORKMANSHIP

The codes and standards discussed in this module were written to ensure that welders consistently make quality welds. Although many weldments will be examined and tested, due to time and cost not all will be. Nevertheless, quality workmanship is expected in every weld a welder makes.

8.1.0 Typical Site Organization

The welder should be able to work with appropriate site representatives to ensure that quality work is achieved. To do this, the site organizational structure needs to be understood. If quality problems arise, the welder needs to follow the appropriate chain of command to eliminate the problems. *Figure 31* shows a typical site organizational structure.

Note

Your site may differ in its organizational structure. Check with your immediate supervisor to find out the structure for your site.

8.2.0 Chain of Command

A welder should always follow the site chain of command. However, there may be instances

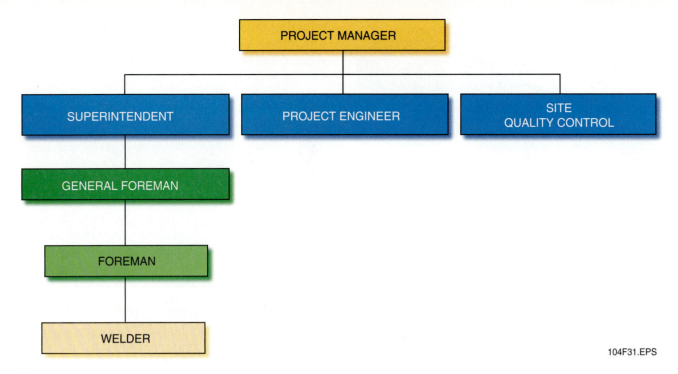

Figure 31 ◆ Typical site organizational structure.

when the welder should bypass the chain of command. Examples of such instances are:

- You have been directed to perform an unsafe act. If you cannot resolve the matter with your immediate supervisor, it is your responsibility to go to the general foreman, superintendent, project manager, or safety engineer.
- You have been directed to perform a weld that requires certification, and you are not certified to perform it. If you cannot resolve the matter with your immediate supervisor, it is your responsibility to go to the general foreman, superintendent, project manager, or quality engineer.

Note

Always try to resolve problems with your immediate supervisor before bypassing the chain of command.

Summary

Quality is everyone's responsibility. If the work being done cannot be defined as quality work, it reflects on all those involved in the process. One essential trait of a craftsperson is a sense of quality workmanship. The craftsperson is generally closest to the work and will therefore have a major impact on product quality. Keeping quality in mind as you perform each step of your job will help you identify and correct small problems before they become major problems. This will make everyone's job easier and instill a sense of pride in what has been accomplished.

Safety, quality, production — each has a cost of its own. While on the project, each of these items should also have proper guidelines. At the completion of the project, when all records for safety, cost, planning, scheduling, and effectiveness have been evaluated, the papers are usually filed away and never accessed again, while quality remains for all eyes to see indefinitely. Quality is, perhaps, the major reason for repeat business; how well the craftsperson performed will be noticed long after the project has been completed.

1. The specifications for acceptable ferrous (Part A) and nonferrous (Part B) base metals are contained in Section _____ of the *American Society of Mechanical Engineers (ASME) Boiler and Pressure Vessel Code (BPVC).*
   a. I
   b. II
   c. V
   d. IX

2. The minimum requirements for the design, materials, fabrication, erection, testing, and inspection of various types of piping systems are covered by _____.
   a. *ASME B31*
   b. *AWS D1.1*
   c. *AWS B1.11*
   d. *API Std. 1104*

3. Which of the following elements defined in a WPS is considered an essential variable?
   a. amperage
   b. travel speed
   c. electrode size
   d. joint design

4. Incomplete joint penetration is a welding _____.
   a. porosity
   b. inclusion
   c. defect
   d. discontinuity

5. The type of porosity in which the weld exhibits a localized grouping of pores is known as _____.
   a. uniformly scattered porosity
   b. clustered porosity
   c. linear porosity
   d. piping porosity

Refer to the following figure to answer Questions 6 through 8.

(A)

(B)

(C)

(D)

104E01.EPS

Match a picture to each discontinuity listed below.

6. _____ Surface slag inclusion

7. _____ Linear porosity

8. _____ Toe crack

9. One method of reducing weld metal cracking is to use _____ electrodes.
   a. fast-fill
   b. fill-freeze
   c. low-hydrogen
   d. fast-freeze

10. The test in which a vacuum box is used is done to check for _____.
    a. inclusions
    b. porosity
    c. leaks
    d. incomplete fusion

Refer to the following figure to answer Question 11.

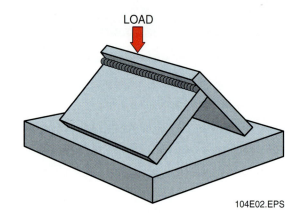

LOAD

104E02.EPS

11. The test method shown above is the _____.
    a. nick-break test
    b. fillet weld break test
    c. impact test
    d. hardness test

12. A typical American Welding Society (AWS) welder qualification test is a V-groove weld with metal backing in the 3G and 4G positions using an _____ electrode.
    a. F4
    b. F3
    c. F2
    d. F1

13. An ASME welder qualification test that qualifies a welder to make all types of welds is to weld pipe in the _____ position.
    a. 1F
    b. 6G
    c. 4F
    d. 4G

14. To properly prepare a test specimen, it is important to _____.
    a. water-quench the specimen while it is hot
    b. grind it widthwise to produce a notch effect
    c. remove any bead reinforcement on either the face or root side of the weldment
    d. make sure the edges are sharp

15. If you are directed to perform a weld for which you are not qualified, you should _____.
    a. bypass the chain of command
    b. report directly to the safety engineer as soon as possible
    c. try to resolve the matter with your immediate supervisor
    d. perform the weld to the best of your ability

Scott Esmeier

Vice President of Production
Schuff Steel Company
Phoenix, AZ

Scott attended Mountainview High School in Mesa, Arizona, where he became involved in the welding program at East Valley Institute of Technology. In his last year of high school, Scott worked in a welding shop earning school credit. After graduation, he went to work as a laborer for Able Steel Fabricators in Mesa. He worked his way up to plant superintendent during fourteen years with Able. Scott then left Able and went to work for Schuff Steel Company in Phoenix. Hired as a general foreman, he has worked his way up to vice president of production after eight years with the company.

How did you become interested in the welding industry?
My family got me started. My dad was a boilermaker. In high school I took every shop class I could because I really didn't have any ambitions to go to college. I knew I wanted to be a craftsman in the trades. I decided on welding after I tried some different classes.

What do you think it takes to be a success in your trade?
Focus. That's one of the main things. You've got to set long-term goals for yourself. And you need consistency, not just in the welding trade, but in any trade. If you're consistent in what you do and you have ambition, there are doors open to you. All you have to do is knock on them. Anyone who's willing to learn can be taught.

What are some of the things you do in your job?
I oversee all scheduling and the complete operations of two different plants, with 350 employees who work directly under me.

What do you like most about your job?
Every day is different. It's never boring, because you're not only dealing with the physical aspects of a welding and fabrication shop, but you're dealing with the people aspects. Every day is a challenge.

What would you say to someone entering the trade today?
Get into it early. Get into it through a school program so that you have a good understanding of what you're getting into. It's not as glamorous as some people make it out to be, but a person can make a very good living doing it. Get into some training programs and really understand the processes. There are some good shadowing programs out there. You have to try it to find out if welding is what you really want to do.

Trade Terms Introduced in This Module

Defect: A discontinuity or imperfection that renders a part of the product or the entire product unable to meet minimum acceptable standards or specifications.

Discontinuity: A change or break in the shape or structure of a part that may or may not be considered a defect, depending on the code.

Embrittled: Metal that has been made brittle and that will tend to crack with little bending.

Hardenable materials: Metals that have the ability to be made harder by heating and then cooling.

Homogeneity: The quality or state of being homogeneous (having a uniform structure or composition throughout).

Inclusion: Foreign matter introduced into and remaining in a weld.

Laminations: Cracks in the base metal formed when layers separate.

Procedure qualification: The demonstration that welds made following a specific process can meet prescribed standards.

Procedure qualification record (PQR): The document containing the results of the nondestructive and destructive testing required to qualify a welding procedure specification (WPS).

Radiograph: Photograph made by passing X-rays or gamma rays through an object and recording the variations in density on photographic film.

Underbead crack: Crack in the base metal near the weld but under the surface.

Example Procedure Qualification Record

QW-483 SUGGESTED FORMAT FOR PROCEDURE QUALIFICATION RECORD (pqr)
(See QW-200.2, Section IX, ASME Boiler and Pressure Vessel Code)
Record Actual Conditions Used to Weld Test Coupon.

Company Name _____

Procedure Qualification Record No. _____ Date _____

WPS No. _____

Welding Process(es) _____

Types (Manual, Automatic, Semi-Auto.) _____

JOINTS (QW-402)

Groove Design of Test Coupon

(For combination qualifications, the deposited weld metal thickness shall be recorded for each filler metal or process used.)

BASE METALS (QW-403)

Material Spec. _____

Type or Grade _____

P-No. _____ to P-No. _____

Thickness of Test Coupon _____

Diameter of Test Coupon _____

Other _____

_____

_____

_____

FILLER METALS (QW-404)

SFA Specification _____

AWS Classification _____

Filler Metal F-No. _____

Weld Metal Analysis A-No. _____

Size of Filler Metal _____

Other _____

Weld Metal Thickness _____

POSITION (QW-405)

Position of Groove _____

Weld Progression (Uphill, Downhill) _____

Other _____

_____

PREHEAT (QW-406)

Preheat Temp. _____

Interpass Temp. _____

Other _____

POSTWELD HEAT TREATMENT (QW-407)

Temperature _____

Time _____

Other _____

_____

_____

GAS (QW-408)

| | Percent Composition | | |
	Gas(es)	(Mixture)	Flow Rate
Shielding	_____	_____	_____
Trailing			
Backing	_____	_____	_____

ELECTRICAL CHARACTERISTICS (QW-409)

Current _____

Polarity _____

Amps _____ Volts _____

Tungsten Electrode Size _____

Other _____

_____

TECHNIQUE (QW-410)

Travel speed _____

String or Weave Bead _____

Oscillation _____

Multipass or Single Pass (per side) _____

Single or Multiple Electrodes _____

Other _____

_____

_____

_____

This form (E00007) may be obtained from the Order Dept., ASME, 22 Law Drive, Box 2300, Fairfield, NJ 07007-2300

(Source ASME B31.1, SEC IX)

104A01.TIF

QW-483 (Back)

Tensile Test (QW-150)

PQR No. _____

Specification No.	Width	Thickness	Area	Ultimate Total Load lb	Ultimate Unit Stress psi	Type of Failure & Location

Guided-Bend Tests (QW-160)

Type and Figure No.	

Toughness Tests (QW-170)

Specimen No.	Notch Location	Notch Type	Test Temp.	Impact Values	Lateral Exp.		Drop Weight	
					% Shear	Mils	Break	No Break

Fillet-Weld Test (QW-180)

Result - Satisfactory: Yes _____ No _____ Penetration into Parent Metal: Yes _____ No _____

Macro - Results _____

Other Tests

Type of Test _____

Deposit Analysis _____

Other _____

- -

Welder's Name _____ Clock No. _____ Stamp No. _____

Tests Conducted by: _____ Laboratory Test No. _____

We certify that the statements in this record are correct and that the test welds were prepared, welded, and tested in accordance with the requirements of Section IX of the ASME Code.

Manufacturer _____

Date _____ By _____

(Detail of record of tests are illustrative only and may be modified to conform to the type and number of tests required by the Code.)

Additional Resources

This module is intended to present thorough resources for task training. The following reference works are suggested for further study. These are optional materials for continued education rather than for task training.

AWS B1.10: Guide for the Nondestructive Inspection of Welds, 1999. Miami, FL: The American Welding Society.

AWS B1.11: Guide for the Visual Inspection of Welds, 2000. Miami, FL: The American Welding Society.

Modern Welding, A. D. Althouse, C. H. Turnquist, W. A. Bowditch, K. E. Bowditch, 2000. Tinley Park, IL: The Goodheart-Willcox Company, Inc.

Welding Handbook, 2001. Miami, FL: The American Welding Society.

Shielded Metal Arc Welding – Equipment and Setup

Course Map

This course map shows all of the modules in the AWS Entry Level Welder – Phase 1 curriculum. The suggested training order begins at the bottom and proceeds up. Skill levels increase as you advance on the course map. The local Training Program Sponsor may adjust the training order.

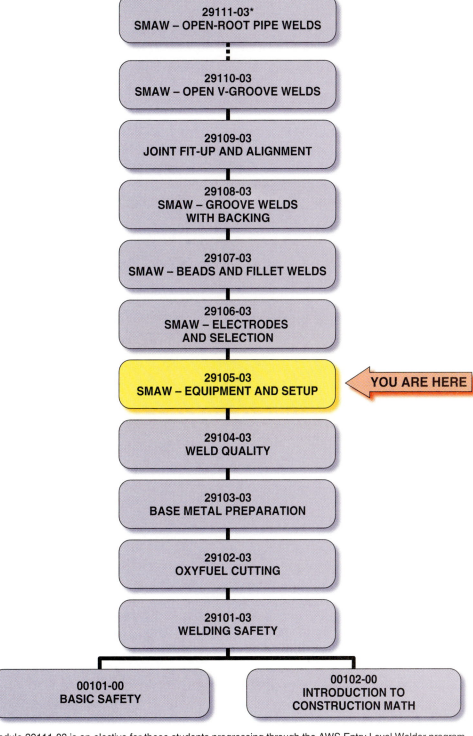

AWS ENTRY LEVEL WELDER—PHASE 1

29111-03*
SMAW – OPEN-ROOT PIPE WELDS

29110-03
SMAW – OPEN V-GROOVE WELDS

29109-03
JOINT FIT-UP AND ALIGNMENT

29108-03
SMAW – GROOVE WELDS
WITH BACKING

29107-03
SMAW – BEADS AND FILLET WELDS

29106-03
SMAW – ELECTRODES
AND SELECTION

29105-03
SMAW – EQUIPMENT AND SETUP   ◀ YOU ARE HERE

29104-03
WELD QUALITY

29103-03
BASE METAL PREPARATION

29102-03
OXYFUEL CUTTING

29101-03
WELDING SAFETY

00101-00
BASIC SAFETY

00102-00
INTRODUCTION TO
CONSTRUCTION MATH

*Module 29111-03 is an elective for those students progressing through the AWS Entry Level Welder program.

29105CMAP.EPS

Figures

Table

Shielded Metal Arc Welding – Equipment and Setup

Objectives

When you have completed this module, you will be able to do the following:

1. Identify and explain shielded metal arc welding (SMAW) safety.
2. Identify and explain welding electrical current.
3. Identify and explain arc welding machines.
4. Explain setting up arc welding equipment.
5. Set up a machine for welding.
6. Identify and explain tools for weld cleaning.

Prerequisites

Before you begin this module, it is recommended that you successfully complete the following: Modules 00100-00 through 29105-03.

Required Trainee Materials

1. Pencil and paper
2. Appropriate personal protective equipment

1.0.0 ◆ INTRODUCTION

Shielded metal arc welding (SMAW) is often referred to as stick welding. The welding process can use either **alternating current (AC)** or **direct current (DC)**, with adjustments made on the welding machine to regulate the level of current. The **electrodes**, also called rods, used are typically coated with a clay-like mixture of fluorides, carbonates, oxides, metal alloys, and binding materials to provide a gas shield for the weld puddle. This coating hardens and covers the weld during its cooling cycle to protect the weld from the atmosphere, which can cause weld degradation, and is then chipped or removed from the weld to

105F01.EPS

Figure 1 ◆ Application of SMAW.

expose the finished, shiny weld. *Figure 1* shows an SMAW welding machine properly connected to a blade on a bulldozer. The workpiece clamp is connected to the bottom edge of the blade, directly below the point of welding.

This module provides an overview of the different types of equipment used for SMAW, characteristics of the currents associated with SMAW, and some of the accessories necessary to properly connect the SMAW equipment. Upon completion of this module, you will have a better understanding of SMAW equipment and how to properly connect the equipment for welding.

2.0.0 ◆ WELDING CURRENT

The **arc** that is produced during SMAW is current that jumps across the gap from the tip of the electrode to the surface of the workpiece, as long as

Ground Clamp Location

If the welder in *Figure 1* needs to weld in a location other than the blade on the dozer, he or she should relocate the workpiece clamp. This will ensure that the welding current does not pass through the hydraulic cylinders or any other contacting surfaces that may be damaged or destroyed by arcing.

the workpiece clamp from the SMAW lead is attached to the workpiece. The current originates in the transformer or generator of the welding machine, is routed through one welding lead and then through the workpiece, and finally returns to the transformer or generator through the other welding lead. If the electrode tip is touched and held against the workpiece instead of allowing for a gap between the two, the electrode will pass the electricity directly to the workpiece without creating an arc. Maximum available current from the welding machine will be passed through the circuit. No welding will take place in this situation because an arc across the gap is needed to heat the workpiece and the electrode to a molten state. *Figure 2* shows an arc being struck using an SMAW process.

WARNING!

Never remove your face shield if the electrode is stuck or frozen to the workpiece. You must free the electrode with your face shield in place because the electrode will flash when it breaks free.

3.0.0 ◆ TYPES OF WELDING CURRENT

SMAW power supplies are designed to produce an output that has a nearly constant current or a nearly constant **voltage**. The currents may be either AC or DC. The type of current used depends on the type of welding to be done and/or the welding equipment available.

3.1.0 AC Welding Current

Alternating current (AC) in welding machines is derived from either a transformer-type welding machine or an alternator-type machine. The transformer-type machine changes high-voltage, low-current commercial AC power to low-voltage, high-current power required for welding. An alternator-type machine uses an electric or fuel-driven motor to turn a rotor inside a number of electromagnets to produce the current needed for welding.

AC current alternates between positive and negative values only because the voltage changes from positive to negative values. Referring to *Figure 3*, which is an actual meter display of 120 volts AC, the top half of the cycle of the voltage path is considered positive. The current follows the voltage. In the bottom half of the cycle, the voltage is considered negative, with the current continuing to follow the voltage. In one complete

105F02.EPS

Figure 2 ◆ Striking an arc.

Striking an Arc

Two general methods of striking the initial arc are:

- Scratching the electrode on the workpiece
- Tapping the electrode on the workpiece

The scratching method is generally used by most beginners and even by many experienced welders. However, it is usually not acceptable for code or finish-product welding unless the burn trails from scratching are within the area covered by the weld. Some of the newer inverter power sources do not activate the open-circuit voltage until after an electrode touches a surface. With these machines, a scratching motion must be accomplished in order to activate the required open-circuit voltage and start the arc.

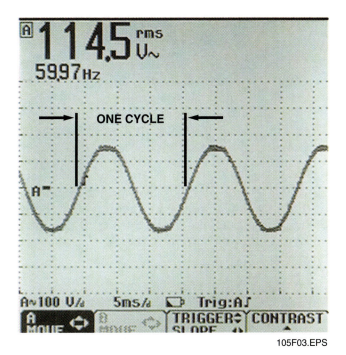

105F03.EPS

Figure 3 ◆ AC current waveform.

cycle, the wave crosses the axis or zero point twice. The peaks or extreme high or low points on the wave indicate the maximum level of the voltage, either positive (on the top) or negative (on the bottom). The number of cycles completed in one second is called the frequency. Frequency varies depending on the number of poles (magnetic field coils) installed in the generator or alternator during the generation stages of the power and the speed of the rotor rotation. The standard frequency in the United States is set at 60 cycles per second.

3.2.0 DC Welding Current

Direct current (DC) is electrical current that has no frequency. It does not travel from positive to negative and negative to positive within the same operation. The waveform for DC power resembles a straight line; therefore, it is not really a wave. Remember, frequency is associated with AC power only because frequency refers to the change in direction during one cycle, and DC power does not change direction during its flow. Direct current in welding is derived either from a transformer-rectifier-type system or from a generator. In a transformer-rectifier-type system, AC voltage first is reduced (transformed) to a lower voltage level through the application of a transformer and then converted (rectified) into DC voltage. Most gasoline-powered SMAW welding machines produce DC current through a generator. SMAW units that

Voltage Level

One cycle of our standard 60 cycles per second AC power occurs in .0167 second, with the corresponding rapid changes in voltage levels unnoticeable during the operation of machinery or in the arc of a welder. What is the voltage level of the AC power at the point where the waveform crosses the center axis?

plug into 60-cycle AC power generally produce DC current through the application of a transformer and a rectifier.

3.2.1 Polarity

Polarity only applies to DC current. Polarity in welding is determined by the way the welding leads are connected either to the welding machine or to the workpiece. The universally accepted theory is that DC current always travels from the negative pole to the positive pole, with the minus-marked terminal on the welding machine indicating the negative pole and the plus-marked terminal indicating the positive pole. In welding or in any other DC circuit, you cannot change that process. You can, however, change the path the current takes in going from the negative pole to the positive pole. In welding, this can be done simply by changing the welding lead connections at the machine or, in newer machines, by changing a switch position on the machine. Most, if not all, welding leads are configured with one end of one lead holding an electrode or rod and the other lead equipped on one end with a clamping device to attach to the workpiece. Both of the other ends of these two cables typically are equipped with ring-type or plug-in connectors for attaching the leads to the welding machine. When the electrode holder lead is connected to the plus or positive terminal on the welding machine and the lead with the clamping device is connected to the minus or negative terminal on the welding machine, direct current electrode positive (DCEP) polarity is established. If these leads are manually switched at the welding machine terminals or through the operation of the manual polarity switch located on the machine, the electrode holder lead becomes the negative lead and the workpiece clamping device lead becomes the positive lead. This is referred to as direct current electrode negative (DCEN) polarity. *Figure 4* shows DCEP and DCEN hookups.

CAUTION

On machines that have polarity switches, do not switch the polarity during operation. This may damage the machine.

DCEN generates more heat in the workpiece than at the electrode and is used for flat-position welding. When DCEP is used, more heat is generated at the electrode than in the workpiece, which results in deep, narrow weld penetration. DCEP also tends to deposit weld material at a faster rate and is recommended for out-of-position welding.

4.0.0 ◆ CHARACTERISTICS OF WELDING CURRENT

The voltage in a welding machine, whether the machine is connected to commercial power or driven by an engine, is reduced or stepped down by a transformer called a **step-down transformer**. The step-down transformer reduces the voltage level and raises the current level. The higher current allows the arc to occur with a voltage potential that doesn't carry a high risk of electrocution. However, electrical shock in welding remains a hazard that can be reduced or eliminated only by using proper personal protective equipment and applying safe welding procedures.

4.1.0 Voltage

Voltage is the measure of the electromotive force or pressure that causes current to flow in a circuit. Two types of voltage are associated with welding current: open-circuit voltage and operating voltage. Open-circuit voltage is the voltage present when the welding machine is on but no arc is present. For SMAW, open-circuit voltage is usually between 50 and 100 volts.

Properties of an Arc

An arc produces a temperature in excess of 6,000°F at the tip of an electrode. This extreme heat is what melts the base metal and electrode, producing a pool of molten metal often referred to as a crater. As the crater solidifies behind the weld, it makes the bond.

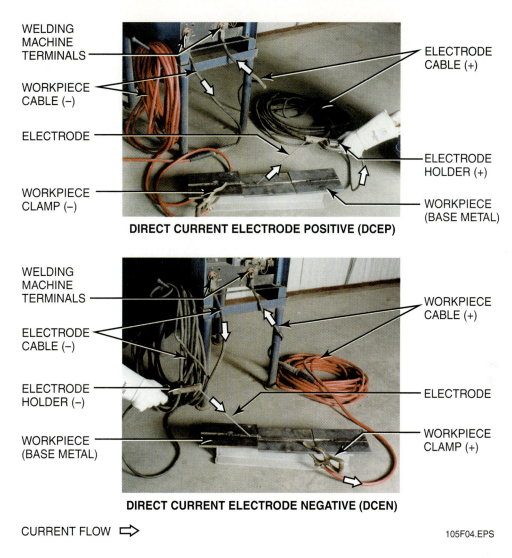

DIRECT CURRENT ELECTRODE POSITIVE (DCEP)

WELDING MACHINE TERMINALS

WORKPIECE CABLE (–)

ELECTRODE

WORKPIECE CLAMP (–)

ELECTRODE CABLE (+)

ELECTRODE HOLDER (+)

WORKPIECE (BASE METAL)

DIRECT CURRENT ELECTRODE NEGATIVE (DCEN)

WELDING MACHINE TERMINALS

ELECTRODE CABLE (–)

ELECTRODE HOLDER (–)

WORKPIECE (BASE METAL)

WORKPIECE CABLE (+)

ELECTRODE

WORKPIECE CLAMP (+)

CURRENT FLOW ⇨

105F04.EPS

Figure 4 ◆ DCEP and DCEN hookups.

Operating voltage, or arc voltage, is the voltage measured after the arc is struck. This voltage is lower than open-circuit voltage and usually measures between 18 and 45 volts. The higher open-circuit voltage is required to establish the arc because the air gap between the electrode and the work has high resistance to current flow. Once the arc is established, less voltage is needed, and the welding machine automatically compensates by lowering the voltage.

This characteristic of an SMAW machine is called variable-voltage, constant-current power and is shown as a volt-ampere output curve in *Figure 5*. Operating voltage is dependent on the current range selected. However, SMAW voltage can be altered somewhat by a change in the arc gap.

>
>
> **Note**
> Voltage can be separately adjusted on welding machines designed for gas metal arc welding (GMAW) or flux cored arc welding (FCAW). These machines will be explained in the next level.

4.2.0 Amperage

Amperage is a measurement of the electric current in a circuit. The unit of measurement for current is the ampere (amp, or A). The number of amps produced by the welding machine determines the amount of heat available to melt the work and the electrode. The current is increased

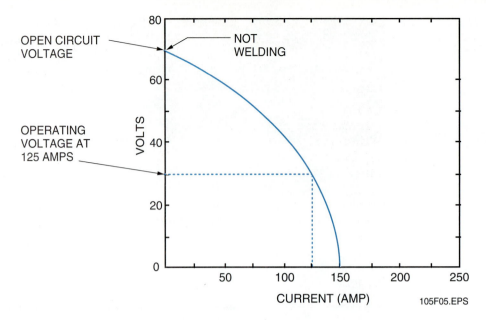

Figure 5 ◆ Volt-ampere output curve.

or decreased according to the size of electrode being used and the position in which the welding is being performed.

5.0.0 ◆ SMAW MACHINE CLASSIFICATIONS

SMAW welding machines are classified by the type of welding current they produce: AC, DC, or AC/DC.

An SMAW welding machine classified as an AC machine will produce only AC welding voltage and current, and a DC machine will produce only DC welding voltage and current. If a welding machine produces both AC and DC welding voltage and current, it is classified as an AC/DC welding machine.

Welding machines that produce only DC welding voltage and current can be further classified by the characteristics of the welding current they produce. If the welding current varies between the higher open-circuit voltage and the lower operating voltage, it is classified as a variable-voltage constant-current power source. This type of welding machine is used for SMAW and gas tungsten arc welding (GTAW). If open-circuit and operating voltage are always the same, the machine is classified as a constant-voltage or constant-potential DC welding machine. This type of machine is used for GMAW and FCAW. *Figure 6* shows the difference in outputs between variable-voltage and constant-voltage welding machines.

6.0.0 ◆ SMAW MACHINE TYPES

Several different types of basic SMAW welding machines are available to produce the current necessary for welding. They are explained in the following sections of this module and include:

- Transformers
- Transformer rectifiers
- Engine-driven generators and alternators

Transformers, transformer rectifiers, and electric motor generators all require electrical power from commercial power lines to operate. This high-voltage, low-amperage current from commercial power lines coming into the welding machine is called the **primary current**. The primary current first flows from the power lines to an electrical circuit breaker or disconnecting fuse box. From the disconnecting device, the current flows to a receptacle. The disconnecting device can be used to shut off the flow of current to the receptacle. The welding machine is plugged into the receptacle. The primary current required for welding machines can be 120- or 240-volt single-phase or 480-volt three-phase.

WARNING!
Coming into contact with the primary current of a welding machine can cause electrocution. Ensure that welding machines are properly grounded.

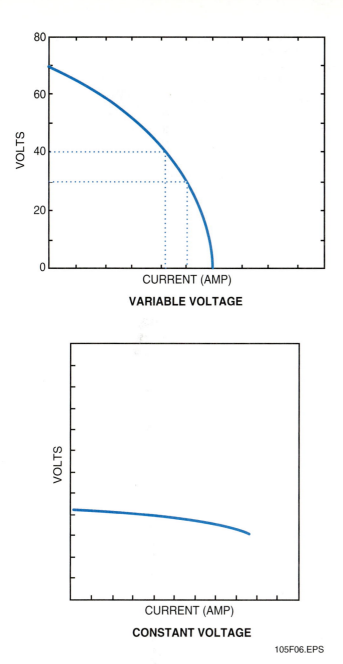

VARIABLE VOLTAGE

CONSTANT VOLTAGE

105F06.EPS

Figure 6 ◆ Variable-voltage and constant-voltage current curves.

Figure 7 shows a typical welding machine circuit.

6.1.0 Transformer Welding Machines

Transformer welding machines without rectifiers produce AC welding current. They use a voltage step-down transformer, which converts high-voltage, low-amperage current from commercial power lines to low-voltage, high-amperage welding current. The primary power required for a transformer welder can be 240-volt single-phase

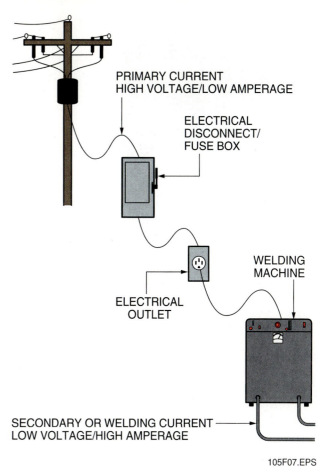

105F07.EPS

Figure 7 ◆ Welding machine primary circuit.

or 480-volt three-phase. Special light-duty transformer welding machines used for sheet metal work are designed to be plugged into a 120-volt outlet. However, most light-duty transformer welding machines operate more efficiently on 240-volt primary power. Heavy-duty industrial transformer welders require 480-volt three-phase primary power.

Transformer welders are not as common as other types of welding machines on the job site, but they are used for special jobs. A transformer welder has an ON/OFF switch, an amperage control, and terminals for connecting the electrode lead and the workpiece lead. *Figure 8* shows a typical transformer welding machine.

6.2.0 Transformer-Rectifier Welding Machines

A rectifier is a device that converts AC current to DC current. A transformer-rectifier welding machine uses a transformer to reduce the primary voltage to welding voltage and then rectifies the

Figure 8 ◆ Transformer welding machine.

105F09.EPS

Figure 9 ◆ Transformer-rectifier welding machine.

current to change it from AC to DC. Transformer-rectifier welding machines can be designed to produce AC and DC welding current or DC current only. Transformer rectifiers that produce both AC and DC welding current are usually intended for lighter duty than those that produce DC current only. Depending on their size, transformer-rectifier welding machines may require 240-volt single-phase power, 240-volt three-phase power, or 480-volt three-phase power.

WARNING!

Coming into contact with the primary circuit of a welding machine can cause electrocution. Ensure that you take proper safety precautions to avoid contact.

Transformer rectifiers can also be designed to produce variable-voltage constant-current or constant-voltage (constant-potential) DC welding current. Some are even designed to provide both

variable and constant voltage; those are referred to as multi-process power sources since they can be used for any welding process. Refer to the manufacturer's documentation to determine how the machine is marked.

Transformer rectifiers have an ON/OFF switch and an amperage control. The welding cables (electrode lead and workpiece lead) are connected to terminals. They often have selector switches to select DCEP or DCEN or to select AC welding current if it is available. If there is no selector switch, the cables must be manually changed on the machine terminals to select the type of current desired. *Figure 9* shows a typical transformer-rectifier welding machine.

Multiple transformer rectifiers are available grouped into a single cabinet, called a pack. When the pack contains six welding machines, it is called a six-pack; when it contains eight welding machines, it is called an eight-pack (*Figure 10*).

The pack in *Figure 10* has eight individual welding machines. The pack has ON/OFF push buttons and a red power-on indicator. Each individual welding machine has its own amperage control and welding cables. In addition, each welding machine has a receptacle for a remote

AC/DC Machines

Transformer-rectifier welding machines that can provide either AC or DC current do not utilize the rectifier circuit when welding with AC current. In this situation, the welding machine functions like a transformer welding machine.

Figure 10 ◆ Eight-pack welding machine.

current control. When the remote current control is plugged in, it can be used to adjust the amperage from a remote location without going back to the pack. The pack unit also has 120-volt receptacles for power tools and lights.

Packs are mounted on skids and have lifting eyes so they can be moved easily. They require only one primary power connection, which is normally 480-volt three-phase. Packs are common on construction sites and existing facilities during rebuilds when many welding machines are required.

6.3.0 Engine-Driven Generator and Alternator Machines

Welding machines can also be powered by gasoline or diesel engines. The engine can be connected to a generator or to an alternator. Engine-driven generators produce DC welding current. Engine-driven alternators produce AC current, which is then fed through a rectifier to produce DC welding current.

To produce welding current, the generator or alternator must turn at a required number of revolutions per minute (rpm). Engines that power alternators and generators have governors to con-trol the engine speed. Most governors have a welding speed switch. The switch can be set to idle the engine when no welding is taking place. When the electrode is touched to the base metal, the governor will automatically increase the speed of the engine to the required rpm for welding. After about 15 seconds of no welding, the engine will automatically return to idle. The switch can also be set to enable the engine to run continuously at welding speed. *Figure 11* shows an example of an engine-driven generator welding machine.

Engine-driven generators and alternators often have an auxiliary power unit to produce 120-volt AC current for lighting, power tools, and other electrical equipment. When 120-volt AC current is required, the engine-driven generator or alternator must run continuously at welding speed.

Engine-driven generators and alternators have engine controls and welding current controls. The engine controls vary with the type and size but normally include:

- Starter
- Voltage gauge
- Temperature gauge
- Fuel gauge
- Hour meter

Inverter Power Sources

Inverter power sources increase the frequency of the incoming primary power. This provides a smaller, lighter power supply with a faster response time and much more waveform control. Inverter power sources are used where limited space and portability are important factors. The controls on these machines vary according to their size and application. An AC/DC unit with a 300A capability that weighs approximately 100 pounds is shown here. It has extensive waveform control for SMAW as well as for GMAW, GTAW, and FCAW.

105P0501.EPS

Engines

The size and type of engine used depends on the size of the welding machine. Single-cylinder engines are used to power small rectifier alternators. Four- and six-cylinder engines are used to power larger generators.

Engine-driven generators and alternators have an amperage control. To change polarity, there may be a polarity switch, or you may have to manually change the welding cables at the welding current terminals.

Many engine-driven generators are mounted on trailers, which makes them portable so they can be used in the field where electricity is not available for other types of welding machines. The disadvantage is that engine-driven generators and alternators are costly to purchase, operate, and maintain because of the added cost of the drive unit.

7.0.0 ◆ SMAW MACHINE RATINGS

The size of a welding machine is determined by the amperage output of the machine at a given duty cycle. The duty cycle of a welding machine is based on a ten-minute period. It is the percentage of ten minutes that the machine can continuously produce its rated amperage without overheating. For example, a machine with a rated output of 300A at 60% duty cycle can deliver 300A of welding current for six minutes out of every ten (60% of ten minutes) without overheating.

Figure 11 ◆ Engine-driven machine.

The duty cycle of a welding machine will be 10%, 20%, 30%, 40%, 60%, or 100%. A welding machine having a duty cycle of 10% to 40% is considered a light-duty to medium-duty machine. Most industrial, heavy-duty machines for manual welding have a 60% duty cycle. Machines designed for automatic welding operations have a 100% duty cycle.

With the exception of 100% duty cycle machines, the maximum amperage that a welding machine will produce is always higher than its rated capacity. A welding machine rated 300A at 60% duty cycle will generally put out a maximum of 375A to 400A. However, since the duty cycle is a function of its rated capacity, the duty cycle will decrease as the amperage is raised over 300A. Welding at 375A with a welding machine rated at 300A with a 60% duty cycle will lower the duty cycle to about 30%. If welding continues for more than three out of ten minutes, the machine will probably overheat.

If the amperage is set below the rated amperage, the duty cycle increases. Setting the amperage at 200A for a welding machine rated at 300A with a 60% duty cycle will increase the duty cycle to 100%. *Figure 12* shows the relationship between amperage and duty cycle.

Welding Machine Circuit Breakers

Most welding machines have a heat-activated circuit breaker that will shut off the machine automatically when it overheats. The machine cannot be turned on again until it has cooled.

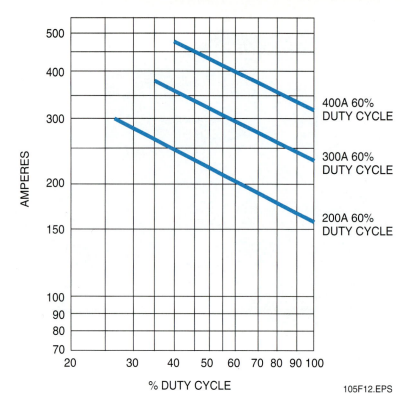

Figure 12 ◆ Amperage and duty cycle.

8.0.0 ◆ WELDING CABLE

Cables that carry welding current are designed for maximum strength and flexibility. The conductors inside the cable are made of fine strands of copper wire. The copper strands are covered with layers of rubber reinforced with nylon or dacron cord. *Figure 13* shows a cutaway section of welding cable.

The size of a welding cable is based on the number of copper strands it contains. Large-diameter cable has more copper strands and can carry more welding current. Typically the smallest welding cable size is No. 8 and the largest is No. 4/0 (four aught).

When selecting welding cable size, the amperage load and the distance the current will travel must be taken into account. The longer the distance the current has to travel, the larger the cable must be to reduce voltage drop and heating caused by electrical resistance. When selecting welding cable, use the rated capacity of the welding machine for the cable amperage requirement. To determine the distance, measure both the electrode and ground leads and add the two lengths together. To identify the welding cable size required, refer to a recommended welding cable size table furnished by most welding cable manufacturers. *Table 1* shows typical welding cable sizes for various welding currents and cable distances.

8.1.0 Welding Cable End Connections

Welding cables must be equipped with the proper end connections. End connections used on welding cables include lugs, quick disconnects, ground clamps, and electrode holders.

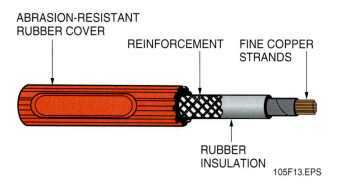

Figure 13 ◆ Welding cable.

 CAUTION

If the end connection is not tightly secured to the cable, the connection will overheat. An overheated connection will cause variations in the welding current and permanent damage to the connector and/or cable. Repair connectors that overheat.

Table 1 Welding Cable Sizes

Machine Size In Amperes	Duty Cycle (%)	Recommended Cable Sizes for Manual Welding				
		Copper Cable Sizes for Combined Lengths of Electrodes Plus Ground Cable				
		Up to 50 Feet	50–100 Feet	150 Feet	150–200 Feet	200–250 Feet
100	20	#8	#4	#3	#2	#1
180	20	#5	#4	#3	#2	#1
180	30	#4	#4	#3	#2	#1
200	50	#3	#3	#2	#1	#1/0
200	60	#2	#2	#2	#1	#1/0
225	20	#4	#3	#2	#1	#1/0
250	30	#3	#3	#2	#1	#1/0
300	60	#1/0	#1/0	#1/0	#2/0	#3/0
400	60	#2/0	#2/0	#2/0	#3/0	#4/0
500	60	#2/0	#2/0	#3/0	#3/0	#4/0
600	60	#3/0	#3/0	#3/0	#4/0	* * *
650	60	#3/0	#3/0	#4/0	* *	* * *

* * Use Double Strand of #2/0

* * * Use Double Strand of #3/0

8.1.1 Lugs and Quick Disconnects

Lugs are used at the end of the welding cable to connect the cable to the welding machine current terminals. Lugs come in various sizes to match the welding cable size and are mechanically crimped onto the welding cable. These lugs are normally ringed to prevent the welding lead from pulling loose from the stud connector on the machine.

Quick disconnects are also mechanically connected to the cable ends. They are insulated and serve as cable extensions for splicing two lengths of cable together. Quick disconnects are connected or disconnected with a half twist. When using quick disconnects, take care to ensure that they are tightly connected to prevent overheating or arcing in the connector. They must be kept clean of dirt, sand, mud, and other contaminants in order to assure a good connection. *Figure 14* shows typical lugs and quick disconnects used as welding cable connectors.

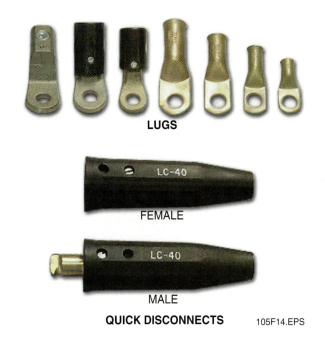

LUGS

FEMALE

MALE

QUICK DISCONNECTS   105F14.EPS

Figure 14 ◆ Lugs and quick disconnects.

8.1.2 Workpiece Clamps

Workpiece clamps establish the connection between the end of the workpiece and the workpiece lead. Workpiece clamps are mechanically connected to the welding cable and come in a variety of shapes and sizes. The size of a workpiece clamp is the rated amperage that it can carry without overheating. When selecting a workpiece clamp, be sure it is rated at least as high as the rated capacity of the power source on which it will be used. *Figure 15* shows examples of workpiece clamps.

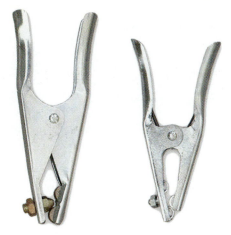

LOW-CURRENT CLAMPS

HIGH-CURRENT CLAMPS

105F15.EPS

Figure 15 ◆ Workpiece clamps.

HEAD ROTATES ON HANDLE TO RELEASE COLLET-TYPE ELECTRODE CLAMP OR TO ADJUST CLAMP FOR ELECTRODE SIZE

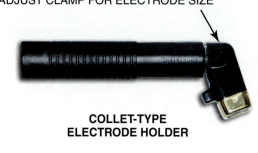

COLLET-TYPE ELECTRODE HOLDER

LEVER-OPERATED JAW

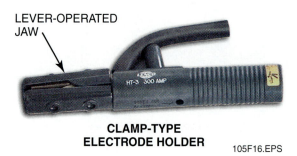

CLAMP-TYPE ELECTRODE HOLDER

105F16.EPS

Figure 16 ◆ Electrode holders.

8.1.3 Electrode Holders

Electrode holders grasp the electrode and provide the electrical contact between the electrode and the end of the welding cable. Electrode holders are also mechanically connected to the welding cable and come in a variety of styles and sizes. The size of an electrode holder is the rated amperage it will carry without overheating. The physical size of an electrode holder is also proportional to its amperage capacity. A small 200A electrode holder will be light and easy to handle but will not hold large electrodes that require more than 200A. When selecting an electrode holder, select the size for the amperage and electrodes being used. *Figure 16* shows two styles of electrode holders.

9.0.0 ◆ SMAW EQUIPMENT SETUP

In order to safely and efficiently weld using the SMAW process, the equipment must be properly selected and set up. The following subsections explain the steps for selecting and setting up SMAW equipment for welding.

9.1.0 Selecting the Proper SMAW Equipment

To select the proper SMAW machine, the following factors must be considered:

- *The welding process* – As previously shown, SMAW requires a variable-voltage power source with a constant current. Other processes may require constant-voltage power sources, but we are focusing on SMAW in this module.
- *The type of welding current (AC or DC)* – Most SMAW processes use DC current.
- *The maximum amperage required* – Consider the composition of the material, its thickness, and other material properties to determine the amperage requirements of the machine.
- *The primary power requirements* – If AC power with the proper receptacle for the welding machine is not available, then a portable or machine-driven machine must be used.

9.2.0 Locating a Welding Machine

The welding machine should be located near the work to be performed. Select a site where the machine will not be in the way but will be pro-

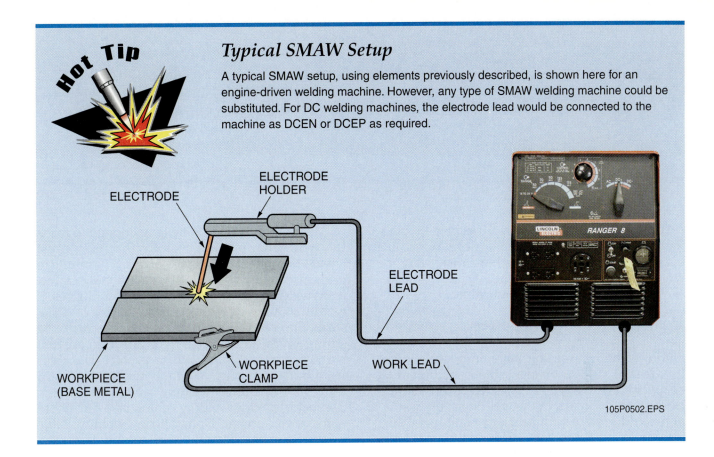

Typical SMAW Setup

A typical SMAW setup, using elements previously described, is shown here for an engine-driven welding machine. However, any type of SMAW welding machine could be substituted. For DC welding machines, the electrode lead would be connected to the machine as DCEN or DCEP as required.

Hot Tip

ELECTRODE

ELECTRODE HOLDER

ELECTRODE LEAD

WORK LEAD

WORKPIECE CLAMP

WORKPIECE (BASE METAL)

RANGER 8

105P0502.EPS

---

tected from welding or cutting sparks. There should be good air circulation to keep the machine cool, and the environment should be free from explosive or corrosive fumes and as free as possible from dust and dirt. Welding machines have internal cooling fans that will pull these materials into the welding machine if they are present. The site should be free of standing water or water leaks. If an engine-driven generator or alternator is used, locate it so it can be easily refueled and serviced.

WARNING!
Never run an engine-driven generator or alternator indoors because it can cause carbon monoxide poisoning.

There should also be easy access to the site so the machine can be started, stopped, and adjusted as needed. If the machine is to be plugged into an outlet, be sure the outlet has been properly installed by a licensed electrician to ensure it is grounded. Also, be sure to identify the location of the electrical disconnect for the outlet before plugging the welding machine into it.

9.3.0 Moving a Welding Machine

Large engine-driven generators are mounted on a trailer frame and can be easily moved by a pickup truck or tractor using a trailer hitch. Other types of welding machines may have a skid base or may be mounted on steel or rubber wheels. When moving welding machines that are mounted on wheels, use care. Some machines are top-heavy and may fall over in a tight turn or if the floor or ground is uneven or soft.

WARNING!
If a welding machine starts to fall over, do not attempt to hold it. Welding machines are very heavy. You can be severely injured if a welding machine falls on you.

Many welding machines have lifting eyes. Lifting eyes are usually provided to move machines mounted on skids. Before lifting any welding machine, refer to the equipment specifications for the weight. Be sure the lifting device and tackle are rated at more than the weight of the machine. When lifting a welding machine, always use a

Moving Welding Machines on Wheels

It is not uncommon on industrial construction projects to move all welding machines using cranes or cherry pickers (small hydraulic boom equipment), even those on trailers or machines with wheels, because of the limited access for vehicular traffic in these facilities.

shackle. Never attempt to lift a machine by placing the lifting hook in the machine eye. Before lifting or moving a welding machine, make sure the welding cables are coiled and secure. *Figure 17* shows how to lift a welding machine.

9.4.0 Stringing Welding Cable

Before stringing welding cable, inspect the cable for damage. Cuts or breaks in the insulation must be repaired with electrical tape before the welding cable is used. Breaks in the welding cable insulation could arc to equipment or to any metal surface the welding contacts. This will damage the surface and cause additional damage to the welding cable. If the welding cable cannot be repaired, it should be discarded.

Welding cables should be long enough to reach the work area but not so long that they must always be coiled. Welding cables must be strung to prevent tripping hazards or damage from traffic. Keep welding cables out of walkways and aisles. If welding cables must cross a walkway or aisle, string the cables overhead. If the cables cannot be strung overhead, use boards or ramps to prevent workers from tripping and to protect the cables from both foot and vehicular traffic. *Figure 18* shows how to protect welding cables with a ramp.

When stringing welding cables, take care not to damage equipment. Do not string welding cable over instrumentation wires and tubing. These can be easily damaged by the weight or movement of the cables. If there are moving parts, keep the welding cables well away from nip points. If the equipment suddenly starts, the welding cables could be pulled into the equipment. When there is new construction or rebuild work taking place, equipment such as cable trays, conduit, piping, and pipe hangers may not be permanently attached. The weight of welding cables could cause these items to collapse. Check such equipment before stringing welding cables.

When stringing welding cable overhead or between floors, use rope to tie off the cables. The weight of the welding cables could damage con-

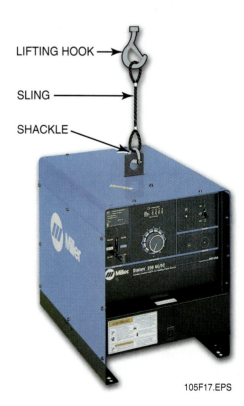

LIFTING HOOK

SLING

SHACKLE

105F17.EPS

Figure 17 ◆ Lifting a welding machine.

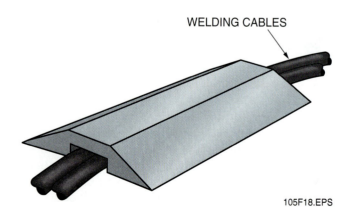

WELDING CABLES

105F18.EPS

Figure 18 ◆ Protecting welding cables.

nections or pull an individual holding onto the electrode holder off balance. Rope is relatively nonconductive and nonabrasive, and it will support the weight of the welding cable without damaging it.

9.5.0 Locating the Workpiece Clamp

The workpiece clamp must be properly located to prevent damage to surrounding equipment. If the welding current travels through any type of bearing, seal, valve, or contacting surface, it could cause severe damage from arcing, which would require that the item be replaced. Carefully check the area to be welded and position the workpiece clamp so the welding current will not pass through any contacting surface. If in doubt, ask your supervisor for assistance before proceeding.

CAUTION

Welding current can severely damage bearings, seals, valves, and contacting surfaces. Position the workpiece clamp to prevent welding current from passing through them.

Welding current passing through electrical or electronic equipment will cause severe damage. Disconnect the workpiece lead at the battery to protect the electrical system before welding on any type of mobile equipment. If welding is per-

formed near the battery, remove the battery. Batteries produce hydrogen gas, which is extremely explosive. A welding spark could cause the battery to explode.

WARNING!

Do not weld near batteries. A welding spark could cause a battery to explode, showering the area with battery acid.

Before welding on or near electronic or electrical equipment or cabinets, contact an electrician. The electrician will isolate the system to protect it.

CAUTION

The slightest spark of welding current can destroy electronic or electrical equipment. Have an electrician check the equipment and, if necessary, isolate the system before welding.

Workpiece clamps must never be connected to pipes carrying flammable or corrosive materials. The welding current could cause overheating or sparks, resulting in an explosion or fire.

The workpiece clamp must make good electrical contact when it is connected. Dirt and paint will inhibit the connection and cause arcing,

Trailer Mud Flaps

A welder was given the job of reattaching a mud flap to the back of a refrigerated trailer. The loaded trailer was parked right next to the welding machine, and it looked to be a very simple job. Because it was so close, the employee reached over and connected the workpiece clamp to the front of the trailer and then pulled the cable with the electrode holder to the back end. After cleaning and positioning the bracket, the mud flap was welded to the trailer. The entire job took less than 15 minutes to complete.

When the customer returned to pick up the trailer, he noticed that the refrigeration unit was off and there was no power at the display panel. A service technician determined that the microprocessor on the refrigeration unit had failed and needed to be replaced. After further investigation, it was determined that the pressure and temperature sensors on the refrigeration unit had provided a path for the welding current to get to the microprocessor. The current was too high for the microprocessor and burned out some of its components. As a result, the welding shop was required to pay more than $1,000 in parts and labor to replace the microprocessor.

The Bottom Line: If the welder had isolated the electronic components from the welding circuit, or at least had located the workpiece clamp close to the weld joint, then the damage would not have occurred.

resulting in overheating of the workpiece clamp. Dirt and paint also affect the welding current and can cause defects in the weld. Clean the surface before connecting the workpiece clamp. If the workpiece clamp is damaged and does not close securely onto the surface, change it.

9.6.0 Energizing Electrically Powered Welding Machines

Electrically powered welding machines are plugged into an electrical outlet. The electrical requirements (primary current) will be on the equipment specification tag displayed prominently on the machine. Most machines will require single-phase 240-volt current or three-phase 480-volt current. Machines requiring single-phase 240-volt power will have a three-prong plug. Machines requiring three-phase 480-volt power will have a four-prong plug. *Figure 19* shows grounded electrical plugs for welding machines.

If a welding machine does not have a power plug, an electrician must connect the machine. The electrician will add a plug or will hard-wire the machine directly into a disconnecting device.

WARNING!

Never use a welding machine until you identify the electrical disconnect. In the event of an emergency, you must be able to quickly turn off the power to the welding machine at the disconnect to prevent injury or even death.

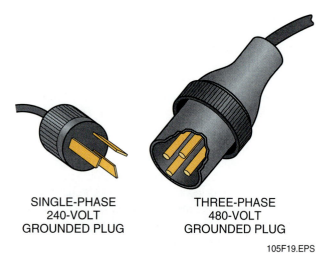

SINGLE-PHASE 240-VOLT GROUNDED PLUG
THREE-PHASE 480-VOLT GROUNDED PLUG

105F19.EPS

Figure 19 ◆ Grounded electrical plugs.

9.7.0 Starting Engine-Driven Welding Machines

Before welding can take place with an engine-driven welding machine, the engine must be checked and then started. As with a car engine, the engine powering the welding machine must also have routine maintenance performed.

9.7.1 Prestart Checks

Many facilities will have prestart checklists that must be completed and signed prior to starting and operating an engine-driven welding machine. If your site has such a checklist, you must complete and sign it. If your site does not have a prestart checklist, perform the following checks before starting the engine.

- Check the oil level using the engine oil dipstick. If the oil is low, add the proper amount of appropriate grade oil. Do not overfill.
- Check the coolant level in the radiator if the engine is liquid-cooled. If the coolant level is low, add coolant. Do not remove the radiator cap on a hot engine.

CAUTION

Do not add plain water to radiators that contain antifreeze. Antifreeze not only protects radiators from freezing in cold weather, it also has rust inhibitors and additives to aid in cooling. If the antifreeze is diluted, it will not function properly. If the weather turns cold, the system may freeze, causing damage to the radiator, engine block, and water pump.

- Check the fuel level. The unit may have a fuel gauge or a dipstick. If the fuel is low, add the correct fuel, diesel or gasoline, to the fuel tank. The type of fuel required should be marked on the fuel tank. If not, contact your supervisor to verify the fuel required and have the tank marked.

CAUTION

Adding gasoline to a diesel engine or diesel to a gasoline engine will cause severe engine problems and can cause a fire hazard. Always be sure to add the correct type of fuel to the fuel tank.

- Check the battery water level unless the battery is sealed. Add water if the battery water level is low.

WARNING!

Battery cell vapors can be explosive. Use extreme care when working with or around a serviceable battery. Avoid exposing the battery to sparks or arcs of any type. Many serious accidents are reported annually due to batteries exploding in workers' faces, caused by accidental arcing or sparking.

- Check the electrode holder to be sure it is not touching the workpiece. If the electrode holder is touching the workpiece, it will arc and overheat the welding system when the welding machine is started.
- Open the fuel shutoff valve if the equipment has one. If there is a fuel shutoff valve, it will be located in the fuel line between the fuel tank and the carburetor.
- Record the hours from the hour meter if the equipment has one. An hour meter records the total number of hours the engine runs. This information is used to determine when the engine needs to be serviced. The hours will be displayed on a gauge similar to an odometer.
- Clean the unit. Use a compressed air hose to blow off the engine and the generator or alternator. Use a rag to remove heavier deposits that cannot be removed with the compressed air.

CAUTION

Never attempt to clean or blow off the engine or areas near the engine while the engine is running. Drive belts and other moving and rotating engine or equipment parts may accelerate loosened debris into eyes or other body parts, possibly causing severe injury.

Note

Clean the engine and equipment after each use.

9.7.2 Starting the Engine

Most engines will have an ON/OFF ignition switch and a starter. They may be combined into a key switch similar to the ignition switch on a car. To start the engine, turn on the ignition switch and press the starter. Release the starter when the engine starts. The engine speed will be controlled by the governor. If the governor switch is set for idle, the engine will slow to an idle after a few seconds. If the governor is set to welding speed, the engine will continue to run at the welding speed.

Small engine-driven welding machines may have an ON/OFF switch and a pull rope. These are started by turning on the ignition switch and pulling the cord, which is similar to starting a lawn mower. These engines do not have a battery.

Engine-driven welding machines should be started about five to ten minutes before they are needed for welding. This will allow the engine to warm up before a welding load is placed on it.

WARNING!

When you start the welding machine, always make sure the welding electrode and/or electrode holder are secure and not touching any conductive surface or worker.

WARNING!

Engine cowlings, or covers, should be closed and latched prior to starting the machine to protect workers from accidental contact with moving parts.

9.7.3 Stopping the Engine

If no welding is required for 30 or more minutes, stop the engine by turning off the ignition switch. If you are finished with the welding machine for the day, also close the fuel valve if there is one. Allow the machine to cool before storing leads inside the engine compartment.

9.7.4 Preventive Maintenance

Engine-driven welding machines require regular preventive maintenance to keep the equipment operating properly. Most sites will have a preventive maintenance schedule based on the hours that the engine operates. In severe conditions, such as a very dusty environment or cold weather, maintenance may have to be performed more frequently.

Responsibility for performing preventive maintenance will vary from site to site. Check with your supervisor to determine who is responsible for performing preventive maintenance.

When performing preventive maintenance, follow the manufacturer's guidelines in the equipment manual. Typical tasks to be performed as part of a preventive maintenance schedule include:

- Changing the oil
- Changing the gas filter
- Changing the air filter
- Checking/changing the antifreeze
- Greasing the undercarriage
- Repacking the wheel bearings
- Replacing worn drive belts
- Maintaining/replacing the spark plugs
- Maintaining the battery
- Cleaning surfaces

10.0.0 ◆ TOOLS FOR CLEANING WELDS

Even though some of the hand tools used to clean welds were presented in the tools section, it is worthwhile to mention some of these tools again and to explain their functions. Hand tools such as files, chipping hammers, and wire brushes, as well as small power tools such as flux chippers and scalers, are common tools used by welding personnel to prepare surfaces prior to welding and also to clean welds after they are completed.

10.1.0 Hand Tools

It is extremely important that the base metals be thoroughly cleaned and prepared prior to welding. Dirty or corroded surfaces are very difficult, if not impossible, to weld properly. Likewise, before any additional weld beads can be applied over previous beads, the previous pass must be cleaned of any slag that was deposited along with the weld bead. Many mechanical methods have been developed to assist in these cleaning and removing processes, including sandblasting and ultrasonic cleaning. However, many welders prefer simpler manual procedures involving tools such as files, wire brushes, and chipping hammers. *Figure 20* shows some of these popular hand tools used by welders to prepare base metals prior to welding and to remove slag after the bead is laid.

105F20.EPS

Figure 20 ◆ Typical hand tools used in welding.

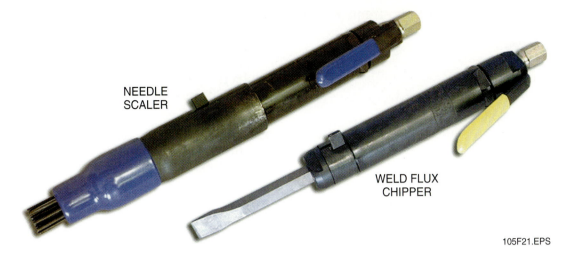

NEEDLE SCALER

WELD FLUX CHIPPER

105F21.EPS

Figure 21 ◆ Weld flux chipper and pneumatic needle scaler.

10.2.0 Pneumatic Cleaning and Slag Removal Tools

Weld slag chippers and needle scalers *(Figure 21)* are pneumatically powered. They are used by welders to clean surfaces and to remove slag from cuts and welds. Weld slag chippers and needle scalers are also excellent for removing paint or hardened dirt but are not very effective for removing surface corrosion. Weld slag chippers have a single chisel, and needle scalers have about 18 to 20 blunt steel needles approximately 10" long. Most weld slag chippers can be converted to needle scalers with a needle scaler attachment.

Summary

Shielded metal arc welding is often referred to as stick welding and remains one of the most widely used forms of welding in the industry because of its simplicity and relatively low cost. However, it is not as foolproof as some of the new technology now used in the welding industry. Proper selection and setup of equipment and accessories are necessary in order to achieve a reliable weld while maintaining a safe working environment.

This module described the welding currents associated with SMAW, the various types of SMAW machines, and some of the accessories needed to connect SMAW equipment. The steps in selecting and setting up SMAW equipment were also discussed. SMAW is the basis on which all other welding processes have been developed, and understanding its process is mandatory in achieving success in the other processes.

Review Questions

1. SMAW is often referred to as _____.
   a. AC welding
   b. stick welding
   c. gap welding
   d. resistive welding

2. During SMAW welding, current jumps across the _____ from the tip of the electrode to the surface of the workpiece.
   a. workpiece clamp
   b. gap
   c. electrode holder
   d. molten metal

3. The type of current derived from either a transformer-type welding machine or an alternator-type machine is _____.
   a. direct current
   b. DC voltage
   c. alternating current
   d. AC or DC current

4. An AC waveform crosses the axis or zero point _____ in one complete cycle.
   a. once
   b. twice
   c. three times
   d. four times

5. The number of cycles completed in one second is called the _____.
   a. polarity
   b. travel
   c. frequency
   d. current value

6. Electrical current that has no frequency is known as _____.
   a. DC current
   b. AC current
   c. AC voltage
   d. 120 volts AC

7. _____ only applies to DC current.
   a. Frequency
   b. Polarity
   c. Amperage
   d. Current flow

8. The voltage in a welding machine is reduced or stepped down by a(n) _____.
   a. voltage limiter
   b. generator
   c. alternator
   d. transformer

9. _____ voltage is present when the welding machine is on but no arc is present.
   a. Open-circuit
   b. Operating
   c. Zero
   d. Minimum

10. If welding current varies between higher open-circuit voltage and lower operating voltage, the current is classified as a _____ power source.
    a. constant-voltage
    b. constant-current
    c. fluctuating-current
    d. varying-current

11. Most light-duty transformer welding machines operate more efficiently on _____ primary power.
    a. 480-volt three-phase
    b. 120-volt single-phase
    c. 220 volts AC
    d. DC

12. A(n) _____ changes alternating current to direct current.
    a. transformer
    b. rectifier
    c. alternator
    d. generator

13. Governors on engine-driven generator and alternator machines directly control the _____ of the machine.
    a. temperature
    b. speed
    c. rectifier
    d. primary power

14. The duty cycle of a welding machine is based on a _____ period.
    a. 10-minute
    b. 15-minute
    c. 20-minute
    d. 30-minute

15. Welding cable lugs are generally _____ to prevent the lead from pulling loose from the welding machine terminal.
    a. slotted
    b. soldered in place
    c. ringed
    d. insulated

Bennett B. (Ben) Grimmett, P.E.

Department–Engineering and NDE
Senior Engineer
Edison Welding Institute (EWI)

Born in Amherst, Ohio, Ben Grimmett was one of four brothers. His father was a maintenance mechanic at Ford Motor Company, performing welding both on the job and at home. Though he had only a sixth-grade education, Ben's father was "one of the cleverest individuals I have ever known," according to Ben. After taking a vocational welding and fabrication program in high school, Ben attended Ohio State University and earned a B.S. in welding engineering. Now senior engineer at Edison Welding Institute, Ben recently earned his master's degree in business administration.

How did you become interested in the welding industry?
As a kid, I remember watching my dad build a swing set out of old pipes and chains. As he was welding the pieces together, I remember watching through the welding hood and being amazed by the process and the associated phenomena–the noises, the sparks, the bright lights. I was quickly hooked on it, even at that young age. It was fascinating to a 10-year-old boy. Since then, welding has always been and will remain an extremely interesting science and technology to me. Welding is truly one of the most under-appreciated technologies in the world.

I was also fortunate enough to have an older engineer take an active interest in my career when I was at Terra Technical College. Jim Harris, who works as a senior staff engineer at Ashland Chemical, became my mentor and helped guide me at a time in my life when I needed someone to help me find direction. Jim continues to be an advisor, confidante, and one of my best friends.

What do you think it takes to be a success in your trade?
Make solutions, not excuses. If you make a mistake, accept it as a learning experience. Nobody wants to hear an excuse. What people want to hear is how you are going to rectify the situation.

If you are not experiencing some failure in life, then you simply aren't doing anything. You need to be able to make decisions, accept responsibility for those decisions, and be willing to see them through to completion. Nothing in the world is more annoying than an engineer who does not make decisions.

Find a mentor. A mentor gives you someone to discuss your goals in life with and who can offer advice on how to achieve them. Mentoring is a two-way relationship. A mentor takes an active, vested interest in seeing you succeed, is never jealous, and takes great satisfaction in sharing your success.

What are some of the things you do in your job?
I evaluate and develop welding methods, testing procedures, training materials and programs, and welding solutions to unique problems and situations. I also develop cost estimates, proposals, and scopes-of-work for contractor research and engineering consulting. I even serve as an expert witness for court cases involving equipment failures.

What do you like most about your job?
I like the variety, the depth of the projects, and the ability to work with some of the most talented people in welding today.

What would you say to someone entering the trade today?
Be able to roll up your sleeves and talk to the person on the floor, then be able to turn around and talk to the CEO of the company–in language both the floorperson and the senior manager understand. Do not be afraid to challenge yourself and take some risks to get noticed. I have had some absolutely stunning fiascos in my life, but each one provided a unique learning opportunity and a chance to evaluate my principles.

Always think long term and big picture, but be able to see the trees and the forest so that you can communicate effectively. Also, be able to sit back and laugh at yourself.

If you are not enjoying your job, find something else. You are going to end up spending a whole bunch of time working; you might as well enjoy it. If you can find someone to mentor you, grab that opportunity and make the most of it. To have someone take an active interest in your life and help guide you is a benefit that cannot be matched by any college or work experience.

Have fun; you only get one chance in this world, so make the most of it. Don't be afraid of failure.

Trade Terms Introduced in This Module

Alternating current (AC): Electrical current that reverses its flow at set intervals.

Amperage: A measurement of the rate of flow of electric current.

Arc: The flow of electricity through an air gap or gaseous space.

Direct current (DC): Electrical current that flows in one direction only.

Electrode: Terminal point to which electricity is brought in the welding operation and from which the arc is produced to do the welding.

Polarity: The direction of flow of electrical current in a direct current welding circuit.

Primary current: Electrical current received from conventional power lines used to energize welding machines.

Step-down transformer: An electrical device that uses two wire coils of different sizes to convert a high-voltage, low-current power source into a lower voltage, higher current output.

Voltage: A measurement of electromotive force or pressure that causes current to flow in a circuit.

Additional Resources

This module is intended to present thorough resources for task training. The following reference works are suggested for further study. These are optional materials for continued education rather than for task training.

OSHA Standard 1926.351, Arc Welding and Cutting.
The Procedure Handbook of Arc Welding, 1994. Cleveland, OH: The Lincoln Electric Company.

CONTREN™ LEARNING SERIES—USER UPDATES

The NCCER makes every effort to keep these textbooks up-to-date and free of technical errors. We appreciate your help in this process. If you have an idea for improving this textbook, or if you find an error, a typographical mistake, or an inaccuracy in NCCER's Contren™ textbooks, please write us, using this form or a photocopy. Be sure to include the exact module number, page number, a detailed description, and the correction, if applicable. Your input will be brought to the attention of the Technical Review Committee. Thank you for your assistance.

Instructors—If you found that additional materials were necessary in order to teach this module effectively, please let us know so that we may include them in the Equipment/Materials list in the Instructor's Guide.

Write: Curriculum Revision and Development Department
National Center for Construction Education and Research
P.O. Box 141104, Gainesville, FL 32614-1104

Fax: 352-334-0932

E-mail: curriculum@nccer.org

Craft _____ Module Name _____

Copyright Date _____ Module Number _____ Page Number(s) _____

Description _____

_____

_____

_____

(Optional) Correction _____

_____

_____

(Optional) Your Name and Address _____

_____

_____

Shielded Metal Arc Welding – Electrodes and Selection

Course Map

This course map shows all of the modules in the AWS Entry Level Welder – Phase 1 curriculum. The suggested training order begins at the bottom and proceeds up. Skill levels increase as you advance on the course map. The local Training Program Sponsor may adjust the training order.

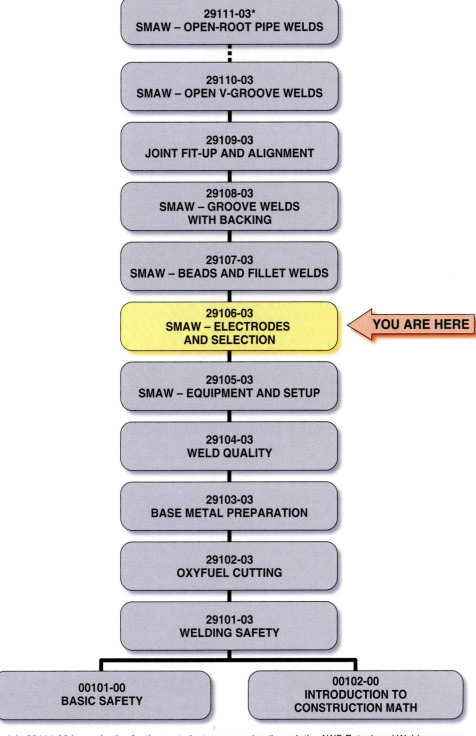

AWS ENTRY LEVEL WELDER—PHASE 1

29111-03*
SMAW – OPEN-ROOT PIPE WELDS

29110-03
SMAW – OPEN V-GROOVE WELDS

29109-03
JOINT FIT-UP AND ALIGNMENT

29108-03
SMAW – GROOVE WELDS
WITH BACKING

29107-03
SMAW – BEADS AND FILLET WELDS

29106-03
SMAW – ELECTRODES
AND SELECTION ◄ YOU ARE HERE

29105-03
SMAW – EQUIPMENT AND SETUP

29104-03
WELD QUALITY

29103-03
BASE METAL PREPARATION

29102-03
OXYFUEL CUTTING

29101-03
WELDING SAFETY

00101-00
BASIC SAFETY

00102-00
INTRODUCTION TO
CONSTRUCTION MATH

*Module 29111-03 is an elective for those students progressing through the AWS Entry Level Welder program.

29106CMAP.EPS

Figures

Tables

Shielded Metal Arc Welding – Electrodes and Selection

Objectives

When you have completed this module, you will be able to do the following:

1. Identify factors that affect electrode selection.
2. Explain the American Welding Society (AWS) and the American Society of Mechanical Engineers (ASME) filler metal classification system.
3. Identify different types of filler metals.
4. Explain the storage and control of filler metals.
5. Explain filler metal traceability requirements and how to use applicable code requirements.
6. Identify and select the proper electrode for an identified welding task.

Prerequisites

Before you begin this module, it is recommended that you successfully complete Modules 00101-00 through 29105-03.

Required Trainee Materials

1. Pencil and paper
2. Appropriate personal protective equipment

1.0.0 ◆ INTRODUCTION

Shielded metal arc welding (SMAW) cannot take place without electrodes. SMAW has often been referred to as stick welding, and the electrodes have been referred to as stick electrodes. The welder must be able to distinguish different types of electrodes and select the correct electrode for the job. In addition, the welder must know how to handle and store electrodes, some of which require special considerations.

This module explains the electrode classification system, electrodes, and the considerations for selecting the correct electrode for the job. It also explains how to properly store electrodes in compliance with the appropriate welding codes.

2.0.0 ◆ SHIELDED METAL ARC WELDING ELECTRODES

A shielded metal arc welding electrode has a wire core coated with **flux.** The wire core transfers the welding current from the electrode holder to the work. The arc at the end of the electrode melts the metal core, the flux coating, and the base metal at a temperature exceeding 6,000°F. The wire core mixes with the melted base metal, forming the majority of the weld. As the flux coating is melted, it:

- Gives off a gaseous shield of carbon dioxide that protects the molten weld by pushing away harmful oxygen and hydrogen present in the atmosphere
- Acts as a fluxing agent to clean and deoxidize the molten metal
- Stabilizes the arc and reduces spatter
- Forms a slag covering over the weld to further protect it from the atmosphere and slow the cooling rate

Figure 1 shows an illustration of an electrode in use.

Some flux coatings also have powdered metal and **alloying** elements in them. The powdered metal adds filler metal and helps improve the appearance of the weld. The alloying elements change the chemical composition and strength of the deposited weld metal. Alloys are normally added to electrodes so the deposited weld metal will match the base metal chemical composition and strength.

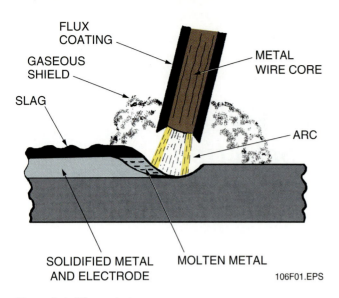

Figure 1 ◆ Electrode in use.

3.0.0 ◆ AWS FILLER METAL SPECIFICATION SYSTEM

The American Welding Society (AWS) writes specifications for welding consumables, such as electrodes, filler rods, and fluxes, used with all the various welding processes. There are a series of different specifications identified as *AWS A5.XX*. These specifications are used in all major welding codes and accepted in all industries except where other approvals are specified, such as the American Bureau of Shipping (ABS), the U.S. Coast Guard, the Federal Highway Administration, Military Specifications (MIL), and Lloyds of London. The American Society of Mechanical Engineers (ASME) Boiler Code electrode specifications closely follow the AWS specifications.

The purpose of the specifications is to set standards that all manufacturers must follow when manufacturing welding consumables. This ensures consistency for the user regardless of which company manufactured the product. The specifications set standards for the:

- Classification system, identification, and marking
- Chemical composition of the deposited weld metal
- Mechanical properties of the deposited weld metal

Some examples of AWS specifications that pertain to shielded metal arc welding electrodes are:

- *A5.1* – Specification for Carbon Steel Covered Arc-Welding Electrodes
- *A5.3* – Specification for Aluminum and Aluminum Alloy Electrodes for Shielded Metal Arc Welding

- *A5.4* – Specification for Stainless Steel Welding Electrodes for Shielded Metal Arc Welding
- *A5.5* – Specification for Low-Alloy Steel Electrodes for Shielded Metal Arc Welding
- *A5.6* – Specification for Covered Copper and Copper Alloy Electrodes
- *A5.11* – Specification for Nickel and Nickel-Alloy Welding Electrodes for Shielded Metal Arc Welding
- *A5.15* – Specification for Welding Electrodes and Rods for Cast Iron
- *A5.16* – Specification for Titanium and Titanium Alloy Welding Electrodes and Rods
- *A5.21* – Specification for Bare Electrodes and Rods for Surfacing
- *A5.24* – Specification for Zirconium and Zirconium Alloy Welding Electrodes and Rods

The most common electrodes are discussed in the *A5.1* and *A5.5* classifications. The electrodes described in the *A5.1* classification are used to weld carbon (mild) steel. The electrodes in the *A5.5* classification are used to weld low-alloy, high-strength steels. These two classifications will be explained later in this module.

Note

The classification number is generally followed by a hyphen and a two-digit number, such as *A5.1-91* or *A5.5-96*. The hyphen and number indicate the year that the classification was last revised. These documents are normally on a five-year revision cycle. Always use the latest version.

3.1.0 Classification System

The classification system used to identify SMAW *A5.1* carbon steel electrodes and *A5.5* low-alloy electrodes uses the prefix E followed by four or five numbers, such as E6010 or E10018. This is printed on each electrode near the end where the electrode holder grasps the electrode. *Figure 2* shows electrode markings.

The meanings of the markings are as follows:

- *E* stands for electrode to signify use with electric welding.
- The first two digits, or three if 100 or over, multiplied by 1,000 designate the minimum tensile strength in pounds per square inch (psi) of the deposited weld metal:
  - E6010 has a minimum tensile strength of 60,000 psi.
  - E7018 has a minimum tensile strength of 70,000 psi.

Sizing and Packaging of SMAW Electrodes

SMAW electrodes are sized according to the wire core diameter and are commonly packaged in 50-pound metal cans, as shown in the photo.

106P0601.EPS

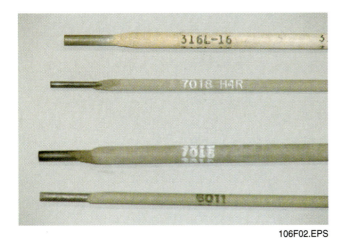

106F02.EPS

Figure 2 ◆ Electrode markings.

– E12018 has a minimum tensile strength of 120,000 psi.
• The third or fourth digit indicates the possible welding positions:
  – *1* can be used to weld in all positions.
  – *2* can be used to weld in flat and horizontal positions only.
  – *4* can be used to weld in flat, horizontal, overhead, and downhill vertical positions.
• The fourth or fifth digit indicates special characteristics of the flux coating and the type of welding current required, such as alternating current (AC), direct current electrode negative (DCEN), or direct current electrode positive (DCEP).

Figure 3 shows the electrode marking system.

Table 1 shows the characteristics of mild steel electrodes covered by the *A5.1* specification.

The low-alloy steel electrodes that are covered by *A5.5* have a suffix in addition to the standard classification numbers. The suffix has a letter and a number. The letter indicates the chemical composition of the deposited weld metal, and the number indicates the composition of the chemical classification. *Figure 4* shows the marking on an electrode from the *A5.5* classification.

The meanings of the suffix letters are:

- *A* – carbon-molybdenum alloy steel
- *B* – chromium-molybdenum alloy steel
- *C* – nickel steel alloy
- *D* – manganese-molybdenum alloy steel
- *G* – other alloys with minimal elements

The meaning of the numbers varies by alloy type but always indicates the percentages of the various alloying elements, such as:

- Carbon (C)
- Manganese (Mn)
- Phosphorus (P)
- Sulphur (S)
- Silicon (Si)

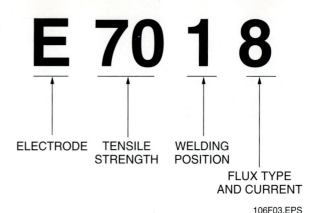

106F03.EPS

Figure 3 ◆ Electrode marking system.

Determining Tensile Strength

What factor determines the selection of an electrode with the proper tensile strength? How is this tensile strength determined?

Table 1 Mild Steel Electrode Characteristics

AWS Class	Flux Coating	Current Requirements	Position	Characteristics
EXX 10	CELLULOSE SODIUM	DCEP	All	DEEP PENETRATION, FLAT BEADS
EXX 20			FLAT, HOR. FILLET	
EXX 11	CELLULOSE POTASSIUM	AC, DCEP	All	DEEP PENETRATION, FLAT BEADS
EXX 12	TITANIA SODIUM	AC, DCEN	All	MEDIUM PENETRATION
EXX 13	TITANIA POTASSIUM	AC, DCEP, DCEN	All	SHALLOW PENETRATION
EXX 14	TITANIA IRON POWDER	AC, DCEP, DCEN	All	MEDIUM PENETRATION, FAST DEPOSIT
EXX 24			FLAT, HOR. FILLET	
EXX 15	LOW-HYDROGEN SODIUM	DCEP	All	MODERATE PENETRATION
EXX 16	LOW-HYDROGEN POTASSIUM	AC, DCEP	All	MODERATE PENETRATION
EXX 27	IRON POWDER, IRON OXIDE	AC, DCEP, DCEN	FLAT, HOR. FILLET	MEDIUM PENETRATION
EXX 18	IRON POWDER LOW-HYDROGEN	AC, DCEP	All	SHALLOW TO MEDIUM PENETRATION
EXX 28			FLAT, HOR. FILLET	

Figure 4 ◆ Marking on a low-alloy, high-strength electrode.

- Nickel (Ni)
- Chromium (Cr)
- Molybdenum (Mo)
- Vanadium (V)

Figure 5 shows the electrode marking system for low-alloy, high-strength electrodes.

Figure 6 shows examples of various electrodes available listed by alloy content.

3.2.0 Manufacturers' Classification

All electrodes manufactured in the United States must have the AWS classification number printed on them as well as on the container in which they are shipped. In addition, electrode manufacturers generally print their own unique classification name or number on the electrode and/or on the container in which the electrodes are shipped. Manufacturers may also make more than one electrode within the same AWS classification. For example, Lincoln manufactures two electrodes with the AWS classification E6010. One is called Fleetweld 5P, and the other is called Fleetweld 5P+.

Lincoln recommends Fleetweld 5P for general fabrication and maintenance welding and states that it is good for pipe. However, Lincoln states that Fleetweld 5P+ is the best choice for pipe. The major difference between the two electrodes is the mechanical properties of the deposited weld metal. Even though both fall into the E6010 classi-

fication, 5P+ has a hardness lower than 5P. When 5P+ is used for the root pass, this lower hardness results in improved resistance to **heat-affected zone** cracking. Also, the 5P+ has slightly higher tensile strength, yield, and elongation.

When there is more than one electrode in an AWS classification and you are unsure about which electrode to use, check with your supervisor.

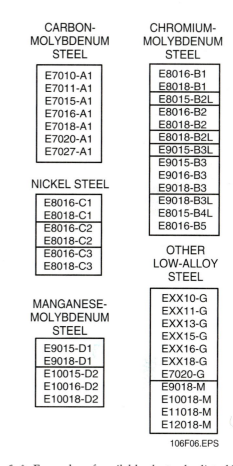

Figure 6 ◆ Examples of available electrodes listed by alloy.

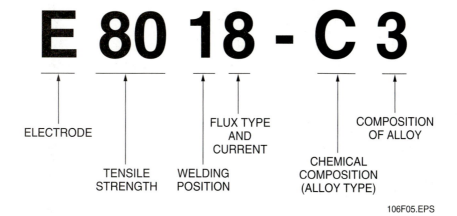

Figure 5 ◆ Example of marking system.

Material Safety Data Sheet (MSDS) for Electrodes

Every box or can of welding electrodes must contain a copy of the MSDS for that product, similar to this one for an E7018 electrode. The complete MSDS for another type of electrode is shown in the *Appendix*.

HAYNES
International

SAFETY DEPARTMENT
1020 WEST PARK AVENUE
P.O. BOX 9013
KOKOMO, INDIANA 46904-9013
INFORMATION: 765-456-6614

MATERIAL SAFETY DATA SHEET

HAYNES INTERNATIONAL, INC.
Welding Consumables

MSDS IDENTIFICATION NUMBER **H1072-4** This replaces H1072-3	PREVIOUS REVISION DATE 01/07/97 DATE REVISED 07/31/99	EMERGENCY PHONE NUMBERS HAYNES: 765-456-6894 CHEMTREC: 800-424-9300 (24-hour contact for Health & Transportation Emergencies)

This Material Safety Data Sheet (MSDS) provides information on a specific group of manufactured metal products. Since these metal products share a common physical nature and constituents, the data presented are applicable to all alloys identified. This document was prepared to meet the requirements of OSHA's Hazard Communication Standard, 29 CFR 1910.1200 and the Superfund Amendments and Reauthorization Act of 1986 Public Law 99-949.

Ingredients:	CAS No.	Wt.%	TLV mg/m³	PEL mg/m³
Iron	7439-89-6	15	10*	10*
Limestone and/or calcium carbonate	1317-65-3	10	10	15
Fluorides (as F)	7789-75-5	<5	2.5	2.5
Silicates and other binders	1344-09-8	<5	10*	10*
Manganese and/or manganese alloys and compounds (as Mn)***	7439-96-5	<5	0.2	1.0(c)
Titanium dioxides (as Ti)***	13463-67-7	<5	10	10
Silicon and/or silicon alloys and compounds (as Si)	7454-21-3	1	10*	10*
Mineral silicates	1332-58-7	0.5	5**	5**
Aluminum oxide and/or Bauxite***	1344-28-1	<0.5	10	10
Carbon steel core wire	7439-89-6	60	10*	10*

106UA0601.EPS

Table 2 Manufacturers' and AWS Classifications for Common Mild-Steel Electrodes

Manufacturer	AWS Classification				
	E6010	E6011	E6013	E7014	E7018
Hobart	Pipemaster 60	335A, 335C	447A, 447C	14A	718, 718C
Airco	Pipecraft	6011, 6011C	6013, 6013D	7014	Easy Arc 7018MR
ESAB	AP100, SW-610, SW-10P	SW-14	SW-15	SW-15P	Atom Arc 7018
Lincoln	Fleetwood 5P, 5P+	Fleetwood 35, 35LS, 180	Fleetwood 37, 57	Fleetwood 47	Jetweld LH-70
McKay	6010M	6011	6013	7014	—

Manufacturers' Welding Guides

Most electrode manufacturers publish paperback welding guides that include information on their electrodes such as recommended amperage settings and uses. These booklets can be obtained through your local distributor for free or at a nominal charge.

Table 2 shows some examples of manufacturers' and AWS classifications for common mild-steel electrodes.

3.3.0 Electrode Sizes

Mild steel electrodes for SMAW are available in standard sizes, usually ranging from ³⁄₃₂" to ¼". They are typically packaged in 50-pound cans, as shown in *Figure 7*. The amperage application for which the electrodes are used is proportional to the diameter size of the electrode. The number of electrodes in a 50-pound can varies with the diameter of the can and the electrodes. The length of the electrode may vary from manufacturer to manufacturer. *Table 3* shows a table, taken from the McKay® electrode catalog, listing some mild steel electrodes and the number of electrodes in one pound. A 50-pound can would have 50 times this number as its contents. For example, there would be 750 McKay-type 6013, ⅛" diameter electrodes in a 50-pound can. As noted in the second row from the top of the table, these electrodes are 14" in length.

4.0.0 ◆ SELECTING ELECTRODES

Because so many different electrodes are available, the welder must be able to distinguish between them and select the correct electrode for the job. The following sections will explain how to select electrodes.

SIDE VIEW

TOP VIEW

106F07.EPS

Figure 7 ◆ 50-pound can of electrodes.

4.1.0 Electrode Groups

Electrodes are often classified into one of four groups based on their general characteristics. The groups are:

- Fast-freeze
- Fast-fill
- Fill-freeze
- Low-hydrogen

Table 3 Number of Electrodes per Pound

McKay Type	Diameter: 3/32" (2.4mm) Length: 14"	1/8" (3.2mm) 14"	5/32" (4.0mm) 14"	3/16" (4.8mm) 14"	7/32" (5.6mm) 18"	1/4" (6.4mm) 18"
6010 PM	30	17	12	8	—	—
6011, Soft-Arc 6011	25	15	11	7	—	—
6013	25	15	10	7	—	—
7014	24	13	9	6	—	—
7024	—	10	7	—	4	2
7018 XLM, Soft-Arc 7018-1	32	15	10	7	—	3
7018 AC, Millennia 7018	32	15	10	7	4	3

4.1.1 Fast-Freeze Electrodes

Fast-freeze electrodes are all-position electrodes that provide deep penetration. However, the arc has a lot of fine weld spatter. The weld bead is flat with distinct ripples and a light slag coating that can be difficult to remove. The arc is easy to control, making fast-freeze electrodes good for vertical and overhead welding. Electrodes in the fast-freeze group include:

- E6010
- E7010
- E6011

Typical applications for fast-freeze electrodes are:

- Vertical-up and overhead plate welding
- Pipe welding (cross-country and in-plant)
- Welds made on galvanized, plated, or painted surfaces
- Joints requiring deep penetration, such as square butts
- Sheet metal welds

WARNING!

Many surface coatings, such as galvanized coatings, give off very toxic fumes when heated, cut, or welded. Fumes from certain types of electrodes may also be hazardous. Breathing these fumes can cause damage to the lungs and other organs. Identify any hazard for the coatings with which you are working by referring to the MSDS for that particular coating. Also, read the MSDS contained in the electrode package. Take the necessary precautions to protect yourself.

4.1.2 Fast-Fill Electrodes

Fast-fill electrodes have powdered iron in the flux, which gives them high deposition rates but requires that they only be used for flat welds and horizontal fillet welds. These electrodes have shallow penetration with excellent weld appearance and almost no spatter. The heavy slag covering is easy to remove. Electrodes in the fast-fill group include:

- E6027
- E7024
- E7028 (also in the low-hydrogen group)

Typical applications for fast-fill electrodes are:

- Production welds on plate more than 1/4" thick or thicker
- Flat and horizontal fillets and lap welds

4.1.3 Fill-Freeze Electrodes

Fill-freeze electrodes, also called fast-follow electrodes, are all-position electrodes that are commonly used for downhill or flat welding. They have medium deposition rates and penetration and are excellent for sheet metal. The weld bead ranges from smooth and ripple-free to even with distinct ripples. The medium slag covering is easy to remove.

Electrodes in the fill-freeze group include:

- E6012
- E6013
- E7014

Typical applications for fill-freeze electrodes are:

- Downhill fillet or lap welds
- Sheet metal lap and fillet welds
- Joints having poor fit-up in the flat position
- General purpose welding

4.1.4 Low-Hydrogen Electrodes

Low-hydrogen electrodes are designed for welding high-sulphur, phosphorus, and medium- to high-carbon steels, which have a tendency to develop porosity and cracks under the weld bead. These problems are caused by hydrogen that is absorbed during welding. Low-hydrogen electrodes are available with fast-fill and fill-freeze characteristics. They produce dense, X-ray-quality welds with excellent **notch toughness** and **ductility.** Electrodes in the low-hydrogen group include:

- E7016
- E7018
- E7028
- E7048

Typical applications for low-hydrogen electrodes are:

- X-ray-quality welds
- Welds requiring high mechanical properties
- Crack-resistant welds in medium- to high-carbon steels
- Welds that resist hot-short cracking in phosphorus steels
- Welds that minimize porosity in sulphur steels
- Welds in thick or highly restrained mild or alloy steels to minimize the danger of weld cracking
- Welds in alloy steels requiring a strength of 70,000 psi or more
- Multiple-pass welds in all positions

106F08.EPS

Figure 8 ◆ Storing low-hydrogen electrodes.

If hydrogen is present in the weld zone, it will create defects. The greatest source of hydrogen is water moisture, which is composed of oxygen and hydrogen. Low-hydrogen electrodes are carefully manufactured to eliminate moisture in the flux coating. If low-hydrogen electrodes are exposed to the atmosphere, they will start to absorb moisture and will no longer be low-hydrogen. For this reason, all low-hydrogen electrodes are shipped in **hermetically sealed** metal containers. Once opened, the electrodes must be stored in a heated oven. *Figure 8* shows low-hydrogen electrodes properly stored. Guidelines for the storage and handling of low-hydrogen electrodes are covered later in this module.

CAUTION

The welding codes establish strict guidelines for the storage, handling, and redrying of low-hydrogen electrodes. These procedures will be referenced in your site quality standards and welding procedure specifications (WPSs). Severe penalties will result from violation of these standards. Improperly storing these electrodes will lead to poor-quality welds. Be sure you read and understand your site standards before working with low-hydrogen electrodes.

4.2.0 Electrode Selection Considerations

Many factors enter into the proper selection of electrodes for the job. The available welding equipment and its current ranges limit the choice of electrodes, and base metal properties play an important role in the selection. Also, the position in which the weld must be made and design factors affect the choice of electrodes for a particular application. Taken from Hobart's welding catalog, *Table 4* indicates position, polarity, and common usages for some of the common steel electrodes, which are listed by AWS classification.

4.2.1 Welding Procedure Specification

If there is a WPS for the weld to be made, the filler metal (electrode) will be specified. It will be given as an AWS classification and/or manufacturer's standard and will sometimes include a manufacturer's name. The filler metal specified in the WPS must be used.

Table 4 Common Steel Electrode Selection Factors

AWS Class	Position	Polarity	Usage
E6011	All	AC, DCEN, DCEP	Good for dirty, rusty steel
E6013	All	AC, DCEN, DCEP	Good for thin steel and poor fit-up
E7014	All	AC, DCEN, DCEP	Good for thin steel
E7018	All	AC, DCEN, DCEP	Good for high deposition; good for medium- or high-carbon steels

DCEN – DC Electrode Negative (Straight Polarity)
DCEP – DC Electrode Positive (Reverse Polarity)

4.2.2 Base Metal Type

The filler metal should match the base metal's chemical composition and mechanical properties. The chemical composition and mechanical properties of the base metal can be obtained by referring to the base metal mill specifications. The mill specifications can then be compared to various tables that list base metal and filler metal acceptability. These tables are available in a number of codes, such as the *AWS Structural Welding Code D1.1, Annex M, Code-Approved Base Metals and Filler Metals*, and the *ASME Boiler and Pressure Vessel Code, Section II, Parts A, B, and C, Material Specifications*. Tables are also available from the electrode or base metal manufacturers. If the filler metal and base metal are not compatible, the weld produced will be defective.

A weld made on mild steel with a high-tensile electrode such as E12018 will be weaker than one made with E7018. The high-tensile weld will be so strong that when stress is placed on the weld, it will not give, transferring all of the stress to the heat-affected zone along the weld. This will cause cracking. A weld made with E7018 will give along with the base metal, distributing the stress across the weld, making a stronger joint. The opposite occurs when mild steel electrodes are used to weld high-strength steel. All the stress occurs in the weld because the surrounding base metal is so much stronger, causing the weld to fail prematurely.

A simple test can be used to determine electrode and base metal compatibility. Using the base metal to be welded, fit a T-joint together. Run a short weld on one side of the T-joint. After the weld cools, strike the vertical section of the T-joint opposite the weld. If the base metal and electrode are compatible, the weld should break. This occurs because the small weld is the weakest point. To prevent the weld from breaking, all you have to do is add more weld. If the electrode and base metal are not compatible, the weld will tear out of the base metal at the toe along the heat-affected zone. Regardless of how much weld you apply, it will not hold if the weld breaks out in the heat-affected zone. *Figure 9* shows the base metal compatibility test.

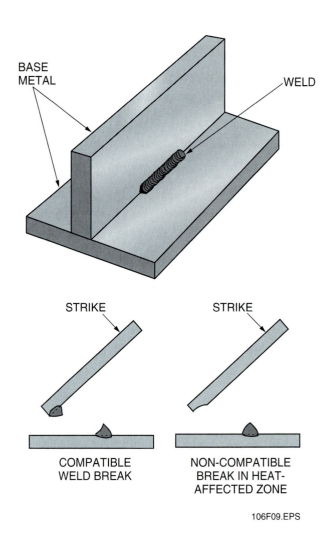

Figure 9 ◆ Base metal compatibility test.

4.2.3 Base Metal Thickness

Normally, an electrode that is smaller in diameter than the thickness of the metal to be welded is used. For sheet metal, a ³⁄₃₂" E6013 electrode may be recommended because it has shallow penetration. For very thick sections, always use a low-hydrogen electrode for its ductile qualities. To minimize joint preparation, a deep penetration E6010 or E7010 electrode can be used for the root pass.

4.2.4 Base Metal Surface Condition

When possible, the base metal surface should always be cleaned before welding. Surface corrosion, plating, coatings, and paint should be removed. If this is not possible, use a deep penetrating electrode, such as an E6010.

WARNING!

Many surface coatings, such as galvanized coatings, give off very toxic fumes when heated, cut, or welded. Breathing these fumes can cause damage to the lungs and other organs. Identify any hazard for the coatings with which you are working by referring to the MSDS for that particular coating. Take the necessary precautions to protect yourself.

4.2.5 Welding Position

The position of welding is important because some electrodes, such as E7024 and E7028, can only be used in the flat and horizontal fillet weld positions. In addition, welding in vertical and overhead positions should never be attempted with electrodes larger than ³⁄₁₆", except for low-hydrogen electrodes. The limitation for low-hydrogen electrodes is ⁵⁄₃₂". As a general rule, the joint should be positioned for flat welding so you can use the largest electrode possible.

4.2.6 Joint Design

The type of joint to be welded is an important consideration when selecting an electrode. When T- and lap joints can be positioned flat, they are usually welded with E7024 or E7028 electrodes. Butt joints that need deep penetration should be welded with E6010 electrodes. If the joint has poor fit-up, use a shallow penetrating electrode such as an E6012 or E6013. If low-hydrogen electrodes are used for open root joints, back gouging may be required. To eliminate the need for back gouging

and to make the root pass easier to run, use metal backing or put the root in with an E6010 electrode and then fill out with a low-hydrogen electrode.

4.2.7 Welding Current

The most common type of welding current is DC. Because all electrodes can be run on DC, there is no problem when DC current is available. Welding machines that can produce only AC current cannot use E6010 or E7010 electrodes. Although most other electrodes will run on AC, they will run better using DC current.

The amount of current produced by the welding machine is also a consideration. Running large diameter electrodes that require high amperages near the maximum output of the welding machine will cause the welding machine to overheat and cut out. Check the welding machine current type, rated output, and duty cycle before selecting the electrode size.

5.0.0 ◆ FILLER METAL STORAGE AND CONTROL

Electrodes used for shielded metal arc welding must be dry. The dry mineral flux coatings on electrodes start to absorb moisture as soon as they are exposed to the atmosphere. In as little as one-half hour, some electrodes can pick up enough moisture from the atmosphere to cause weld defects. Some of these defects, such as underbead cracking, may not be visible to the naked eye.

Low-hydrogen electrodes pick up moisture faster than any other type of electrode. Also, because of the nature of the base metal welded with low-hydrogen electrodes, a smaller amount of moisture will cause defects. For these reasons, special care must be used when handling and storing low-hydrogen electrodes.

5.1.0 Code Requirements

All welding codes require electrodes with a low-hydrogen coating to receive special handling. They must be shipped in hermetically sealed containers. Immediately after opening the sealed containers, low-hydrogen electrodes must be stored in special ovens at a temperature of 250°F (120°C). When low-hydrogen electrodes are removed from the oven, they can only be exposed to the atmosphere for a designated period of time. The period of time varies with the classification of the electrode but generally ranges from one-half hour to four hours. The electrodes must then be returned to the oven and generally require redrying at higher temperatures. The exact procedures for handling filler metal at your site are typically written out in a quality standard. These procedures must be followed. If you are unsure of the requirements at your site, check with your supervisor.

CAUTION

The consequences of not following your site's standards on electrode handling and storage may be severe. Be sure you understand your site's standards before handling electrodes. Improper handling of electrodes can cause weld defects. The information provided in this module is of a general nature. Your site's particular quality standards must be followed.

5.2.0 Receiving Filler Metal

Inspect all filler metal as it is received. All low-hydrogen electrodes (EXX15, 16, or 18), stainless steel electrodes, and nickel alloy coated electrodes must be supplied in hermetically sealed containers. If the seal has been broken, reject the electrodes. Mark them *Not to Be Used* and notify your supervisor.

All non-low-hydrogen electrodes must be shipped in moisture-resistant containers. Inspect these containers for damage. The electrodes can be used if their coatings are not damaged and they have not come into direct contact with water or oil. Mark electrodes that cannot be used and notify your supervisor.

Each container must be marked with at least the following information:

- ASME, AWS, or other electrode specification number and electrode classification designation
- Electrode size
- Weight

If the containers are not properly marked, mark them *Not to Be Used* and notify your supervisor.

5.3.0 Storing Filler Metal

All filler metal must be stored in a warm (40°F minimum), dry storage area. Filler metals are not to be stored directly on the floor. Electrodes that are stored on the floor will form **condensation.** Use pallets or shelves to store electrodes off the floor. They must also be stored in such a manner that they will not be damaged. Once low-hydrogen, stainless steel, or nickel alloy electrode containers have been opened, the electrodes must immediately be stored in an oven.

5.4.0 Storage Ovens

Electrode storage ovens are typically electrical and are controlled by a thermostat. Storage ovens for large shops can hold several hundred pounds of electrodes and can reach maximum temperatures of more than 800°F. This maximum temperature is too high for most electrodes. Always follow the electrode manufacturer's guidelines for storing electrodes in ovens. Smaller portable ovens that are designed to be used on the job may hold only 25 or 30 pounds of electrodes. These ovens have a maximum temperature of 300°F. Depending on the WPS or job quality standards, portable ovens may be required to hold electrodes at the point of use. The electrodes remaining in the portable ovens after performing the task must be returned to the holding oven. *Figure 10* shows an electrode storage oven.

Storage ovens should be set at about 300°F (minimum temperature 250°F). The oven must remain on at all times. Never unplug an oven or plug it into an outlet that could accidentally be shut off.

5.4.1 Exposure Times

The time that a low-hydrogen, stainless steel, or nickel alloy electrode can be exposed to the atmosphere is limited and depends on the electrode. It is the welder's responsibility to take out of the oven only the number of electrodes that can be

Electrode Mix-Up

Do not mix electrode types in an oven. If electrodes of various types must be stored in the same oven, keep them organized by placing them on separate shelves.

PORTABLE OVEN

FLOOR- OR TABLE-
MOUNTED OVEN

106F10.EPS

Figure 10 ◆ Electrode storage oven.

used before the maximum exposure time is reached. The exposure time is defined as the period of time that the electrode is not in a heated holding or portable oven.

CAUTION

The following exposure times are general guidelines only. Excessive exposure times will cause weld defects. Check your site quality standards for the actual exposure times for your location.

Typical exposure times are shown in *Table 5.*

Samples of electrodes being used at a site are often collected by the quality control group for moisture testing. Depending on the results of this testing, the maximum exposure times can be adjusted. In addition, if high moisture content is found in the electrodes, the welds made with the electrodes may have to be replaced.

Depending on the site quality standards, electrodes that exceed the maximum exposure time must either be destroyed or dried. Check your site quality standards to see if you can dry the electrodes or if they must be destroyed. Low-hydrogen electrodes that are exposed to water must be destroyed.

Table 5 Typical Electrode Exposure Times

Electrode	Exposure Time– Maximum Hours
E70XX Series	4
E80XX Series	4
E90XX Series	1
E100XX Series	½
E110XX Series	½
E3XX (stainless)	4
Nickel and nickel alloy	4

5.4.2 Drying Electrodes

Electrodes that exceed the maximum exposure time can be dried once. They are dried by placing them in a holding oven at an elevated temperature for a period of time. The amount of time and temperature depend on the nature of the moisture pickup and the electrode. If the electrodes are to be used on welds subject to X-ray inspection, the drying temperatures are typically higher.

Electrodes not subject to X-ray inspection must be brought up to 300°F and held at that temperature for at least one hour. Electrodes subject to X-ray inspection must be brought up to 750°F and held at that temperature for at least one hour.

6.0.0 ◆ FILLER METAL TRACEABILITY REQUIREMENTS

Welding codes and the WPS on a project or job site normally specify the type of electrode to be used. Once a weld is made, there must be a means to trace and document the type of electrode that was used to make the weld or welds in order to verify that the correct welding codes and WPS were followed.

Traceability requirements vary according to client requirements and specifications. On many job sites, these traceability requirements may be found in the quality control documentation mandating the job. One of the more common requirements followed may be found in the *ASME Boiler and Pressure Vessel Code, Section II, Material Specifications*. This section of the ASME code lists acceptable base metal and filler metal combinations.

Note

The traceability requirements in site quality control manuals must be followed. The consequences for violation of the requirements can be severe. Understand traceability requirements before proceeding with a welding job. Verify this information with your supervisor, project engineer, or welding inspector.

Summary

Many different electrodes are produced by many different manufacturers. To ensure quality and uniformity, the American Welding Society has established a classification system. Understanding this classification system and why electrodes are placed in a particular classification is essential for the welder to correctly identify and/or select electrodes for the weld to be made.

Welders must be able to select proper types of filler metals and electrodes to make welds on specific materials. It is also important to understand how to properly store electrodes to prevent damage or deterioration.

Review Questions

1. Welding current is transferred from the electrode holder, through the _____, and into the work.
   a. base metal
   b. flux coating
   c. welding ground
   d. wire core of the electrode

2. The harmful oxygen and hydrogen present in the atmosphere surrounding a molten weld is pushed away from the weld by the _____ created during the welding process.
   a. gaseous shield of carbon dioxide
   b. gaseous shield of nitrogen
   c. welding spatter
   d. heat

3. The purpose of the specifications written by the American Welding Society (AWS) is to set standards that all manufacturers must follow when manufacturing _____.
   a. welding consumables
   b. welding machines
   c. chemicals
   d. metals

4. The hyphen and number to the right of an AWS classification number indicate the _____.
   a. type of electrodes discussed in the classification
   b. year the classification was last revised
   c. number of chapters in the document
   d. year the classification was written

5. All-position electrodes that provide deep penetration belong to the _____ electrode group.
   a. fast-fill
   b. fill-freeze
   c. fast-freeze
   d. low-hydrogen

6. Electrodes designed for welding high-sulphur, phosphorus, and medium- to high-carbon steels belong to the _____ electrode group.
   a. fast-fill
   b. fill-freeze
   c. fast-freeze
   d. low-hydrogen

7. Of all electrodes, the ones that collect moisture faster than any other type are the _____ electrodes.
   a. fast-fill
   b. fill-freeze
   c. fast-freeze
   d. low-hydrogen

8. All filler metal must be stored in a dry storage area at a minimum temperature of _____.
   a. 30°F
   b. 40°F
   c. 50°F
   d. 60°F

9. Electrodes subject to X-ray inspections must be brought up to _____ for at least one hour.
   a. 250°F
   b. 300°F
   c. 500°F
   d. 750°F

10. The traceability requirements of filler metals vary according to the _____ requirements.
    a. base metal
    b. welder type
    c. site and client
    d. state and local

Jeff Hyde

Maintenance and Reliability Engineer
Flint Hills Resources

As a young man, Jeff Hyde learned to weld on his grandparents' farm in southeastern Kansas. Jeff's career demonstrates that life has a lot to offer to someone with ambition and perseverance.

How did you get your start?

As a teenager I spent summers on my grandparents' farm. My grandfather taught me to weld when I was 14. I also learned to repair farm equipment and did a lot of that work, but it wasn't as much fun as welding. When I finished high school, the oil fields were booming. Because of my ability to weld, I got a job as an oil rig welder in Odessa, Texas, and took welding classes at Odessa College at night.

Going from being an oilfield welder to a maintenance engineer is a big jump. How did that happen?

The oil field boom didn't last forever. After a couple of years of welding oil rigs, I decided to join the Navy and learn about electronics as a sonar technician. When my Navy hitch was up, I took advantage of GI education assistance funds to go to college, where I majored in mechanical engineering.

When I finished college, I was hired as a project engineer in the petrochemical industry. Because of my background in welding and electronics, and the knowledge of metallurgy I gained from my welding background, the move into maintenance and reliability engineering was natural.

What do you do in your job?

I review the maintenance needs for plant mechanical equipment and determine the safest, most reliable, and most cost-effective maintenance approach. I also troubleshoot equipment failures and perform failure analysis to determine what caused the equipment to fail. An important part of my job is designing methods to prevent failures from reoccurring.

What do you like about the work you do?

I'm given a lot of freedom to do what I think is necessary in my area of responsibility. Also, the job gives me the opportunity to do the things I really like to do. I have always liked to build things and to tinker with anything new. And I've always enjoyed taking broken machines apart to figure out why they failed and to see if I could fix them.

What do you think it takes to be a success?

To be a success at anything, you can't give up when the going gets tough. During my life, I've encountered many difficulties that could have prevented me from reaching my goals. Any of them would have given me a good excuse to quit, but I didn't let them.

I'm not a genius—ask any of my former teachers. I learned by seeking out the people who could teach me what I needed to know. I always kept my goals in view so that problems wouldn't defeat me.

What advice would you give someone just starting out?

Find something you truly like to do and then do that to the best of your ability. It may take a while, and you may have to try different things before you find what you're looking for. If it takes more training or education, go after it, and don't let anything stand in your way.

Trade Terms Introduced in This Module

Alloy: A metal that has had other elements added to it that substantially change its mechanical properties.

Condensation: The process in which water from the atmosphere forms on a cool surface.

Downhill welding: Welding with a downward progression from top to bottom.

Ductile: Capable of being bent or shaped without breaking.

Flux: Material used to prevent, dissolve, or facilitate the removal of oxides and other undesirable substances on a weld.

Heat-affected zone: The part of the base metal that has been altered but not melted by the heat.

Hermetically sealed: Having an airtight seal.

Low-hydrogen electrode: An electrode specially manufactured to contain little or no moisture.

Notch toughness: The ability of a material to resist breaking at points where stress is concentrated.

Traceability: The ability to verify that a procedure has been followed by reviewing the documentation step by step.

Sample Material Safety Data Sheet

Date:	**4/1/99**	MSDS No.:	**US-M297**
Supersedes:	**3/15/95**		
Trade Name:	**Lincoln 7018-1**		
Sizes:	**All**		

MATERIAL SAFETY DATA SHEET
For Welding Consumables and Related Products
Conforms to Hazard Communication Standard 29CFR 1910.1200 Rev. October 1988

SECTION I - IDENTIFICATION

Manufacturer/ Supplier:
The Lincoln Electric Company
22801 St. Clair Avenue
Cleveland, OH 44117-1199
(216) 481-8100

Product Type: Covered Electrode

Classification: AWS E7018-1

SECTION II - HAZARDOUS MATERIALS (¹)

I M P O R T A N T !
This section covers the materials from which this product is manufactured. The fumes and gases produced during welding with the normal use of this product are covered by Section V; see it for industrial hygiene information.
CAS Number shown is representative for the ingredients listed. All ingredients listed may not be present in all sizes.
(1) The term 'hazardous' in 'Hazardous Materials' should be interpreted as a term required and defined in the Hazards Communication Standard and does not necessarily imply the existence of any hazard.

Ingredients:	CAS No.	Wt.%	TLV mg/m³	PEL mg/m³
Iron	7439-89-6	15	10*	10*
Limestone and/or calcium carbonate	1317-65-3	10	10	15
Fluorides (as F)	7789-75-5	< 5	2.5	2.5
Silicates and other binders	1344-09-8	< 5	10*	10*
Manganese and/or manganese alloys and compounds (as Mn)***	7439-96-5	< 5	0.2	1.0(c)
Titanium dioxides (as Ti)***	13463-67-7	< 5	10	10
Silicon and/or silicon alloys and compounds (as Si)	7440-21-3	1	10*	10*
Mineral silicates	1332-58-7	0.5	5**	5**
Aluminum oxide and/or Bauxite***	1344-28-1	< 0.5	10	10
Carbon steel core wire	7439-89-6	60	10*	10*

Supplemental Information: (*) Not listed. Nuisance value maximum is 10 mg/m³. TLV value for iron oxide is 5 mg/m³.

(c) Values are for manganese fume. STEL (Short Term Exposure Limit) is 3.0 milligrams per cubic meter.

(**) As respirable dust.

(***) Subject to the reporting requirements of Sections 311, 312, and 313 of the Emergency Planning and Community Right-to-Know Act of 1986 and of 40CFR 370 and 372.

SECTION III - FIRE AND EXPLOSION HAZARD DATA

Non Flammable; Welding arc and sparks can ignite combustibles and flammable products. See Z49.1 referenced in Section VI.

(CONTINUED ON SIDE TWO)

106A01.TIF

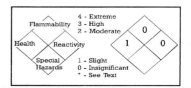

		4 - Extreme
Flammability	3 - High
	2 - Moderate	0
Health	Reactivity	1	0
Special	1 - Slight
Hazards	0 - Insignificant
	* - See Text

SECTION IV - HEALTH HAZARD DATA

Threshold Limit Value: The ACGIH recommended general limit for Welding Fume NOC - (Not Otherwise Classified) is 5 mg/m^3. ACGIH-1987-88 preface states that the TLV-TWA should be used as guides in the control of health hazards and should not be used as fine lines between safe and dangerous concentrations. See Section V for specific fume constituents which may modify this TLV. Threshold Limit Values are figures published by the American Conference of Government Industrial Hygienists. Units are milligrams per cubic meter of air.

Effects of Overexposure: Electric arc welding may create one or more of the following health hazards:
Fumes and Gases can be dangerous to your health. Common entry is by inhalation. Other possible routes are skin contact and ingestion.

Short-term (acute) overexposure to welding fumes may result in discomfort such as metal fume fever, dizziness, nausea, or dryness or irritation of nose, throat, or eyes. May aggravate pre-existing respiratory problems (e.g. asthma, emphysema). Exposure to extremely high levels of fluorides can cause abdominal pain, diarrhea, muscular weakness, and convulsions. In extreme cases it can cause loss of consciousness and death.

Long-term (chronic) overexposure to welding fumes can lead to siderosis (iron deposits in lung) and may affect pulmonary function. Manganese overexposure can affect the central nervous system, resulting in impaired speech and movement. Bronchitis and some lung fibrosis have been reported. Repeated exposure to fluorides may cause excessive calcification of the bone and calcification of ligaments of the ribs, pelvis and spinal column. May cause skin rash. WARNING: This product, when used for welding or cutting, produces fumes or gases which contain chemicals known to the State of California to cause birth defects and, in some cases, cancer. (California Health & Safety Code Section 25249.5 et seq).

Arc Rays can injure eyes and burn skin. *Skin cancer has been reported.*
Electric Shock can kill. If welding must be performed in damp locations or with wet clothing, on metal structures or when in cramped positions such as sitting, kneeling or lying, if there is a high risk of unavoidable or accidental contact with workpiece, use the following equipment: Semiautomatic DC Welder, DC Manual (Stick) Welder, or AC Welder with Reduced Voltage Control.

Emergency and First Aid Procedures: Call for medical aid. Employ first aid techniques recommended by the American Red Cross. IF BREATHING IS DIFFICULT give oxygen. IF NOT BREATHING employ CPR (Cardiopulmonary Resuscitation) techniques. IN CASE OF ELECTRICAL SHOCK, turn off power and follow recommended treatment. In all cases call a physician.

SECTION V - REACTIVITY DATA

Hazardous Decomposition Products: Welding fumes and gases cannot be classified simply. The composition and quantity of both are dependent upon the metal being welded, the process, procedure and electrodes used.

Other conditions which also influence the composition and quantity of the fumes and gases to which workers may be exposed include: coatings on the metal being welded (such as paint, plating, or galvanizing), the number of welders and the volume of the worker area, the quality and amount of ventilation, the position of the welder's head with respect to the fume plume, as well as the presence of contaminants in the atmosphere (such as chlorinated hydrocarbon vapors from cleaning and degreasing activities.)

When the electrode is consumed, the fume and gas decomposition products generated are different in percent and form from the ingredients listed in Section II. Decomposition products of normal operation include those originating from the volatilization, reaction, or oxidation of the materials shown in Section II, plus those from the base metal and coating, etc., as noted above.

Reasonably expected fume constituents of this product would include: Primarily iron oxide and fluorides; secondarily complex oxides of manganese, potassium, silicon and sodium.

Maximum fume exposure guideline and PEL for this product is 5.0 milligrams per cubic meter.

Gaseous reaction products may include carbon monoxide and carbon dioxide. Ozone and nitrogen oxides may be formed by the radiation from the arc.

Determine the composition and quantity of fumes and gases to which workers are exposed by taking an air sample from inside the welder's helmet if worn or in the worker's breathing zone. Improve ventilation if exposures are not below limits. See ANSI/AWS F1.1, F1.2, F1.4, and F1.5, available from the American Welding Society, 550 N.W. LeJeune Road, Miami, FL 33126.

SECTION VI AND VII
CONTROL MEASURES AND PRECAUTIONS FOR SAFE HANDLING AND USE

Read and understand the manufacturer's instruction and the precautionary label on the product. Request Lincoln Safety Publication E205. See American National Standard Z49.1, 'Safety In Welding and Cutting' published by the American Welding Society, 550 N.W. LeJeune Road, Miami, FL, 33126 and OSHA Publication 2206 (29CFR1910), U.S. Government Printing Office, Washington, D.C. 20402 for more details on many of the following:

Ventilation: Use enough ventilation, local exhaust at the arc, or both to keep the fumes and gases from the worker's breathing zone and the general area. Train the welder to keep his head out of the fumes. *Keep exposure as low as possible.*

Respiratory Protection: Use respirable fume respirator or air supplied respirator when welding in confined space or general work area when local exhaust or ventilation does not keep exposure below TLV.

Eye Protection: Wear helmet or use face shield with filter lens shade number 12 or darker. Shield others by providing screens and flash goggles.

Protective Clothing: Wear hand, head, and body protection which help to prevent injury from radiation, sparks and electrical shock. See Z49.1. At a minimum this includes welder's gloves and a protective face shield, and may include arm protectors, aprons, hats, shoulder protection, as well as dark substantial clothing. Train the welder not to permit electrically live parts or electrodes to contact skin . . . or clothing or gloves if they are wet. Insulate from work and ground.

Disposal Information: Discard any product, residue, disposable container, or liner as ordinary waste in an environmentally acceptable manner according to Federal, State and Local Regulations unless otherwise noted.

106A02.TIF

Additional Resources

This module is intended to present thorough resources for task training. The following reference works are suggested for further study. These are optional materials for continued education rather than for task training.

OSHA Standard 1926.351, Arc Welding and Cutting.

The Procedure Handbook of Arc Welding, 1994. Cleveland, OH: The Lincoln Electric Company.

Stick Electrode Welding Guide, 1999. Cleveland, OH: The Lincoln Electric Company.

Craft _____ Module Name _____

Copyright Date _____ Module Number _____ Page Number(s) _____

Description _____

_____

_____

_____

_____

(Optional) Correction _____

_____

_____

(Optional) Your Name and Address _____

_____

_____

_____

Shielded Metal Arc Welding – Beads and Fillet Welds

Course Map

This course map shows all of the modules in the AWS Entry Level Welder – Phase 1 curriculum. The suggested training order begins at the bottom and proceeds up. Skill levels increase as you advance on the course map. The local Training Program Sponsor may adjust the training order.

AWS ENTRY LEVEL WELDER—PHASE 1

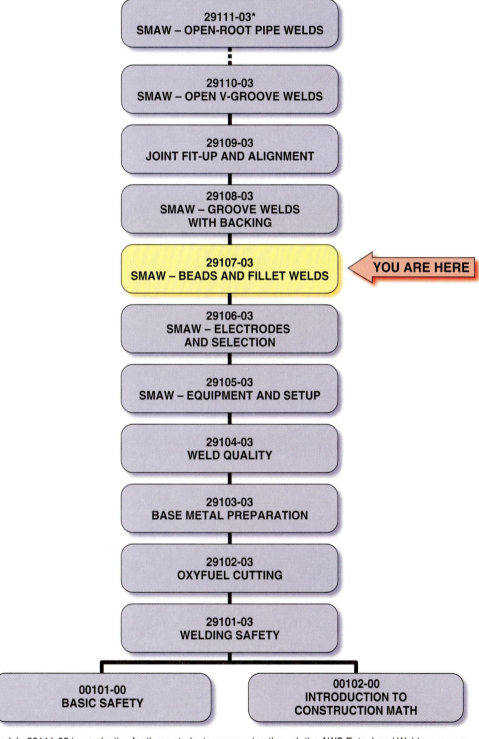

29111-03*
SMAW – OPEN-ROOT PIPE WELDS

29110-03
SMAW – OPEN V-GROOVE WELDS

29109-03
JOINT FIT-UP AND ALIGNMENT

29108-03
SMAW – GROOVE WELDS WITH BACKING

29107-03
SMAW – BEADS AND FILLET WELDS ◄ YOU ARE HERE

29106-03
SMAW – ELECTRODES AND SELECTION

29105-03
SMAW – EQUIPMENT AND SETUP

29104-03
WELD QUALITY

29103-03
BASE METAL PREPARATION

29102-03
OXYFUEL CUTTING

29101-03
WELDING SAFETY

00101-00
BASIC SAFETY

00102-00
INTRODUCTION TO CONSTRUCTION MATH

*Module 29111-03 is an elective for those students progressing through the AWS Entry Level Welder program.

29107CMAP.EPS

Figures

Table

Shielded Metal Arc Welding – Beads and Fillet Welds

Objectives

When you have completed this module, you will be able to do the following:

1. Set up shielded metal arc welding (SMAW) equipment.
2. Describe methods of striking an arc.
3. Properly strike and extinguish an arc.
4. Describe causes of arc blow and wander.
5. Make stringer, weave, and overlapping beads.
6. Make fillet welds in the:
   - Horizontal (2F) position
   - Vertical (3F) position
   - Overhead (4F) position

Prerequisites

Before you begin this module, it is recommended that you successfully complete the following: Modules 00101-00 through 29106-03.

Required Trainee Materials

1. Pencil and paper
2. Appropriate personal protective equipment

1.0.0 ◆ INTRODUCTION

One of the most basic welds is the fillet weld. Before a trainee can make a fillet weld, however, the trainee must be able to set up arc welding equipment, strike an arc, and maintain an arc to run a welding bead. This module will teach trainees how to strike an arc, run beads, and make fillet welds in all positions using **SMAW** electrodes. SMAW is often referred to by the nonstandard name *stick welding*.

2.0.0 ◆ SAFETY SUMMARY

The following is a summary of safety procedures and practices that must be observed while cutting or welding. Keep in mind that this is just a summary. Complete safety coverage is provided in the Level One module *Welding Safety*. If you have not completed that module, do so before continuing. Above all, be sure to wear appropriate protective clothing and equipment when welding or cutting.

2.1.0 Protective Clothing and Equipment

- Always use safety goggles with a full face shield or a helmet. The goggles, face shield, or helmet lens must have the proper light-reducing tint for the type of welding or cutting to be performed. Never directly or indirectly view an electric arc without using a properly tinted lens.
- Wear proper protective leather and/or flame retardant clothing along with welding gloves that will protect you from flying sparks and molten metal as well as heat.
- Wear 8" or taller high-top safety shoes or boots. Make sure that the tongue and lace area of the footwear will be covered by a pant leg. If the tongue and lace area is exposed or the footwear must be protected from burn marks, wear leather spats under the pants or chaps and over the front of the footwear.
- Wear a solid material (non-mesh) hat with a bill pointing to the rear or, if much overhead cutting or welding is required, a full leather hood with a welding face plate and the correct tinted lens. If a hard hat is required, use a hard hat that allows the attachment of rear deflector material and a face shield.
- If a full leather hood is not worn, wear a face shield and snug-fitting welding goggles over

safety glasses for gas welding or cutting. Either the face shield or the lenses of the welding goggles must be an approved shade 5 or 6 filter. For electric arc welding, wear safety goggles and a welding hood with the correct tinted lens (shade 9 to 14).

- If a full leather hood is not worn, wear earmuffs or at least earplugs to protect ears and ear canals from sparks from overhead operations.

2.2.0 Fire/Explosion Prevention

- Never carry matches or gas-filled lighters in your pockets. Sparks can cause the matches to ignite or the lighter to explode, causing serious injury.
- Never perform any type of heating, cutting, or welding until a hot work permit is obtained and an approved fire watch is established during and after the operation. Most work site fires caused by these types of operations are started by cutting torches.
- Never use oxygen to blow off clothing. The oxygen can remain trapped in the fabric for a time. If a spark hits the clothing during this time, the clothing can burn rapidly and violently out of control.
- Make sure that any flammable material in the work area is moved or shielded by a fire-resistant covering. Approved fire extinguishers must be available before any heating, welding, or cutting operations are attempted.
- Never release a large amount of oxygen or use oxygen as compressed air. Its presence around flammable materials or sparks can cause rapid and uncontrolled combustion. Keep oxygen away from oil, grease, or other petroleum products.
- Never release a large amount of fuel gas, especially acetylene. Methane and propane tend to concentrate in and along low areas and can be ignited a considerable distance from the release point. Acetylene is lighter than air but is even more dangerous. When mixed with air or oxygen, it will explode at much lower concentrations than any other fuel.
- To prevent fires, maintain a neat and clean work area and make sure that any metal scrap or slag is cold before disposal.
- Before cutting or welding containers, such as tanks or barrels, check to see if they have contained any explosive, hazardous, or flammable materials, including petroleum products, citrus products, or chemicals that decompose into toxic fumes when heated. As a standard practice, always clean and then fill any tanks or bar-

rels with water or purge them with a flow of inert gas to displace any oxygen.

2.3.0 Work Area Ventilation

- Make sure confined space procedures are followed before conducting any welding or cutting in the confined space.
- Make sure confined spaces are ventilated properly for cutting or welding purposes.
- Never use oxygen in confined spaces for ventilation purposes.
- Always perform cutting or welding operations in a well-ventilated area. Cutting or welding operations involving materials, coatings, or electrodes containing cadmium, mercury, lead, zinc, chromium, and beryllium result in toxic fumes. For cutting or welding of such materials, always use proper area ventilation and wear an approved full face, supplied-air respirator (SAR) that uses breathing air supplied externally of the work area. For occasional, very short-term exposure to fumes from zinc or copper coated materials, a high-efficiency particulate arresting (HEPA)-rated or metal-fume filter may be used on a standard respirator.

3.0.0 ◆ ARC WELDING EQUIPMENT SETUP

Before welding can take place, the area has to be made ready, the welding equipment must be set up, and the metal to be welded must be prepared. The following sections explain how to set up equipment for welding.

3.1.0 Preparing the Welding Area

To practice welding, a welding table, bench, or stand is needed. The welding surface must be steel, and provisions must be made for placing weld coupons out of position. *Figure 1* shows a typical welding station.

To set up the area for welding, follow these steps:

Step 1 Check to be sure the area is properly ventilated. Make use of doors, windows, and fans.

Step 2 Check the area for fire hazards. Remove any flammable materials before proceeding.

Step 3 Check the location of the nearest fire extinguisher. Do not proceed unless the extinguisher is charged and you know how to use it.

107F01.EPS

Figure 1 ◆ Welding station.

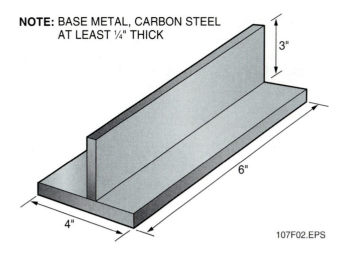

NOTE: BASE METAL, CARBON STEEL AT LEAST ¼" THICK

3"

6"

4"

107F02.EPS

Figure 2 ◆ Fillet weld coupons.

Step 4 Position a welding table near the welding machine.

Step 5 Set up flash shields around the welding area.

3.2.0 Preparing the Weld Coupons

The weld coupons should be carbon steel, ¼" to ¾" thick. Use a wire brush or grinder to remove heavy mill scale or corrosion. Prepare weld coupons to practice the welds indicated as follows:

- *Striking an arc* – The coupons can be any size or shape that is easily handled.
- *Running beads* – The coupons can be any size or shape that is easily handled.
- *Overlapping beads* – The coupons can be any size or shape that is easily handled.
- *Fillet welds* – Cut the metal into 4" × 6" rectangles for the base and 3" × 6" rectangles for the web.

Figure 2 shows the weld coupons for fillet welding.

Steel for practice welding is expensive and difficult to obtain. Every effort should be made to conserve and not waste the material that is available. Reuse weld coupons until all surfaces have been welded on. Weld on both sides of the joint and then cut the weld coupon apart and reuse the pieces. Use material that cannot be cut into weld coupons to practice striking an arc and running beads.

3.3.0 Electrodes

Obtain a small quantity of the electrodes to be used. Electrodes are sometimes referred to by the nonstandard term of rods. For the welding exercises in this module, ⅛" and ⁵⁄₃₂" E6010 and E7018 electrodes will be used. Obtain only the electrodes to be used for a particular welding exercise at one time. Have some type of pouch or rod holder in which to store the electrodes to prevent them from becoming damaged. Never store electrodes loose on a table. They may end up on the floor where they can become damaged or create a tripping hazard. Some type of metal container, such as a pail, must also be available to discard hot electrode stubs.

Note

The electrode sizes are recommendations. Other electrode sizes may be substituted depending on site conditions. Check with your instructor for the electrode size to use.

WARNING!

Do not throw electrode stubs on the floor. They easily roll, and someone could step on one, slip, and fall.

Rod Holders

Several different types of rod holders can be purchased. One type is a leather pouch, and another is a sealable holder that can be used to prevent the electrodes from getting wet.

107P0701.EPS

3.4.0 Preparing the Welding Machine

Identify a welding machine (*Figure 3*) to use and then follow these steps to set it up for welding:

Step 1 Verify that the welding machine can be used for DC (direct current) welding.

Note

An AC (alternating current) welding machine can be used if a DC machine is not available. If an AC welding machine is used, use E6011 and E6013 electrodes.

Step 2 Check to be sure that the welding machine is properly grounded through the primary current receptacle.

Step 3 Check the area for proper ventilation.

Step 4 Verify the location of the primary current disconnect.

Step 5 Set the polarity to direct current electrode positive (DCEP).

Step 6 Connect the clamp of the workpiece lead to the workpiece.

WARNING!

Even though workpiece leads and clamps are sometimes called *ground leads* or *clamps,* they may not actually be grounded. In this case, the full open-circuit voltage of the welding machine may exist between the workpiece lead or clamp and any grounded object.

Step 7 Set the amperage for the electrode type and size to be used. Typical settings are shown in *Table 1.*

PORTABLE 200A ENGINE-DRIVEN AC/DC WELDING MACHINE

DC STICK RANGE CONTROL

AC/DC POLARITY SWITCH

AC/DC AMPERAGE CONTROL

ON/OFF SWITCH

ENGINE START SWITCH

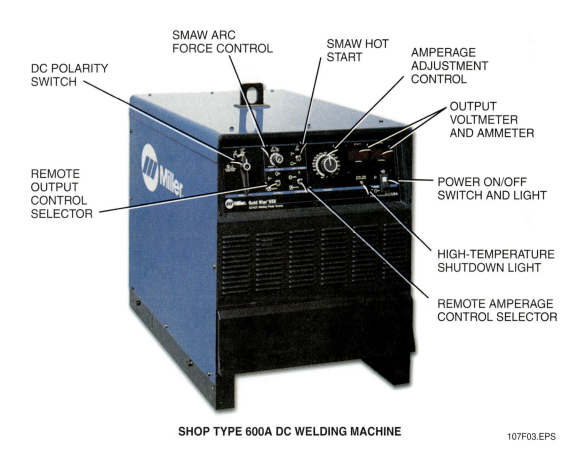

DC POLARITY SWITCH

SMAW ARC FORCE CONTROL

SMAW HOT START

AMPERAGE ADJUSTMENT CONTROL

OUTPUT VOLTMETER AND AMMETER

REMOTE OUTPUT CONTROL SELECTOR

POWER ON/OFF SWITCH AND LIGHT

HIGH-TEMPERATURE SHUTDOWN LIGHT

REMOTE AMPERAGE CONTROL SELECTOR

SHOP TYPE 600A DC WELDING MACHINE

107F03.EPS

Figure 3 ◆ Typical AC/DC and DC welding machines.

Constant-Current Machines

The AC or AC/DC machines desired for SMAW or gas tungsten arc welding (GTAW) are basically constant-current types of machines, some of which are referred to as *droop-current machines* or *droopers*. In a true constant-current machine, the output current varies little over a relatively wide range of circuit voltage. Because of this, raising and lowering the electrode from the workpiece has little effect on the weld quality. In addition, striking an arc is easier with a true constant-current machine because of the high open-circuit voltage. On the other hand, droopers can be used to control the molten weld pool temperature more easily. This is because the current, and resultant weld heat, changes much more with a smaller change in voltage that is caused by raising or lowering the electrode in relation to the workpiece.

Table 1 Amperage Settings for Electrodes

Electrode	Size	Amperage
E6010	⅛"	75A to 130A
E6010	⁵⁄₃₂"	90A to 175A
E6011	⅛"	75A to 120A
E6011	⁵⁄₃₂"	90A to 160A
E7018	⅛"	90A to 150A
E7018	⁵⁄₃₂"	120A to 190A
E6013	⅛"	110A to 150A
E6013	⁵⁄₃₂"	150A to 200A

Note

Amperage recommendations vary by manufacturer, position, current type, and electrode brand. For specific recommendations, refer to the manufacturer's literature for the electrode being used.

Step 8 Check to be sure that the electrode holder is not touching the workpiece lead clamp, the workpiece, or a grounded object.

Step 9 Turn on the welding machine.

4.0.0 ◆ STRIKING AN ARC

The first step in starting a weld is striking the arc. To strike an arc, touch the end of the electrode to the base metal and then quickly raise it to the correct arc length. The general rule is that the arc length should be the diameter of the electrode being used. If a ⅛" electrode is being used, the arc length should be ⅛". The arc length is measured from the end of the electrode core to the base metal. If the electrode has a heavier flux coating, such as E7018 or E6013, the end of the electrode core will be recessed in the coating. For these electrodes, the visible arc length (from the end of the coating to the base metal) should be slightly less to compensate for the part of the arc that is recessed. This will ensure that the actual arc length is correct. *Figure 4* shows arc lengths for a ⅛" rod.

There are two ways to strike an arc: the scratching method and the tapping method.

4.1.0 Scratching Method

The scratching method (*Figure 5*) is the easiest way to strike an arc and is used by trainees who are learning to weld. It is similar to striking a match. The end of the electrode is simply scratched along the base metal to establish an arc.

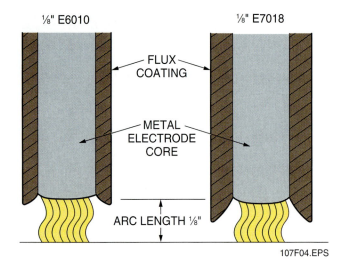

Figure 4 ◆ Arc lengths.

Restriking a Cold Electrode

When attempting to strike an arc using a cold, partially consumed electrode, the cup formed by hardened flux sheath extending beyond the metal core may not allow the core to contact the workpiece to establish an arc. Scratch striking or hard tapping can be used to break off the sheath. Unfortunately, this sometimes causes large pieces of the flux coating to break off the sides, exposing part of the core. Electrodes with missing flux are difficult to strike. If an arc is struck, it will be unstable and will blow or wander until the core is consumed up to where the entire flux sheath cup is reestablished. During this time, poor welds will occur. One method of removing the flux sheath without chipping flux away from the sides of the rod is to draw the end of the rod across the face of a medium-coarse file. If code requirement welding is being performed, electrodes with chipped, missing, or cracked flux coatings must be discarded.

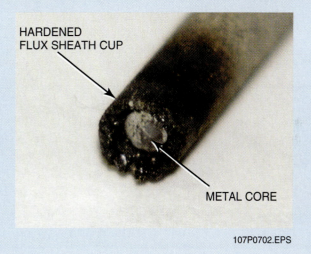

107P0702.EPS

When the arc has been established, the electrode is raised to establish the correct arc length.

4.2.0 Tapping Method

The tapping method (*Figure 6*) is the best method of establishing an arc when using a transformer DC welding machine. It is more difficult and takes more practice to perfect, but as welders develop their skills, it is the method that should be used. The scratching method leaves **arc strikes** on the base metal, which are not allowed by the welding codes unless they occur within the welded area. With the tapping method, the arc is usually established just after the point where the welding should begin (¼" to ½"). The arc is then moved back to the correct starting point as the arc stabilizes. To strike an arc with the tapping method, move the electrode quickly to the base metal, lightly tap it, and then raise the electrode to establish the correct arc length.

Figure 5 ◆ Scratching method of striking an arc.

107F05.EPS

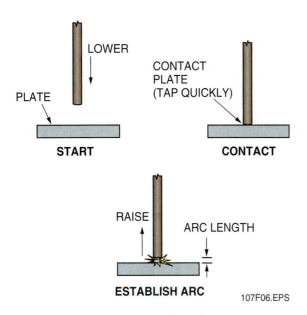

START

LOWER

PLATE

CONTACT

CONTACT
PLATE
(TAP QUICKLY)

ESTABLISH ARC

RAISE

ARC LENGTH

107F06.EPS

Figure 6 ◆ Tapping method of striking an arc.

4.3.0 Practicing Striking and Extinguishing an Arc

Striking the arc correctly is an important step in learning to weld. Practice both methods using ⅛" E6010 or E6011 electrodes until you can strike and maintain an arc consistently. When striking an arc, place the base metal flat on the welding table. To extinguish the arc at the end of a weld, quickly lift the electrode away from the work.

> **NOTE**
>
> Striking an arc with E7018 (low hydrogen) electrodes is more difficult than with most other electrodes because they tend to stick to the base metal more. Practice striking an arc with E7018 after you become more skilled running beads with E6010 or E6011.

Inverter Power Sources

When not actively welding, some inverter power sources used in the SMAW mode deactivate open-circuit voltage at the electrode. Only a very low sensing voltage exists. For these machines, the scratch method must be used to activate the welding voltage and current and start the arc.

Accurate Tap Striking of an Electrode

An accurate method of tap striking an electrode at an exact spot is to rest the electrode on the finger of your gloved, free hand like a pool cue. Angle the rod in the direction of travel. Then move the electrode quickly down and up to strike the arc. This method is feasible if your helmet can be lowered by nodding your head or if you are using an auto-darkening lens helmet.

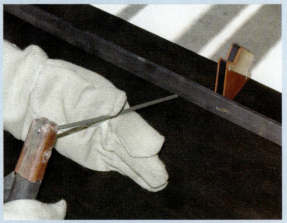

107P0703.EPS

Reusing a Stuck Electrode

Carefully inspect the end of any electrode that has been stuck to the workpiece and freed by bending. If any of the flux coating is missing, cracked, or badly burned, discard the electrode.

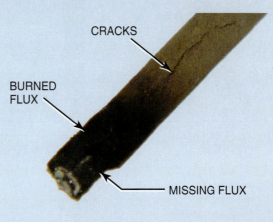

CRACKS

BURNED FLUX

MISSING FLUX

107P0704.EPS

When practicing striking an arc, you will generally experience two problems:

- The end of the electrode welds itself to the base metal. This is caused by keeping the electrode in contact with the base metal too long or trying to maintain too short an arc length. If this occurs, free the electrode by quickly moving the electrode holder from side to side. If this fails, the electrode should be released from the holder and removed from the base metal with pliers. When the electrode is released from the holder, arcing generally occurs between the holder jaws and the electrode. If arcing has occurred, inspect the holder jaws for any heavy pitting or gouging damage and replace them if necessary.
- The arc will break (go out). This is caused by raising the end of the electrode too far above the base metal.

These problems will be eliminated as you gain experience in striking and controlling the arc.

5.0.0 ◆ ARC BLOW

When current flows, especially DC current, strong magnetic fields may be created. The magnetic fields tend to concentrate in corners, in deep grooves, or in the ends of the base metal. When the arc approaches these concentrated magnetic fields, it is deflected. This phenomenon is known as arc blow. In some cases, ferrous metal welding fixtures or jigs that are part of a DC welding current path and subjected to repeated current flow can become magnetized and contribute to arc blow. In AC welding, arc blow is rarely a problem because the magnetic field is constantly reversing at twice the frequency of the primary power source. This effectively cancels any strong magnetic field effects. DC arc blow can cause defects such as excessive weld spatter and porosity. If arc blow occurs, try one or more of these methods to control it (*Figure 7*).

- Change the position of the workpiece lead clamp. This will change the flow of welding current, affecting the way the magnetic fields are created.
- Shorten the arc length. With a shorter arc length, the magnetic field will have less effect on the arc.
- Change the angle of the electrode. The normal electrode angle is 10° to 15° in the direction of travel. Raising the electrode angle toward 90° or, in extreme cases, up to 20° in the opposite direction of travel will compensate for the arc blow.
- If possible, tack-weld the workpieces together at the ends and in the middle.
- Periodically degauss welding fixtures or jigs with an AC-powered degaussing coil or probe.

6.0.0 ◆ STRINGER AND WEAVE BEADS

A stringer bead is a weld bead that is made with very little or no side-to-side motion of the electrode. The width of a stringer bead will vary

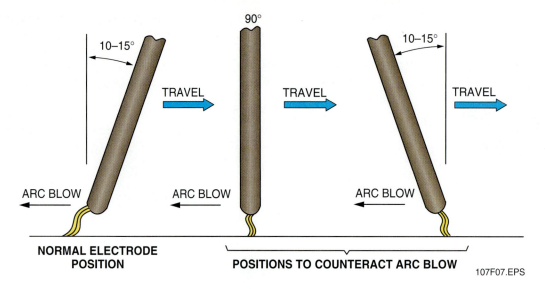

NORMAL ELECTRODE POSITION

POSITIONS TO COUNTERACT ARC BLOW

107F07.EPS

Figure 7 ◆ Controlling arc blow.

Arc Shifting or Wandering

Sometimes when two workpieces are first being joined, the arc may deposit material on one piece only or wander back and forth from one piece to the other. Both of these occurrences are usually the result of too long an arc. If this is the case, the arc will always jump to the closest workpiece. This is especially noticeable with electrodes that are ⅛" or smaller when welding gapped butt or T-joints. If the arc initially deposits material on only one workpiece, it is possible that only one of the workpieces is electrically connected to the workpiece lead from the welder and the other piece is not in the circuit. In this case, weaving the electrode back and forth across the workpiece at the start of the weld will usually bridge the gap between the workpieces and tack the pieces together so that they are both in the circuit. Other solutions are to clamp a piece of metal across both pieces or to perform the welding on a metal tabletop.

with the electrode type. A **weave bead** is a weld bead that is made with a side-to-side motion of the electrode. *Figure 8* shows stringer and weave beads. Outside of welding joints, stringer and weave beads are used for resurfacing or hardsurfacing.

Note
The width of stringer and weave beads is usually specified in the drawing welding symbols, welding code, or welding procedure specification (WPS) being used for the work at your site. Do not exceed the widths specified for the work.

STRINGER BEAD

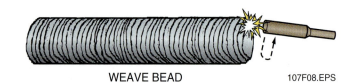

WEAVE BEAD

107F08.EPS

Figure 8 ◆ Stringer and weave beads.

6.1.0 Practicing Stringer Beads with E6010

Practice running stringer beads in the flat position using ⅛" E6010 or E6011 electrodes. After striking the arc, the electrode axis angle should be a 10° to 15° **drag angle** in the direction of travel for the weld and at a zero **work angle** that is perpendicular (90°) to the base metal across the weld. *Figure 9* shows electrode axis angles.

Travel angles used for SMAW and other types of welding are categorized as either drag or **push angles**. An electrode drag angle is when the electrode axis points at the weld bead during the running of a weld bead. A push angle is the opposite condition. It is when the electrode axis points away from the weld bead and toward the direction of travel when running a weld bead. Except for vertical welds, SMAW is usually accomplished with an electrode drag angle. Electrode work angle is the side-to-side tilt of a plane containing the electrode axis from a perpendicular line to the major work surface. The plane is formed by the intersection of the weld axis and the electrode

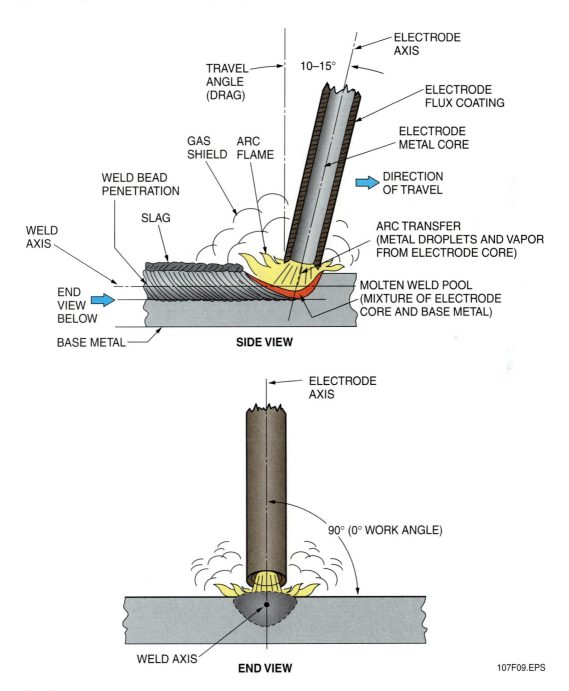

Figure 9 ◆ SMAW characteristics with proper travel and work angle.

axis. Work angles range from 0° to less than 90°. In the case of a single bead being run on a flat plate, the work angle is zero. The work angle for a fillet weld in a T- or corner joint is always taken from a perpendicular line to the non-butting member.

The whipping motion, also called a stepping motion, shown in *Figure 10* can be used when depositing the stringer bead to control the weld puddle. It is performed by moving the electrode up and forward about ¼", then down and backward about ³⁄₁₆". A short pause at the end of the backward travel deposits the weld metal. The exact length of travel can vary. By momentarily lengthening and advancing the arc, the molten weld pool is allowed to cool. Lengthening the arc lowers the arc temperature and reduces metal

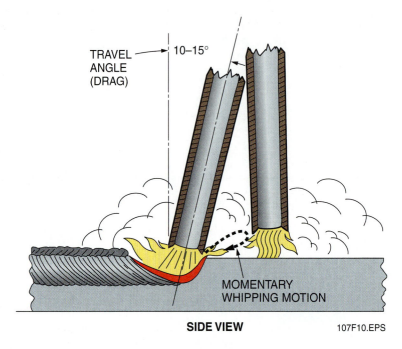

TRAVEL ANGLE (DRAG) — 10–15°

MOMENTARY WHIPPING MOTION

SIDE VIEW   107F10.EPS

Figure 10 ◆ Whipping motion.

Stabilize Your Welding Position, Watch the Immediate Welding Area, and Listen to the Arc

A welder should be in a relaxed, comfortable position when welding. Because of the limited view of the welding area through a dark helmet lens, a beginning welder may sway due to a loss of sense of balance. To counteract this, a welder should sit or lean against something to achieve and maintain a stable and relaxed position. This will reduce fatigue and ensure personal safety.

Beginning welders must learn to view the entire electric arc welding work area through the shaded lens of a helmet and to listen to the sound of the arc. At first, a beginning welder's focus is normally concentrated only on the electric arc because of the difficulty of striking and keeping an arc established. Once beginners have managed to strike and maintain an arc, however, the tremendous light and heat energy generated within the arc is usually fascinating and distracting to beginners for a time. With practice, the beginning welder will gradually be able to shift focus to see all sides of a molten weld pool, the weld buildup at the trailing edge of the pool, the cooling slag cover over the weld buildup, and the adjacent welding area on the workpiece. The welder will become attuned to the correct sound of the arc for the rod and work being welded, and the arc itself may not even be noticed. Once the welder is able to view the entire weld area and listen to the arc, running straight and correct welds is relatively easy.

transfer from the rod core. Advancing the arc pre-heats the base metal ahead of the weld and burns off contaminants.

Carefully observe and listen to the weld as it is being made. If the arc length, speed, angles, and motions are correct, a distinctive frying sound will be heard. Experiment by making minor changes in these factors and notice the effects.

Follow these steps to run stringer beads:

Step 1 Shield your eyes, strike the arc, and move the electrode back slightly to the correct starting point. Then position the electrode for the travel angle and perpendicular to the base metal.

Step 2 Hold the arc in place until the weld puddle widens to about two times the diameter of the electrode.

Step 3 Slowly move the arc forward, keeping a constant arc length. As necessary, use the whipping motion and adjust the forward speed to control buildup and preheat the weld zone.

WARNING!

Always wear face protection to prevent hot slag from hitting your face.

Step 4 Continue to weld until a bead about 2" to 3" long is formed and then break the arc by lifting the electrode straight up quickly. A crater will be left at the point where the arc was broken.

Step 5 Chip the weld slag with a chipping hammer.

Step 6 Clean the bead with a wire brush.

Step 7 Have your instructor inspect the bead for:
- Straightness
- A uniform appearance on the bead face
- A smooth, flat transition with complete fusion at the toes (edges) of the weld
- No porosity
- No undercut at the toes (edges)
- No inclusions
- No cracks
- No overlap

Step 8 Continue welding beads until you can make acceptable welds every time. *Figure 11* shows proper and improper weld beads.

6.2.0 Practicing Stringer Beads with E7018

Continue running stringer beads, this time using ⅛" E7018 electrodes. Use the same electrode angles, but do not whip the electrode. When running all low-hydrogen electrodes, the arc should never leave the weld puddle, and the visible arc should be shorter than with E6010 due to the thicker, longer sheath of the flux coating. Weld defects, such as porosity or hydrogen embrittlement, can occur if the arc leaves the weld puddle or if the arc is too long. The arc can be moved within the weld puddle to control the bead shape.

LASH

To ensure good welding results, remember LASH.

*Length of arc (**L**)* – The distance between the electrode and the base metal (usually one times the electrode diameter).

*Angle (**A**)* – Two angles are critical:

- *Travel angle* – The longitudinal angle of the electrode in relation to the axis of the weld joint

- *Work angle* – The traverse angle of the electrode in relation to the axis of the weld joint

*Speed (**S**)* – Travel speed is measured in inches per minute (IPM). The width of the weld will determine if the travel speed is correct.

*Heat (**H**)* – Controlled by the amperage setting and dependent upon the electrode diameter, base metal type, base metal thickness, and the welding position.

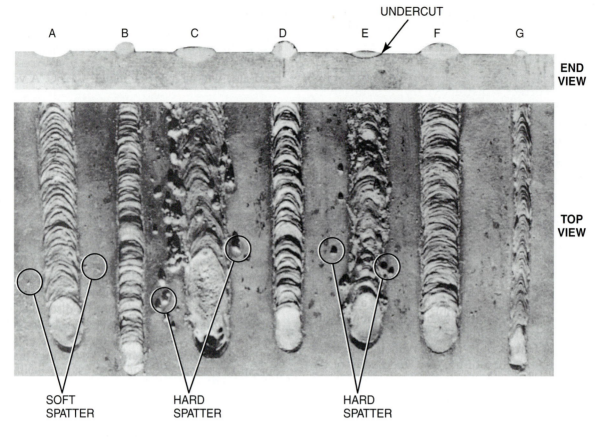

A = Correct current, arc length, and travel speed. Note the easily removed spatter (soft spatter).

B = Current set too low.

C = Current set too high. Note the hard spatter (spatter firmly bonded to base material that must be ground or chiseled off). Note the pointed ends of the bead indicating the weld pool was too hot and cooled too slowly. Impurities are usually trapped in the weld due to the slow cooling.

D = Arc length too short (narrow, high bead caused by arc pressure).

E = Arc length too long. Note the hard spatter and bead undercut of the edges.

F = Travel speed too slow (wide high bead).

G = Travel speed too fast. Note the pointed ends of bead.

107F11.EPS

Figure 11 ◆ Effect of current, arc length, and travel speed on SMAW beads.

6.3.0 Restarting a Weld

A **restart** is the point where one weld bead stops and another starts. Restarts are important because most SMAW welds cannot be made without at least one restart, and an improperly made restart will create a weld defect. A restart must be made so that it blends smoothly into the rest of the weld and does not stand out. The technique for making a restart is the same for stringer and weave beads. Follow these steps to make a restart as shown in *Figure 12*:

Step 1 Just before the previous electrode is used up, quickly increase the weld speed to taper the weld for ¼" to ⅜" and then break the arc.

Step 2 Chip any slag from the tapered section and crater.

Step 3 With a new electrode, restrike the arc in front of the crater and in line with the weld. (The welding codes do not allow arc strikes outside the area to be welded.)

Step 4 Move the electrode back onto the tapered weld section and develop the weld puddle with a slight circular motion, maintaining the correct arc length and electrode angles.

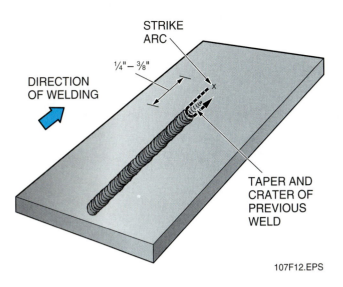

Figure 12 ◆ Making a restart.

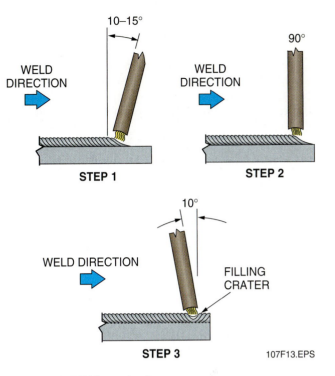

Figure 13 ◆ Weld termination.

Step 5 Start the forward motion as the bead achieves the proper width and continue to weld.

Step 6 Inspect the restart. A properly made restart will blend into the bead, making it hard to detect.

If the restart has undercut, not enough time was spent on the taper or the crater to fill it. If undercut is on one side or the other, use more of a side-to-side motion as you move into the taper. If the restart has a hump, it was overfilled; too much time was spent before resuming the forward motion.

Continue to practice restarts until they are correct. Use the same techniques for making restarts whenever performing SMAW for other types of welds.

6.4.0 Terminating a Weld

Terminations are made at the end of a weld. A termination leaves a crater. When making a termination, the welding codes require that the crater be filled to the full cross section of the weld. This can be difficult because most terminations are at the edge of a plate where welding heat tends to build up, making filling the crater more difficult.

The technique for making a termination is basically the same for all SMAW. Follow these steps to make a termination (*Figure 13*):

Step 1 As you approach the end of the weld, start to stand the electrode up toward 90° and slow the forward travel.

Step 2 Stop forward movement about ⅛" from the end of the plate and slowly angle the electrode to about 10° toward the start of the weld.

Step 3 Move about ⅛" toward the weld and break the arc when the crater is filled.

Step 4 Inspect the termination. The crater should be filled to the full cross section of the weld.

Starting and Stopping Welds with Weld Tabs

Another method of eliminating welding starting and stopping points is to tack-weld starting (run-on) and stopping (run-off) weld tabs to the workpiece. These tabs allow the arc to stabilize and achieve correct penetration of the workpiece at the start and to terminate with a weld crater off the workpiece at the end. They are especially useful on groove welds requiring multiple passes or when low-hydrogen or fast-fill electrodes are used. Because both ends of a multiple-pass weld may taper down, they help control underfill and back-burn at the start and finish of the welding. After the weld tabs are cut off, the weld is of continuous width and penetration across the entire workpiece.

6.5.0 Practicing Weave Beads with E6010

Practice running weave beads in the flat position using ⅛" E6010 or E6011 electrodes. After striking the arc, the electrode angle should be 10° to 15° in the direction of travel and perpendicular (90°) to the base metal.

The weave bead is made by moving the electrode back and forth. Many different patterns can be used to make a weave bead, including circles, crescents, figure-eights, and zigzags. When making a weave bead, use care at the toes to ensure proper tie-in to the base metal. To ensure proper tie-in at the toes, slow down or pause slightly at the edges. Pause longer when using the zigzag motion. The pause at the edges will also flatten out the weld, giving it the proper profile.

Figure 14 shows the weave motions.

Carefully observe and listen to the weld as it is being made. If the arc length, speed, angles, and motions are correct, a distinctive frying sound will be heard. Experiment by making minor changes in these factors and notice the effects.

Follow these steps to run weave beads (*Figure 15*):

Step 1 Strike the arc and position the electrode for a drag angle. Make sure it is perpendicular to the base metal (0° work angle).

Step 2 Hold the arc in place until the weld puddle widens to about two times the diameter of the electrode.

Step 3 Slowly move the arc forward in a weaving motion, keeping a constant arc length.

WARNING!
Always wear face protection to prevent hot slag from hitting your face.

Step 4 Continue to weld until a weave bead about 2" to 3" long is formed, and then break the arc by lifting the electrode straight up. A crater will be left at the point where the arc was broken. Chip any slag from around the crater.

Step 5 Make a restart and continue welding to the end of the plate.

Step 6 Chip the weld slag with a chipping hammer.

Step 7 Clean the bead with a wire brush.

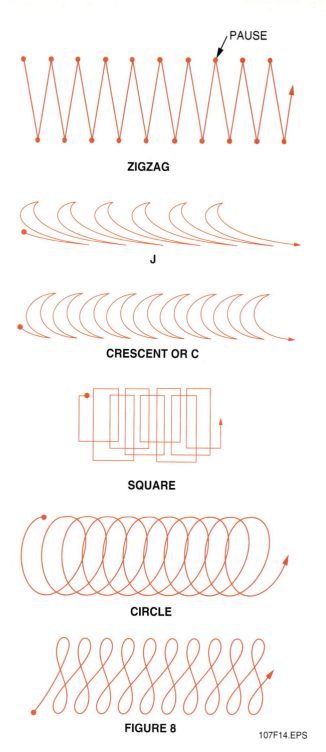

PAUSE

ZIGZAG

J

CRESCENT OR C

SQUARE

CIRCLE

FIGURE 8

107F14.EPS

Figure 14 ◆ Weave motions.

Step 8 Have the instructor inspect the bead. It should show:
- Weld bead straight to within ⅛"
- Uniform appearance on the bead face
- Smooth, flat transition with complete fusion at the toes of the weld
- Crater and restarts filled to the full cross section of the weld

Weave Bead Width

Although the usual width of a straight bead is two to three times the electrode diameter, the bead width for a weave bead can usually be up to, but should not exceed, eight times the electrode diameter.

FILLER
WEAVE
BEAD

107F15.EPS

Figure 15 ◆ Weave bead.

- No porosity
- No undercut at the toes
- No inclusions
- No cracks
- No overlap

Continue welding weave beads until you can make acceptable welds every time.

6.6.0 Practicing Weave Beads with E7018

Repeat running weave beads using ⅛" E7018 electrodes. Use the same electrode angles, but do not whip the electrode. When running low-hydrogen electrodes, the arc should never leave the weld puddle, and the visible arc should be shorter than with E6010. Weld defects, such as porosity or hydrogen embrittlement, can occur if the arc is too long or if it leaves the weld puddle. The arc can be moved within the weld puddle to control the bead shape.

7.0.0 ◆ OVERLAPPING BEADS

Overlapping beads are made by depositing connective weld beads parallel to one another. The parallel beads overlap, forming a flat surface. This is also called *padding*. Overlapping beads are used to build up a surface and to make multi-pass welds. Both stringer and weave beads can be overlapped. Properly overlapped beads, when viewed from the end, will form a relatively flat surface (see *Figure 16*).

7.1.0 Practicing Overlapping Beads with E6010

Follow these steps to weld overlapping stringer beads using ⅛" E6010 electrodes:

Step 1   Mark out a 4" square on a piece of steel.

Step 2   Weld a stringer bead along one edge.

Step 3   Clean the weld.

Step 4   After striking the arc for the next stringer bead, and with the proper travel angle, position the electrode at a work angle of 10° to 15° to the side of the previous bead to obtain proper tie-in. *Figure 17* shows the electrode work angle for overlapping beads.

Step 5   Continue running stringer beads until the 4" square is covered.

Note

The base metal will get very hot as it is built up. Using pliers, cool it by plunging it into water.

Step 6   Continue building layers of stringer beads, one on top of the other, until the technique is perfected.

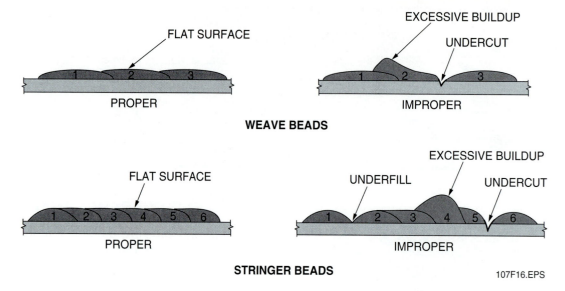

WEAVE BEADS

STRINGER BEADS

107F16.EPS

Figure 16 ◆ Proper and improper overlapping beads.

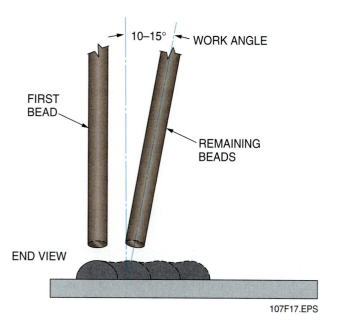

107F17.EPS

Figure 17 ◆ Electrode work angle for overlapping beads.

Continue welding overlapping beads using the weaving technique. Remember to angle the electrode toward the previous bead to obtain a good tie-in.

7.2.0 Practicing Overlapping Beads with E7018

Repeat welding overlapping beads using ⅛" E7018 electrodes. Build a pad using stringer beads and then repeat building the pad using weave beads as indicated in *Figure 18*. Keep a short arc and do not whip the electrode.

SIDE VIEW

END VIEW   107F18.EPS

Figure 18 ◆ Build a pad with E7018 electrodes.

8.0.0 ◆ FILLET WELDS

A fillet weld is a weld that is approximately triangular in cross section and is used with T-, lap, and corner joints. The sizes and locations of fillet welds are given as welding symbols. The two types of fillet welds are convex and concave (see *Figure 19*). A convex fillet weld has its surface

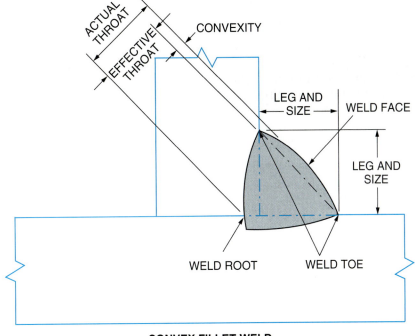

CONVEX FILLET WELD

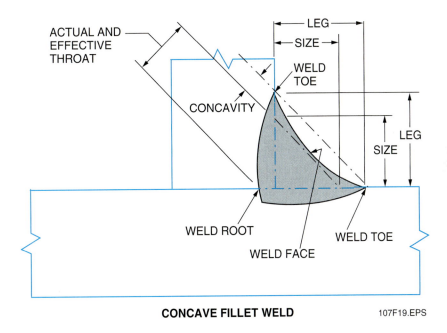

CONCAVE FILLET WELD

107F19.EPS

Figure 19 ◆ Convex and concave fillet welds.

bowed out like the outside surface of a ball. A concave fillet weld has its surface bowed in like the inside surface of a bowl. The terms used to describe a fillet weld are:

- *Weld face* – The exposed surface of the weld
- *Leg* – The distance from the root of the joint to the toe of a fillet weld

- *Weld toe* – The junction between the face of a weld and the base metal
- *Weld root* – The point at which the back of the weld intersects the base metal
- *Size* – The leg lengths of the largest right triangle that can be drawn within the cross section of a fillet weld

- *Actual throat* – The shortest distance from the root of the weld to its face
- *Effective throat* – The minimum distance, minus any convexity, from the root of the weld to its face
- *Theoretical throat* – The distance from the beginning of the joint root (with a zero opening) that is perpendicular to the hypotenuse of the largest right triangle that can be inscribed within the cross section of a fillet weld

As shown in *Figure 20*, fillet welds may be either equal leg or unequal leg. The face may be slightly convex, flat, or slightly concave. Welding codes require that fillet welds have a uniform concave or convex face, although a slightly non-uniform face is acceptable. The convexity of a fillet weld or individual surface bead should not exceed 0.07 times the actual face width or individual surface bead plus 0.06".

Note

Refer to your site's WPS for specific requirements on fillet welds. The information in this module is provided as a general guideline only. The site WPS or quality specifications must be followed for all welds. Check with your supervisor if you are unsure of the specifications for your application.

A fillet weld is unacceptable and must be repaired if the profile has insufficient throat, excessive convexity, excessive undercut, overlap, insufficient leg, or incomplete fusion, as shown in *Figure 21*.

Fillet welds require little base metal preparation except for cleaning the weld area and removing all slag from cut surfaces. Any slag from oxyfuel, plasma arc, or carbon arc cutting will cause porosity in the weld. For this reason, the codes require that all slag be removed prior to welding.

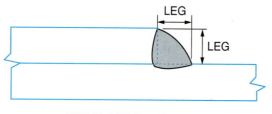

EQUAL LEG FILLET WELD

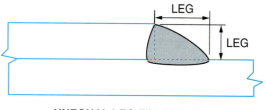

UNEQUAL LEG FILLET WELD

107F20.EPS

Figure 20 ◆ Equal leg and unequal leg fillet welds.

8.1.0 Fillet Weld Positions

The most common fillet welds are made in lap and T-joints. The weld position for plate is determined by the **weld axis** and the orientation of the workpiece. The positions for fillet welding on plate are flat, or 1F (the letter *F* stands for fillet); horizontal, or 2F; vertical, or 3F; and overhead, or 4F. In the 1F and 2F positions, the weld axis can be inclined up to 15°. Any weld axis inclination for the other positions varies with the rotational position of the weld face as specified in AWS standards. *Figure 22* shows the fillet welding positions for plate.

8.2.0 Practicing Horizontal Fillet Welds with E6010 (2F Position)

Practice horizontal fillet (2F) welding by placing multiple-pass fillet welds in a T-joint using ⁵⁄₃₂" E6010 electrodes. When making horizontal fillet welds, pay close attention to the electrode angles and travel speed. For the first bead, the electrode work angle is 45°. The work angle is adjusted for all

Preferred Fillet Weld Contours

In single-pass, weave-bead fillet welds where two workpieces are being joined at an angle (not lap joints), flat or slightly convex faces are usually preferred because weld stresses are more uniformly distributed through the fillet and workpieces.

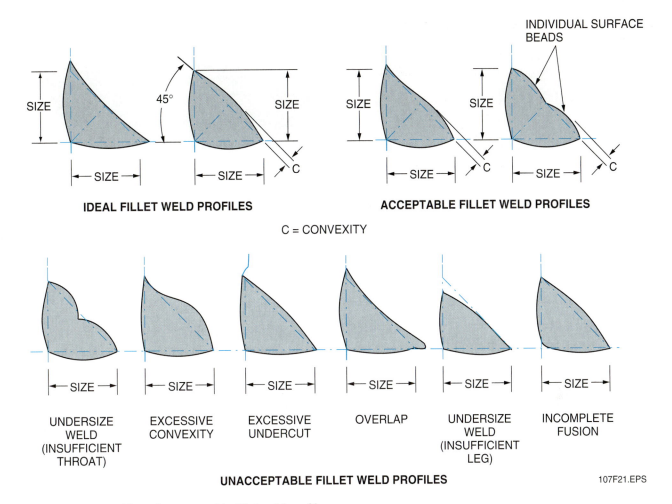

IDEAL FILLET WELD PROFILES   ACCEPTABLE FILLET WELD PROFILES

C = CONVEXITY

INDIVIDUAL SURFACE BEADS

UNACCEPTABLE FILLET WELD PROFILES

| UNDERSIZE WELD (INSUFFICIENT THROAT) | EXCESSIVE CONVEXITY | EXCESSIVE UNDERCUT | OVERLAP | UNDERSIZE WELD (INSUFFICIENT LEG) | INCOMPLETE FUSION |

107F21.EPS

Figure 21 ◆ Acceptable and unacceptable fillet weld profiles.

other welds. Increase or decrease the travel speed to control the amount of weld-metal buildup.

Follow these steps to make a horizontal fillet weld:

Step 1 Tack two plates together to form a T-joint for the fillet weld coupon. Clean the tack welds. *Figure 23* shows the fillet weld coupon.

Step 2 Position the coupon on the welding table.

Step 3 Run the first bead along the root of the joint using an electrode work angle of approximately 45° with a 10° to 15° drag angle. Use a C- or J-weave and push the arc into the root. Keep the root of the joint fusing together or a notch will appear on the leading edge of the bead.

Step 4 Clean the weld by chipping the slag and wire brushing the weld.

Tacking and Aligning Workpieces

When tacking workpieces together, both sides of the workpieces are usually tacked with welds that are about ½" long to position the workpieces and minimize distortion when the final welds are made. After the first tack weld, use a hammer or other tool to align the workpieces side to side and end to end; then tack the opposite side. Tack the far ends of the workpieces in the same manner. Intermediate tack welds can be made every 5" to 6" as necessary to minimize lengthwise distortion.

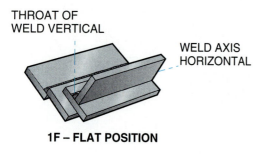

THROAT OF WELD VERTICAL

WELD AXIS HORIZONTAL

1F – FLAT POSITION

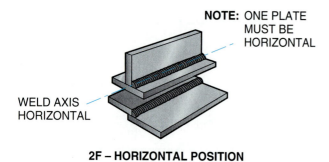

NOTE: ONE PLATE MUST BE HORIZONTAL

WELD AXIS HORIZONTAL

2F – HORIZONTAL POSITION

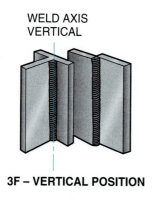

WELD AXIS VERTICAL

3F – VERTICAL POSITION

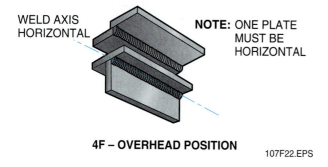

WELD AXIS HORIZONTAL

NOTE: ONE PLATE MUST BE HORIZONTAL

4F – OVERHEAD POSITION

107F22.EPS

Figure 22 ◆ Fillet welding positions for plate.

Step 5 Using a slight oscillation, run the second bead along the bottom toe of the first weld, overlapping about 75% of the first bead. Use the electrode work angle shown in *Figure 24*.

Step 6 Clean the weld.

NOTE: BASE METAL, CARBON STEEL AT LEAST ¼" THICK

107F23.EPS

Figure 23 ◆ Fillet weld coupon.

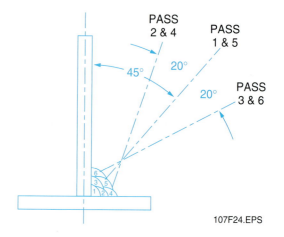

PASS 2 & 4

PASS 1 & 5

PASS 3 & 6

45° 20° 20°

107F24.EPS

Figure 24 ◆ Multiple-pass 2F weld sequences and work angles.

Step 7 Repeat Steps 5 and 6 for each of the remaining bead passes. Run the beads along the toes of the underlying beads and overlap them about 50%.

Step 8 Have the instructor inspect the weld. The weld is acceptable if it has:
- Uniform appearance on the bead face
- Craters and restarts filled to the full cross section of the weld
- Uniform weld size ±⅟₁₆"
- Acceptable weld profile in accordance with the applicable code
- Smooth transition with complete fusion at the toes of the weld
- No porosity
- No undercut
- No overlap
- No inclusions
- No cracks

8.3.0 Practicing Horizontal Fillet Welds with E7018 (2F Position)

Repeat horizontal fillet (2F) welding using ⁵⁄₃₂" E7018 electrodes. Use the same procedure, bead

---

Test Joint Heat Dissipation

In T-joints, the welding heat dissipates more rapidly in the thicker or the non-butting member. On various bead passes, the arc may have to be concentrated slightly more on the thicker or the non-butting member to compensate for the heat loss.

T-Joint Vertical Plate Undercut

The most common defect for T-joints is undercut on the vertical plate of the joint. Use of a J-weave usually eliminates the problem. However, if the problem persists, angling the arc slightly toward the vertical plate and the bead at the top of the weave will force more metal into the bead at the top edge of the weld.

sequence, and electrode angles that were used for the horizontal fillet weld with E6010 electrodes. Use a short arc and do not whip the electrode.

8.4.0 Practicing Vertical Fillet Welds with E6010 (3F Position)

Practice vertical fillet (3F) welding by placing multiple-pass fillet welds in a T-joint using ⅛" E6010 electrodes. Normally, vertical welds are accomplished by welding uphill from the bottom to the top using an electrode push angle (up-angle). Because of the uphill welding and push angle, this type of weld is sometimes called *vertical-up* fillet welding. When vertical welding, either stringer or weave beads can be used. On the job, the site WPS or quality standard will specify which technique to use. Typically, weave beads are used with E6010 electrodes and, for welding

carbon steel, with E7018 electrodes. Stringer beads are generally called for when welding alloy steels with low-hydrogen electrodes.

Note

Check with your instructor to see if you should run stringer beads, run weave beads, or practice both techniques.

Follow these steps to make a vertical-up fillet weld:

Step 1   Tack two plates together to form a T-joint for the fillet weld coupon.

Step 2   Tack-weld the coupon in the vertical position.

Vertical Fillet Welds

When making vertical fillet welds, pay close attention to the electrode angles and travel speed. For the first bead, the electrode work angle is approximately 45°. The work angle is adjusted for all other welds. Increase or decrease the travel speed to control the amount of weld metal buildup.

Step 3 Run the first bead along the root of the joint (starting at the bottom) using an electrode (work) angle of approximately 45° with a 0° to 10° push angle. Use a whipping motion by quickly raising the electrode about ¼" and then dropping it back into the weld puddle. Pause in the weld puddle to fill the crater.

Step 4 Clean the weld by chipping the slag and wire brushing the weld.

Step 5 Run the second bead using a weave technique, such as a C-pattern. Use a slow motion across the face of the weld, pausing at each toe for penetration and to fill the crater. A slight whip can be used to control the puddle when you reach the toe. Adjust the travel speed across the face of the weld to control the buildup. *Figure 25* shows the bead sequence and electrode angles for weave beads (all degrees shown are approximate). In practice, all bead passes run the entire length of the weld.

Step 6 Clean the weld.

Step 7 Continue to run weld beads as shown in *Figure 25*.

Step 8 Clean the weld.

Step 9 Have the instructor inspect the weld. The weld is acceptable if it has:
- Uniform appearance on the bead face
- Craters and restarts filled to the full cross section of the weld
- Uniform weld size ±¹⁄₁₆"
- Acceptable weld profile in accordance with the applicable code
- Smooth transition with complete fusion at the toes of the weld
- No porosity
- No undercut
- No overlap
- No inclusions
- No cracks

8.5.0 Practicing Vertical Fillet Welds with E7018 (3F Position)

Repeat vertical fillet (3F) welding using ⅛" E7018 electrodes. For the root pass and stringer fill

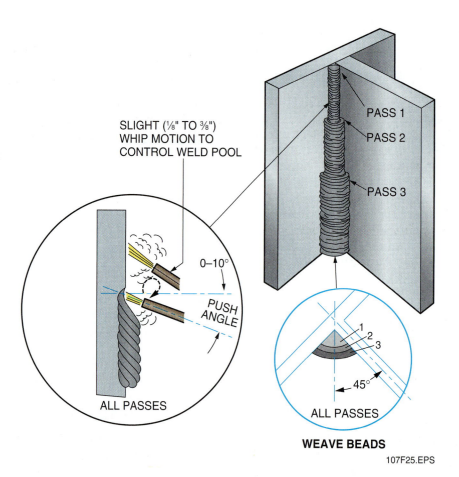

SLIGHT (⅛" TO ⅜") WHIP MOTION TO CONTROL WELD POOL

0–10°

PUSH ANGLE

ALL PASSES

PASS 1

PASS 2

PASS 3

45°

ALL PASSES

WEAVE BEADS

107F25.EPS

Figure 25 ◆ Multiple-pass 3F weld sequences and work angles (E6010).

Vertical Weave Bead

To help control undercut, an alternate pattern such as a small triangular weave can be used in vertical welds. By pausing the rod at the edges, the previous undercut can be filled. This action also creates undercut at the existing weld pool, but that will be filled in by the next weave.

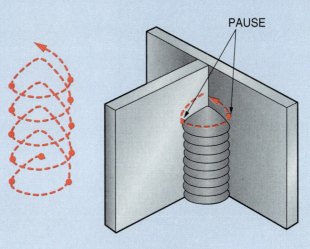

TRIANGULAR WEAVE PATTERN

107UA0701.EPS

beads, use a quick side-to-side motion, moving the electrode about ⅛" without removing the arc from the weld puddle. Pause slightly at each toe to penetrate and fill the crater to prevent undercut. The electrode work angle should be approximately 45° with a 0° to 10° upward push angle. For the remaining stringer beads, use an electrode work angle of approximately ± 20° from 45° as required.

If weave beads are to be used, use the same electrode work angles as the root pass. For the filler beads, move slowly across the face, increasing or decreasing the travel speed to control the buildup. Remember to use a short arc. Do not whip the electrode.

Figure 26 shows the bead pass sequence and electrode angle for weave and stringer beads. All degrees shown are approximate. Again, all bead passes would extend the entire length of the weld.

8.6.0 Practicing Overhead Fillet Welds with E6010 (4F Position)

Practice overhead fillet (4F) welding by welding multiple-pass convex fillet welds in a T-joint using ⅛" E6010 electrodes. When making overhead fillet welds, pay close attention to the electrode angles and travel speed. For the first bead, the electrode work angle is approximately 45°. The work angle

is adjusted for all other welds. Increase or decrease the travel speed to control the amount of weld metal buildup.

Follow these steps to make an overhead fillet weld:

Step 1 Tack two plates together to form a T-joint for the fillet weld coupon.

Step 2 Tack-weld the coupon so it is in the overhead position.

Step 3 Run the first bead along the root of the joint using an electrode angle of approximately 45° with a 10° to 15° drag angle. Use a slight oscillation (circular or side-to-side motion) to tie-in the weld at the toes.

Step 4 Clean the weld.

Step 5 Using a slight oscillation, run the second bead along the bottom toe of the first weld, overlapping about 75% of the first bead. Use the electrode work angle shown in *Figure 27*.

Step 6 Clean the weld.

Step 7 Repeat Steps 5 and 6 for each of the remaining bead passes. Run the beads along the weld toes of the underlying beads and overlap them about 50%.

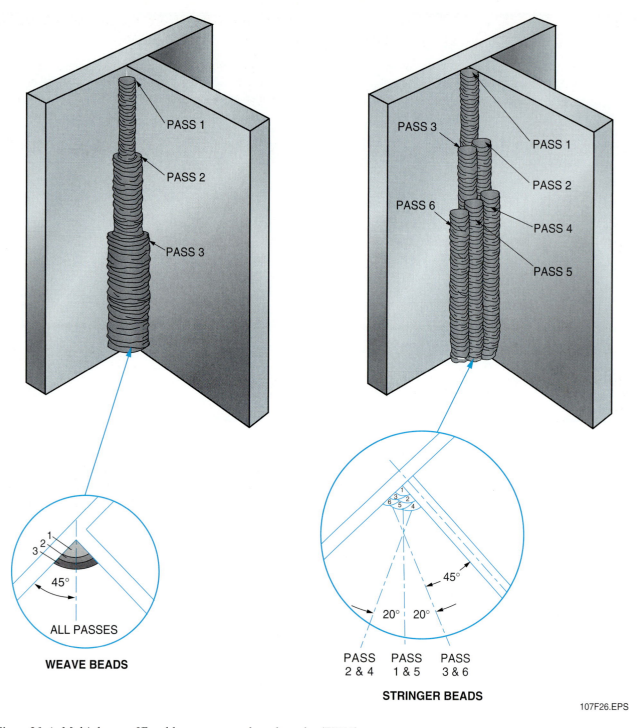

PASS 1
PASS 2
PASS 3

PASS 3
PASS 1
PASS 2
PASS 6
PASS 4
PASS 5

1
3 2
3
45°

ALL PASSES

WEAVE BEADS

1
3 2
6 5 4

45°

20° 20°

PASS
2 & 4

PASS
1 & 5

PASS
3 & 6

STRINGER BEADS

107F26.EPS

Figure 26 ◆ Multiple-pass 3F weld sequences and work angles (E7018).

Step 8   Have the instructor inspect the weld. The weld is acceptable if it has:
- Uniform appearance on the bead face
- Craters and restarts filled to the full cross section of the weld
- Uniform weld size $\pm\frac{1}{16}$"
- Acceptable weld profile in accordance with the applicable code

- Smooth transition with complete fusion at the toes of the weld
- No porosity
- No undercut
- No overlap
- No inclusions
- No cracks

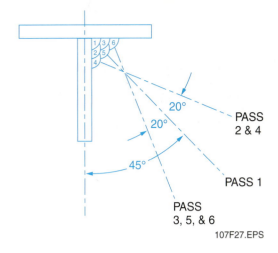

Figure 27 ◆ Multiple-pass 4F weld sequences and work angles.

8.7.0 Practicing Overhead Fillet Welds with E7018 (4F Position)

Repeat overhead fillet (4F) welding using ⅛" E7018 electrodes. Use the same procedure, bead sequence, and electrode angles that were used for the overhead fillet weld with E6010 electrodes. Use a short arc and do not whip the electrode.

Summary

Striking an arc, running stringer and weave beads, and making fillet welds are essential skills a welder must have to perform basic welding jobs and to progress to more difficult welding procedures. It is important to practice these welds until acceptable welds are produced.

These welds can be performed in different positions: flat, vertical, horizontal, and overhead. It is necessary to perform welding with different types of electrodes in different positions.

Review Questions

1. When preparing to practice welding, the welding surface must be _____.
   a. wood
   b. steel
   c. nonferrous
   d. nongrounded

2. When preparing to practice welding fillet welds, cut the metal into _____ for the base.
   a. 3" × 6" rectangles
   b. 4" × 6" rectangles
   c. 4" × 4" squares
   d. 6" × 6" squares

3. When striking an arc, the general rule for arc length is that the arc length should be _____ of the electrode being used.
   a. one-quarter the diameter
   b. one-half the diameter
   c. three-quarters the diameter
   d. the diameter

4. The easiest method of striking an arc is the _____ method.
   a. scratching
   b. grounding
   c. weaving
   d. tapping

5. DC arc blow can cause defects such as excessive weld spatter and _____.
   a. porosity
   b. burn through
   c. increased magnetic fields
   d. decreased magnetic fields

6. A weld bead with very little or no side-to-side motion of the electrode is called a(n) _____ bead.
   a. arc
   b. weave
   c. stringer
   d. whipping

7. After striking the arc when practicing stringer welds, the electrode angle should be 10° to 15° in the direction of travel and _____ to the base metal.
   a. 30°
   b. 45°
   c. 60°
   d. 90°

8. The point where one weld bead stops and another starts is called a _____.
   a. retrace
   b. restart
   c. weave
   d. blend

9. Overlapping beads are made by depositing connective weld beads parallel to one another to form a(n) _____ surface.
   a. flat
   b. undercut
   c. underfill
   d. excessive buildup

10. The part of a fillet weld that can be convex or concave is called the _____.
    a. leg
    b. throat
    c. face
    d. root

Bill Cherry

Manager, Welder Testing & Training
Zachry Construction Corp.
Deer Park, Texas

Bill Cherry was born in San Antonio, Texas, the second of six children. After receiving his high school diploma in 1970, he was hired by a road boring company. He learned to do welding on the job, fixing reamers and other road boring equipment. In 1976 Bill decided to concentrate on welding and took a job as a welder's helper at Zachry Construction in Deer Park. Eventually, Bill became a certified welder and then a certified welding inspector. In his current position at Zachry, Bill trains welders and also performs quality control.

What do you think it takes to be a success in your trade?
In the welding and inspection trades, a person has to continually strive to better himself by learning new processes. I get very bored if I don't have a challenge in my job. Being limited in my capabilities is not very rewarding to me.

What are some of the things you do in your job?
I test welders for job sites, train beginner welders, help experienced welders to upgrade their skills, and track welder continuity. I also furnish weld test coupons to our out-of-state job sites.

What do you like most about your job?
It's a new challenge every day, helping people master their craft.

What would you say to someone entering the trade today?
Don't limit yourself in your endeavors. Having basic minimum skills will not be very rewarding and will provide little or no opportunity for advancement.

Trade Terms Introduced in This Module

Arc blow: The deflection of the arc from its intended path by magnetic forces.

Arc strike: A discontinuity consisting of localized melting of the base metal or finished weld caused by the arc. It occurs as a result of striking the arc outside the area to be welded.

Drag angle: The travel angle when the electrode is pointing in a direction opposite to the welding progression.

Oscillation: A side-to-side motion.

Push angle: The travel angle when the electrode is pointing in the same direction as the welding progression.

Restart: The point in the weld where one weld bead stops and the continuing bead is started.

Rods: A nonstandard term for the welding electrodes used in SMAW.

SMAW: Shielded metal arc welding; sometimes called by the nonstandard name stick welding.

Stringer bead: A type of weld bead made without appreciable weaving motion. With shielded metal arc welding (SMAW), stringer beads are not more than three times the electrode diameter.

Weave bead: A type of weld bead made by transverse oscillation of the electrode.

Weld axis: A straight line drawn through the center of a weld along the length of the weld.

Weld coupon: The metal pieces to be welded together as a test or practice.

Work angle: An angle less than 90° between a line perpendicular to the major workpiece surface and a plane determined by the electrode axis and the weld axis. In a T-joint or corner joint, the line is perpendicular to the non-butting member. The definition of work angle for a pipe weld is covered in a later module.

Performance Accreditation Tasks

The Performance Accreditation Tasks (PATCs) correspond to and support the learning objectives in AWS EG2.0-95, Guide for the Training and Qualification of Welding Personnel: Entry-Level Welder.

Note that in order to satisfy all learning objectives in AWS EG2.0-95, the instructor must also use the PATCs contained in the second level in the Welding Contren™ Learning Series.

PATCs provide specific acceptable critera for performance and help to ensure a true competency-based welding program for students.

The following tasks are designed to evaluate your ability to run beads and fillet welds with SMAW equipment. Perform each task when you are instructed to do so by your instructor. As you complete each task, bring it to your instructor for evaluation. Do not proceed to the next task until instructed to do so by your instructor.

BUILD A PAD WITH E6010 ELECTRODES IN THE FLAT POSITION

Using ⅛" E6010 electrodes, build up a pad of weld metal on carbon steel plate as indicated.

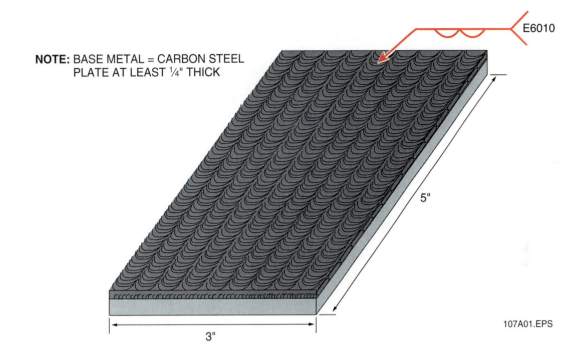

NOTE: BASE METAL = CARBON STEEL PLATE AT LEAST ¼" THICK

E6010

5"

3"

107A01.EPS

Criteria for Acceptance:

- Weld beads straight to within ⅛"
- Uniform rippled appearance on the bead face
- Craters and restarts filled to the full cross-section of the weld
- Face of the pad flat to within ⅛"
- Smooth flat transition with complete fusion at the toes of one bead into the face of the previous bead
- No porosity
- No overlap at weld toes
- No undercut
- No inclusions
- No cracks

BUILD A PAD WITH E7018 ELECTRODES IN THE FLAT POSITION

Using ⅛" E7018 electrodes, build up a pad of weld metal on carbon steel plate as indicated.

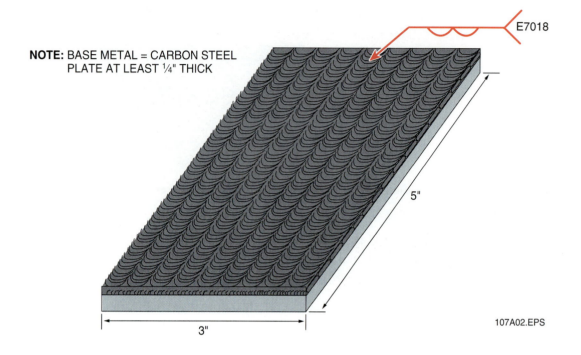

NOTE: BASE METAL = CARBON STEEL PLATE AT LEAST ¼" THICK

E7018

5"

3"

107A02.EPS

Criteria for Acceptance:

- Weld beads straight to within ⅛"
- Uniform rippled appearance on the bead face
- Craters and restarts filled to the full cross-section of the weld
- Face of the pad flat to within ⅛"
- Smooth flat transition with complete fusion at the toes of one bead into the face of the previous bead
- No porosity
- No overlap at weld toes
- No undercut
- No inclusions
- No cracks

HORIZONTAL (2F) FILLET WELD
WITH E6010 ELECTRODES

Using ⅛" E6010 electrodes, make a horizontal fillet weld as indicated.

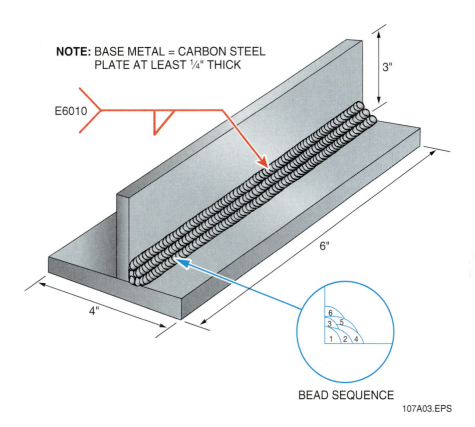

NOTE: BASE METAL = CARBON STEEL
PLATE AT LEAST ¼" THICK

E6010

3"

6"

4"

BEAD SEQUENCE

6
3 5
1 2 4

107A03.EPS

Criteria for Acceptance:

- Uniform rippled appearance on the bead face
- Craters and restarts filled to the full cross-section of the weld
- Uniform weld size, ±¹⁄₁₆"
- Acceptable weld profile in accordance with AWS D1.1
- Smooth transition with complete fusion at the toes of the weld
- No porosity
- No undercut
- No inclusions
- No cracks
- No overlap

HORIZONTAL (2F) FILLET WELD WITH E7018 ELECTRODES

Using ⅛" E7018 electrodes, make a horizontal fillet weld as indicated.

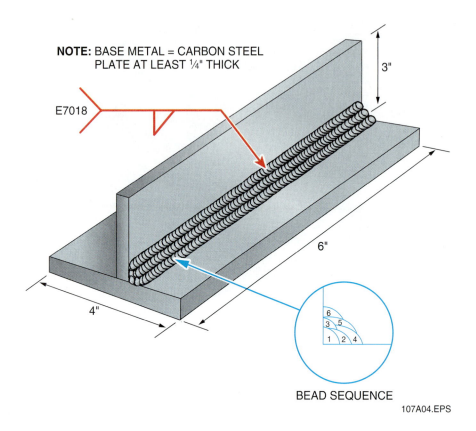

NOTE: BASE METAL = CARBON STEEL
PLATE AT LEAST ¼" THICK

E7018

3"

6"

4"

6
3 5
1 2 4

BEAD SEQUENCE

107A04.EPS

Criteria for Acceptance:

- Uniform rippled appearance on the bead face
- Craters and restarts filled to the full cross-section of the weld
- Uniform weld size, ±⅟₁₆"
- Acceptable weld profile in accordance with AWS D1.1
- Smooth transition with complete fusion at the toes of the weld
- No porosity
- No undercut
- No inclusions
- No cracks
- No overlap

VERTICAL (3F) FILLET WELD
WITH E6010 ELECTRODES

Using ⅛" E6010 electrodes, make a vertical fillet weld as indicated.

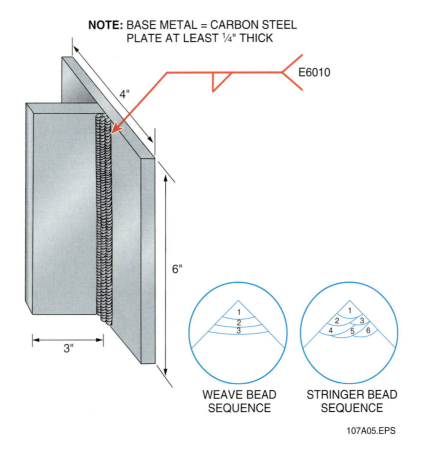

NOTE: BASE METAL = CARBON STEEL
PLATE AT LEAST ¼" THICK

E6010

4"

6"

3"

WEAVE BEAD
SEQUENCE

STRINGER BEAD
SEQUENCE

107A05.EPS

Criteria for Acceptance:

- Uniform rippled appearance on the bead face
- Craters and restarts filled to the full cross-section of the weld
- Uniform weld size, ±¹⁄₁₆"
- Acceptable weld profile in accordance with AWS D1.1
- Smooth transition with complete fusion at the toes of the weld
- No porosity
- No undercut
- No inclusions
- No cracks
- No overlap at weld toes

VERTICAL (3F) FILLET WELD WITH E7018 ELECTRODES

Using ⅛" E7018 electrodes, make a vertical fillet weld as indicated.

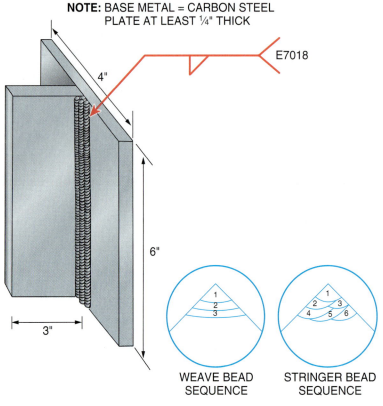

NOTE: BASE METAL = CARBON STEEL PLATE AT LEAST ¼" THICK

E7018

4"

6"

3"

WEAVE BEAD SEQUENCE

STRINGER BEAD SEQUENCE

107A06.EPS

Criteria for Acceptance:

- Uniform rippled appearance on the bead face
- Craters and restarts filled to the full cross-section of the weld
- Uniform weld size, ±1⁄16"
- Acceptable weld profile in accordance with AWS D1.1
- Smooth transition with complete fusion at the toes of the weld
- No porosity
- No undercut
- No inclusions
- No cracks
- No overlap at weld toes

OVERHEAD (4F) FILLET WELD
WITH E6010 ELECTRODES

Using ⅛" E6010 electrodes, make an overhead fillet weld as indicated.

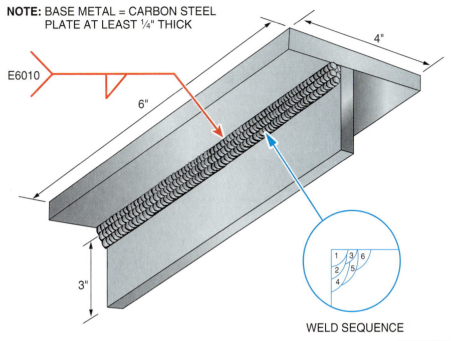

NOTE: BASE METAL = CARBON STEEL PLATE AT LEAST ¼" THICK

E6010

4"

6"

3"

WELD SEQUENCE

107A07.EPS

Criteria for Acceptance:

- Uniform rippled appearance on the bead face
- Craters and restarts filled to the full cross-section of the weld
- Uniform weld size, ±¹⁄₁₆"
- Acceptable weld profile in accordance with AWS D1.1
- Smooth transition with complete fusion at the toes of the weld
- No porosity
- No undercut
- No inclusions
- No cracks
- No overlap

OVERHEAD (4F) FILLET WELD WITH E7018 ELECTRODES

Using ⅛" E7018 electrodes, make an overhead fillet weld as indicated.

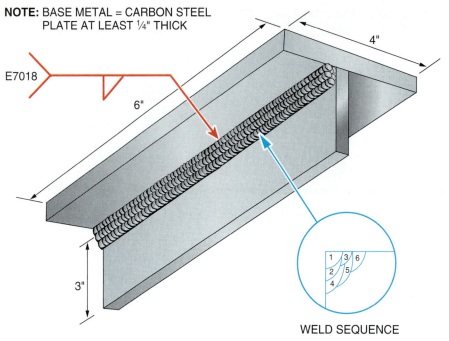

NOTE: BASE METAL = CARBON STEEL PLATE AT LEAST ¼" THICK

E7018

4"

6"

3"

WELD SEQUENCE

107A08.EPS

Criteria for Acceptance:

- Uniform rippled appearance on the bead face
- Craters and restarts filled to the full cross-section of the weld
- Uniform weld size, ±¹⁄₁₆"
- Acceptable weld profile in accordance with AWS D1.1
- Smooth transition with complete fusion at the toes of the weld
- No porosity
- No undercut
- No inclusions
- No cracks
- No overlap

Additional Resources

This module is intended to present thorough resources for task training. The following reference works are suggested for further study. These are optional materials for continued education rather than for task training.

Modern Welding, 1977. A. D. Althouse, C. H. Turnquist, W. A. Bowditch, K. E. Bowditch. Tinley Park, IL: The Goodheart-Willcox Company, Inc.

The Procedure Handbook of Arc Welding, 1994. Cleveland, OH: The Lincoln Electric Company.

Welding Technology, 2001. J. W. Giachino, W. R. Weeks, G. S. Johnson. Homewood, IL: American Technical Publishers, Inc.

CONTREN™ LEARNING SERIES—USER UPDATES

The NCCER makes every effort to keep these textbooks up-to-date and free of technical errors. We appreciate your help in this process. If you have an idea for improving this textbook, or if you find an error, a typographical mistake, or an inaccuracy in NCCER's Contren™ textbooks, please write us, using this form or a photocopy. Be sure to include the exact module number, page number, a detailed description, and the correction, if applicable. Your input will be brought to the attention of the Technical Review Committee. Thank you for your assistance.

Instructors—If you found that additional materials were necessary in order to teach this module effectively, please let us know so that we may include them in the Equipment/Materials list in the Instructor's Guide.

Write: Curriculum Revision and Development Department
National Center for Construction Education and Research
P.O. Box 141104, Gainesville, FL 32614-1104

Fax: 352-334-0932

E-mail: curriculum@nccer.org

Craft _____   Module Name _____

Copyright Date _____   Module Number _____   Page Number(s) _____

Description _____

_____

_____

_____

(Optional) Correction _____

_____

_____

(Optional) Your Name and Address _____

_____

_____

Shielded Metal Arc Welding – Groove Welds with Backing

Course Map

This course map shows all of the modules in the AWS Entry Level Welder – Phase 1 curriculum. The suggested training order begins at the bottom and proceeds up. Skill levels increase as you advance on the course map. The local Training Program Sponsor may adjust the training order.

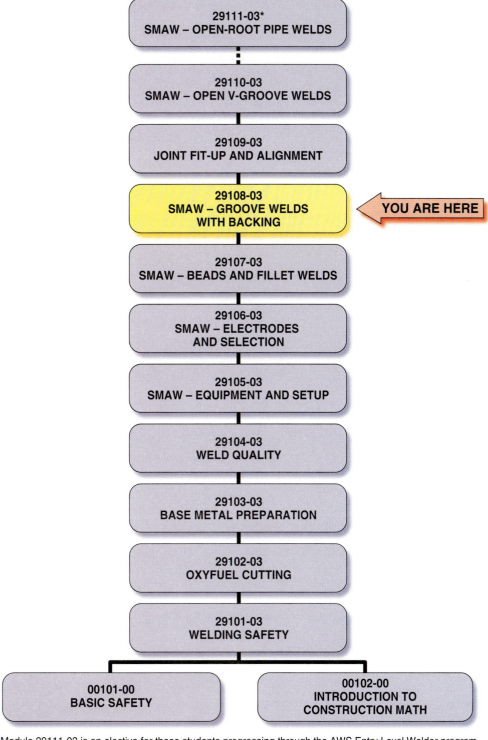

AWS ENTRY LEVEL WELDER—PHASE 1

29111-03*
SMAW – OPEN-ROOT PIPE WELDS

29110-03
SMAW – OPEN V-GROOVE WELDS

29109-03
JOINT FIT-UP AND ALIGNMENT

29108-03
SMAW – GROOVE WELDS
WITH BACKING   ◄ YOU ARE HERE

29107-03
SMAW – BEADS AND FILLET WELDS

29106-03
SMAW – ELECTRODES
AND SELECTION

29105-03
SMAW – EQUIPMENT AND SETUP

29104-03
WELD QUALITY

29103-03
BASE METAL PREPARATION

29102-03
OXYFUEL CUTTING

29101-03
WELDING SAFETY

00101-00
BASIC SAFETY

00102-00
INTRODUCTION TO
CONSTRUCTION MATH

*Module 29111-03 is an elective for those students progressing through the AWS Entry Level Welder program.

29108CMAP.EPS

Figures

Shielded Metal Arc Welding – Groove Welds with Backing

Objectives

When you have completed this module, you will be able to do the following:

1. Identify and explain groove welds.
2. Identify and explain groove welds with backing.
3. Set up shielded metal arc welding (SMAW) equipment for making V-groove welds.
4. Perform SMAW for V-groove welds with backing in the:
   - Flat (1G) position
   - Horizontal (2G) position
   - Vertical (3G) position
   - Overhead (4G) position

Prerequisites

Before you begin this module, it is recommended that you successfully complete the following: Modules 00100-00 through 29107-03.

Required Trainee Materials

1. Pencil and paper
2. Appropriate personal protective equipment

1.0.0 ◆ INTRODUCTION

Different welding applications require different shapes at the point where two metal parts are to be joined. This module explains groove welds, the arc welding equipment used with groove welds, and the different ways to make groove welds.

2.0.0 ◆ GROOVE WELDS

A groove weld is a weld made in an opening in a part or between two parts. Groove welds can be made in all five of the basic types of joints as well as on the surface of a part. There are several different groove styles. The name of the groove is derived from its shape. Some must be prepared; others appear as a result of fitting the parts to be welded together.

2.1.0 Typical Groove Weld Styles

Groove weld styles include the square, V, U, bevel, J, flare V, and flare bevel. *Figure 1* shows typical single groove weld styles.

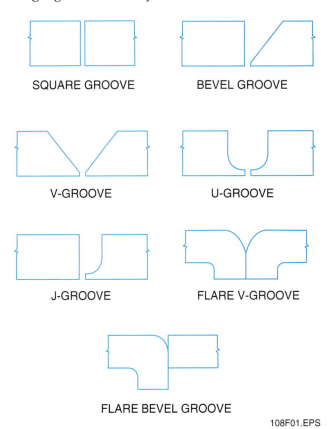

SQUARE GROOVE BEVEL GROOVE

V-GROOVE U-GROOVE

J-GROOVE FLARE V-GROOVE

FLARE BEVEL GROOVE

108F01.EPS

Figure 1 ◆ Single groove weld styles.

Groove Weld Penetration Definitions

If a groove only goes part of the way through the joint, welding in the groove results in what is called a *partial joint penetration (PJP)* weld. If the groove allows for welding the complete thickness of the base metals, it is called a *complete joint penetration (CJP)* weld.

2.2.0 Single and Double Groove Welds

V-, bevel, U-, and J-grooves can be used and welded on one side of a joint or on both sides. If they are on both sides of a joint, they are called *double groove* welds. *Figure 2* shows double groove weld styles.

2.3.0 Groove Weld Terms

The terms used to describe a groove are:

- *Bevel angle* – The angle formed between the prepared edge of a member and a line perpendicular to the surface of the member.
- *Bevel face* – The prepared surface of a bevel edge shape.
- *Bevel radius* – The radius used to form the shape of a J- or U-groove.
- *Depth of bevel* – The distance the joint preparation extends into the base metal (used for partial joint penetration welds).

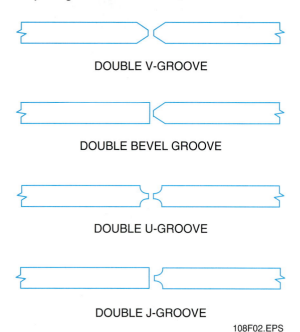

DOUBLE V-GROOVE

DOUBLE BEVEL GROOVE

DOUBLE U-GROOVE

DOUBLE J-GROOVE

108F02.EPS

Figure 2 ◆ Double groove weld styles.

- *Groove angle* – The included (total) angle between the groove faces of a weld groove.
- *Groove face* – Any surface in a weld groove prior to welding. See *bevel face, root face,* and *bevel radius.*
- *Joint root* – The portion of a joint where the members approach closest to each other.
- *Root face* – The portion of the edge of a part to be joined by a groove weld that has not been beveled or grooved.
- *Root opening* – The separation between the members to be joined at the root of the joint.

Figure 3 shows examples of grooves. The terms used to describe a groove weld are:

- *Face reinforcement* – Reinforcement of the weld by excess weld metal on the side from which the weld was made.
- *Groove weld size* – The joint penetration (depth of joint preparation plus root penetration). The size can never be greater than the base metal thickness.
- *Root reinforcement* – Reinforcement of the weld by excess weld metal on the side other than the side from which the weld was made.
- *Root surface* – The exposed surface of a weld opposite the side from which the welding was done.
- *Weld face* – The exposed surface of the weld.
- *Weld root* – The point at which the back of the weld extends the furthest into the weld joint.
- *Weld toe* – The points where the face of the weld metal and the base metal meet.

Figure 4 shows a V-groove weld.

2.4.0 Combination Groove and Fillet Welds

Groove welds are often used in combination with fillet welds to maintain the strength of the welded joint and use less weld metal. *Figure 5* shows various groove and fillet weld combinations.

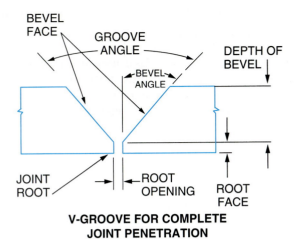

V-GROOVE FOR COMPLETE JOINT PENETRATION

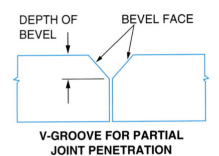

V-GROOVE FOR PARTIAL JOINT PENETRATION

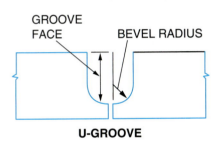

U-GROOVE

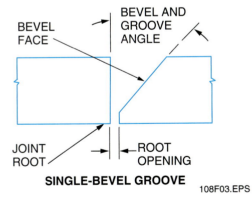

SINGLE-BEVEL GROOVE

108F03.EPS

Figure 3 ◆ Grooves.

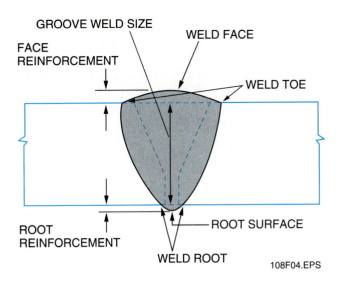

108F04.EPS

Figure 4 ◆ V-groove weld.

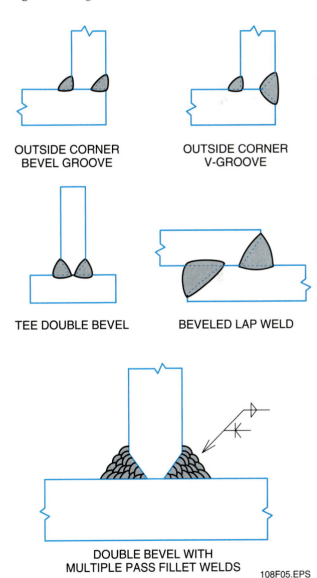

OUTSIDE CORNER BEVEL GROOVE

OUTSIDE CORNER V-GROOVE

TEE DOUBLE BEVEL

BEVELED LAP WELD

DOUBLE BEVEL WITH MULTIPLE PASS FILLET WELDS

108F05.EPS

Figure 5 ◆ Various groove and fillet weld combinations.

2.5.0 Backings

Although single groove welds can be done without backing, single groove welds with backing are used for high-strength, high-quality welds. Backing also makes welding the root easier. Thick metal pieces that are to be joined must be beveled so that the weld will penetrate its full thickness. The simplest form of backing is a strip of metal matching the base metal being welded. Other types of backing include dissimilar metals, ceramic, and coated tapes that can be removed when the welding is completed.

This module will explain the preparation and welding of V-groove welds with metal backing using low-hydrogen electrodes in all positions. Other types of groove welds can be performed in the same manner. Performance demonstrations will require that the welds meet the qualification standards in the applicable code for visual and destructive testing.

> **Note**
>
> The dimensions and specifications in this module are designed to be representative of codes in general and are not specific to any certain code. Always follow the proper codes for your site.

3.0.0 ◆ WELDING EQUIPMENT SETUP

Before welding can take place, the area has to be made ready, the welding equipment must be set up, and the metal to be welded must be prepared. The following sections explain how to set up arc welding equipment for welding.

3.1.0 Safety Practices

The following is a summary of safety procedures and practices that must be observed while cutting or welding. Keep in mind that this is just a summary. Complete safety coverage is provided in the Level One module *Welding Safety*. If you have not completed that module, do so before continuing. Above all, be sure to wear appropriate protective clothing and equipment when welding or cutting.

3.1.1 Protective Clothing and Equipment

- Always use safety goggles with a full face shield or a helmet. The goggles, face shield, or helmet lens must have the proper light-reducing tint for the type of welding or cut-

ting to be performed. Never directly or indirectly view an electric arc without using a properly tinted lens.
- Wear proper protective leather and/or flame retardant clothing along with welding gloves that will protect you from flying sparks and molten metal as well as heat.
- Wear 8" or taller high-top safety shoes or boots. Make sure that the tongue and lace area of the footwear will be covered by a pant leg. If the tongue and lace area is exposed or the footwear must be protected from burn marks, wear leather spats under the pants or chaps and over the front top of the footwear.
- Wear a solid material (non-mesh) hat with a bill pointing to the rear or, if much overhead cutting or welding is required, a full leather hood with a welding face plate and the correct tinted lens. If a hard hat is required, use a hard hat that allows the attachment of rear deflector material and a face shield.
- If a full leather hood is not worn, wear a face shield and snug-fitting welding goggles over safety glasses for gas welding or cutting. Either the face shield or the lenses of the welding goggles must be an approved shade 5 or 6 filter. For electric arc welding or cutting, wear safety goggles and a welding hood with the correct tinted lens (shade 10 to 14).
- If a full leather hood is not worn, wear earmuffs or at least earplugs to protect ear canals from sparks.

3.1.2 Fire/Explosion Prevention

- Never carry matches or gas-filled lighters in your pockets. Sparks can cause the matches to ignite or the lighter to explode, causing serious injury.
- Never perform any type of heating, cutting, or welding until a hot work permit is obtained and an approved fire watch is established. Most work site fires caused by these types of operations are started by cutting torches.
- Never use oxygen to blow off clothing. The oxygen can remain trapped in the fabric for a time. If a spark hits the clothing during this time, the clothing can burn rapidly and violently out of control.
- Make sure that any flammable material in the work area is moved or shielded by a fire-resistant covering. Approved fire extinguishers must be available before any heating, welding, or cutting operations are attempted.
- Never release a large amount of oxygen or use oxygen as compressed air. Its presence around flammable materials or sparks can cause rapid

AWS ENTRY LEVEL WELDER – TRAINEE MODULE 29108-03

and uncontrolled combustion. Keep oxygen away from oil, grease and other petroleum products.

- Never release a large amount of fuel gas, especially acetylene. Methane and propane tend to concentrate in and along low areas and can be ignited a considerable distance from the release point. Acetylene is lighter than air but is even more dangerous than methane. When mixed with air or oxygen, it will explode at much lower concentrations than any other fuel.
- To prevent fires, maintain a neat and clean work area and make sure that any metal scrap or slag is cold before disposal.
- Before cutting or welding containers such as tanks or barrels, check to see if they have contained any explosive, hazardous, or flammable materials including petroleum products, citrus products, or chemicals that decompose into toxic fumes when heated. As a standard practice, always clean and then fill any tanks or barrels with water or purge them with a flow of inert gas to displace any oxygen.

3.1.3 Work Area Ventilation

- Make sure confined space procedures are followed before conducting any welding or cutting in the confined space.
- Make sure confined spaces are ventilated properly for cutting or welding purposes.
- Always perform cutting or welding operations in a well-ventilated area. Cutting or welding operations involving zinc or cadmium materials or coatings result in toxic fumes. For long-term cutting or welding of such materials, always wear an approved full face, supplied-air respirator (SAR) that uses breathing air supplied outside of the work area. For occasional, very short-term exposure, a HEPA-rated or metal-fume filter may be used on a standard respirator.
- Never use oxygen in confined spaces for ventilation purposes.

3.2.0 Preparing the Welding Area

To practice welding, a welding table, bench, or stand is needed. The welding surface must be steel, and provisions must be made for placing welding coupons out of position. *Figure 6* shows a typical welding station.

To set up the area for welding, follow these steps:

Step 1 Check to be sure the area is properly ventilated. Use doors, windows, and fans.

108F06.EPS

Figure 6 ◆ Welding station.

Step 2 Check the area for fire hazards. Remove any flammable materials before proceeding.

Step 3 Check the location of the nearest fire extinguisher. Do not proceed unless the extinguisher is charged and you know how to use it.

Step 4 Position a welding table near the welding machine.

Step 5 Set up flash shields around the welding area.

3.3.0 Preparing the Weld Coupons

If possible, the weld coupons should be carbon steel ⅜" thick to conform with AWS limited-thickness test coupon requirements. If this size is not readily available, ¼"- to ¾"-thick steel can be used for practice welds.

Clean the steel plate prior to welding by using a wire brush or grinder to remove heavy mill scale or corrosion.

For each weld coupon, you will need two pieces ⅜" × 3" × 7" and one piece ¼" × 1" × 8".

Prepare weld coupons to practice the following welds:

- *Running beads* – The coupons can be any size or shape that is easily handled.

Beveling Guide

When no track cutter is available, use a piece of angle iron to guide your cutting torch in making the bevel.

108P0801.EPS

- *V-groove welds with backing* – Cut the metal into 3" × 7" rectangles, with the 7" lengths beveled at approximately 22½°. Cut backing strips at least 1" wide by 8" long. The backing strips must be the same base metal type as the beveled strips.

Follow these steps to prepare the V-groove with metal backing weld coupon:

Step 1   Check the bevel. There should be no land and no slag, and the bevel angle should be approximately 22½°.

Step 2   Center the beveled strips on the backing strip with a ¼" gap and tack-weld them in place. Place the tack welds on the back side of the joint in the lap formed by the backing strip and the beveled plate. Use three to four ½" tack welds on each beveled plate. Be sure the backing strip is tight against the beveled plates. *Figure 7* shows the V-groove with metal backing weld coupon.

Backing Strip

When the backing strip is ½" to 1" longer at each end of the weld groove, it allows the welder to start and stop the bead outside the weld groove.

Figure 8 shows the alternate horizontal weld coupons. Degrees shown are approximate.

Note

Check with your instructor on whether to use the standard V-groove preparation or the alternate preparation for horizontal weld coupons.

3.4.0 Electrodes

Obtain a small quantity of the electrodes to be used. The welding exercises in this module use $\frac{3}{32}$", $\frac{1}{8}$", and $\frac{5}{32}$" E7018 electrodes.

Note

The electrode sizes are recommendations. Other electrode sizes may be substituted depending on site conditions. Check with your instructor for the electrode sizes to use.

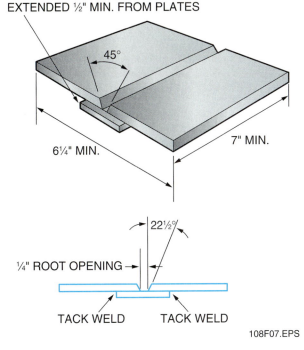

BACKING STRIP 1" MIN. WIDTH
EXTENDED ½" MIN. FROM PLATES

45°

6¼" MIN.   7" MIN.

22½°

¼" ROOT OPENING

TACK WELD   TACK WELD

108F07.EPS

Figure 7 ◆ V-groove with metal backing weld coupon.

When welding in the horizontal position, an alternate joint preparation can be used. The alternate joint has one plate beveled at approximately 45° and the other plate at approximately 90°. The plates are positioned with the beveled plate above the 90° plate and a ¼" root opening.

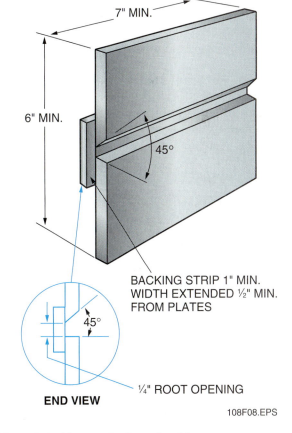

NOTE: BASE METAL MILD STEEL

7" MIN.

6" MIN.

45°

BACKING STRIP 1" MIN.
WIDTH EXTENDED ½" MIN.
FROM PLATES

45°

END VIEW

¼" ROOT OPENING

108F08.EPS

Figure 8 ◆ Alternate horizontal weld coupons.

Hot Tip

Conserving Material

Steel for practice welding is expensive and difficult to obtain. Every effort should be made to conserve and not waste the material that is available. Reuse weld coupons until all surfaces have been welded on by cutting the weld coupon apart and reusing the pieces. Use material that cannot be cut into weld coupons to practice running beads.

Keep these general rules in mind when running low-hydrogen electrodes:

- Do not whip low-hydrogen electrodes.
- Maintain a short arc because a low-hydrogen weld relies on its slag to protect the molten metal. It does not have a heavy gaseous shield.
- Remove all slag between passes.
- Remove only a small number of electrodes at a time from the oven because low-hydrogen electrodes pick up moisture quickly.
- Restart by striking the arc ahead of the crater, move quickly back into the crater, and then proceed as usual. This technique welds over the arc strike, eliminating porosity and giving a smoother restart.
- When running stringer beads, use a slight side-to-side motion with a slight pause at the weld toe to tie in and flatten the bead.
- When running weave beads, use a slight pause at the weld toe to tie in and flatten the bead.
- Do not use low-hydrogen electrodes with chipped or missing flux.
- Set the amperage in the lower portion of the suggested range for better puddle control when welding vertically.

Obtain only the electrodes to be used for a particular welding exercise at one time. Have some type of pouch or rod holder in which to store the electrodes to prevent them from becoming damaged. Never store electrodes loose on a table. They may end up on the floor where they are a tripping hazard and can become damaged. Some type of metal container such as a pail must also be available to discard hot electrode stubs.

WARNING!
Do not throw electrode stubs on the floor. They roll easily and someone could step on one, slip, and fall.

Hot Tip

LASH

To ensure good welding results, remember LASH.

Length of arc (L) – The distance between the electrode and the base metal (usually one times the electrode diameter).

Angle (A) – Two angles are critical:

- *Travel angle* – The longitudinal angle of the electrode in relation to the axis of the weld joint
- *Work angle* – The traverse angle of the electrode in relation to the axis of the weld joint

Speed (S) – Travel speed is measured in inches per minute (IPM). The width of the weld will determine if the travel speed is correct.

Heat (H) – Controlled by the amperage setting and dependent upon the electrode diameter, base metal type, base metal thickness, and the welding position.

3.5.0 Preparing the Welding Machine

Identify a welding machine to use and then follow these steps to set it up for welding:

Step 1   Verify that the welding machine can be used for DC (direct current) welding.

Step 2   Check to be sure that the welding machine is properly grounded through the current receptacle.

Step 3   Check the area for proper ventilation.

Step 4   Verify the location of the current disconnect.

Step 5   Set the polarity to direct current electrode positive (DCEP).

Step 6   Connect the clamp of the workpiece lead to the workpiece.

Step 7   Set the amperage for the electrode type and size to be used. Typical settings are:

Electrode	Size	Amperage
E7018	$\frac{3}{32}$"	50A to 110A
E7018	$\frac{1}{8}$"	90A to 150A
E7018	$\frac{5}{32}$"	120A to 190A

Note

Amperage recommendations vary by manufacturer, position, current type, and electrode brand. For specific recommendations, refer to the manufacturer's recommendations for the electrode being used.

Step 8   Check to be sure the electrode holder is not grounded.

Step 9   Turn on the welding machine.

Figure 9 shows typical DC and AC/DC welding machines.

4.0.0 ◆ V-GROOVE WELDS WITH BACKING

The V-groove with backing is a common groove weld normally made with low-hydrogen electrodes. The backing can be dissimilar material, steel, or a backing weld run with E7018. Depending on the code or procedure, low-hydrogen elec-

trodes may be used for the root pass on an open root weld. However, if low-hydrogen electrodes are used on an open root, back gouging and back welding may be required. Using a steel backing eliminates the need for back gouging and is a fast and effective way to prepare a joint for welding. It also makes the root pass easier to run.

The V-groove with steel backing is also the standard AWS qualification test for plate welding. The most common test is the limited-thickness qualification (up to ¾"), which requires the use of ⅜" carbon steel plate and welding with low-hydrogen electrodes.

4.1.0 Groove Weld Positions

Groove welds can be made in all positions. The weld position for plate is determined by the weld and the orientation of the workpiece. Groove weld positions for plate are flat, or 1G (groove); horizontal, or 2G; vertical, or 3G; and overhead, or 4G. In the 1G and 2G positions, the weld axis can be inclined up to 15°. Any weld axis inclination for the other positions varies with the rotational position of the weld face as specified in AWS standards. *Figure 10* shows the groove weld positions.

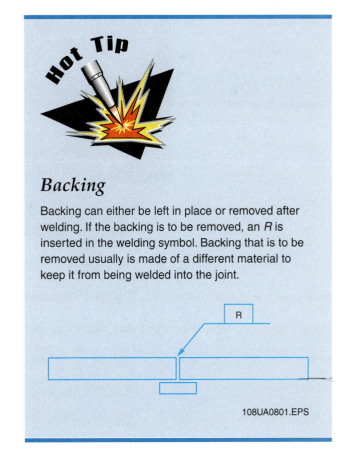

Backing

Backing can either be left in place or removed after welding. If the backing is to be removed, an *R* is inserted in the welding symbol. Backing that is to be removed usually is made of a different material to keep it from being welded into the joint.

108UA0801.EPS

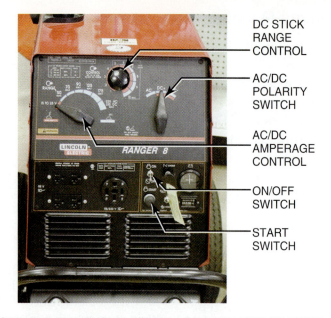

PORTABLE 200A ENGINE-DRIVEN AC/DC WELDING MACHINE

DC STICK RANGE CONTROL

AC/DC POLARITY SWITCH

AC/DC AMPERAGE CONTROL

ON/OFF SWITCH

START SWITCH

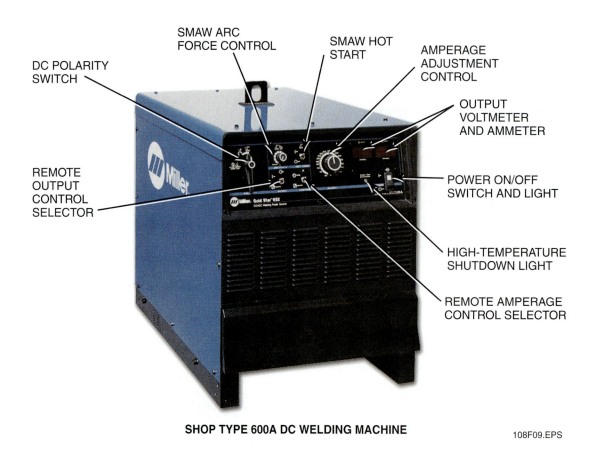

SMAW ARC FORCE CONTROL

SMAW HOT START

AMPERAGE ADJUSTMENT CONTROL

DC POLARITY SWITCH

OUTPUT VOLTMETER AND AMMETER

REMOTE OUTPUT CONTROL SELECTOR

POWER ON/OFF SWITCH AND LIGHT

HIGH-TEMPERATURE SHUTDOWN LIGHT

REMOTE AMPERAGE CONTROL SELECTOR

SHOP TYPE 600A DC WELDING MACHINE

108F09.EPS

Figure 9 ◆ Typical DC and AC/DC welding machines.

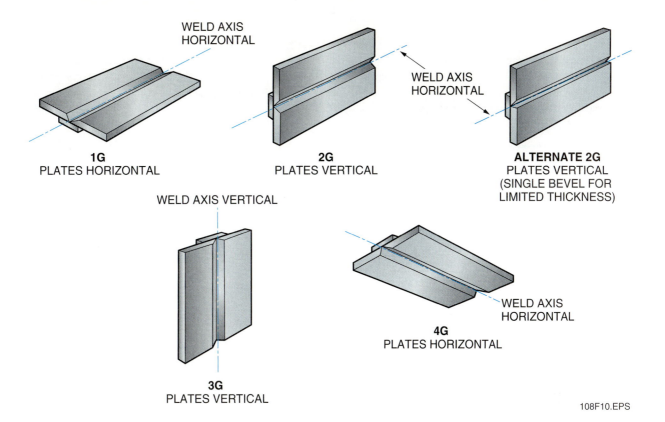

Figure 10 ◆ Groove weld positions for plate.

108F10.EPS

4.2.0 Acceptable and Unacceptable Groove Weld Profiles

Groove welds should be made with slight reinforcement and a gradual transition to the base metal at each toe. Groove welds must not have excessive reinforcement, underfill, undercut, or overlap. If a groove weld has any of these defects it must be repaired.

> **Note**
>
> Refer to your site's welding procedure specifications (WPSs) for specific requirements on groove welds. The information in this module is provided as a general guideline only. The site WPS or quality specifications must be followed for all welds. Check with your supervisor if you are unsure of the specifications for your application.

Figure 11 shows acceptable and unacceptable groove weld profiles.

5.0.0 ◆ SMAW OF V-GROOVE WELDS WITH BACKING

This section of the module explains how to perform the following V-groove welds using a metal backing strip:

- Flat welds
- Horizontal welds
- Vertical welds
- Overhead welds

5.1.0 Practicing Flat V-Groove Welds with Backing (1G Position)

Practice flat V-groove welds using ³⁄₃₂", ⅛", or ⁵⁄₃₂" E7018 electrodes. Stringer beads or weave beads can be used. When using weave beads, keep the electrode at a 0° work angle with a 10° to 15° drag angle. When using stringer beads, use a 10° to 15° drag angle, but adjust the electrode work angle to tie in the weld on one side or the other as needed. Pay particular attention at the termination of the weld to fill the crater. If the coupon gets too hot between passes, you may have to cool it in water.

Follow these steps to practice V-groove welds with metal backing in the flat position:

Step 1   Tack-weld the practice coupon together following the example given earlier (see *Figure 12*).

Step 2   Position the weld coupon flat on the welding table.

Step 3   Run the root pass using ³⁄₃₂", ⅛", or ⁵⁄₃₂" E7018 electrodes. *Figure 13* shows the root, filler, and cover pass sequences and work angles. Degrees are approximate.

Step 4   Clean the weld.

Step 5   Run the filler and cover passes, cleaning between each bead, to complete the weld using ³⁄₃₂", ⅛", or ⁵⁄₃₂" E7018 electrodes. Use stringer or weave beads as directed by your instructor.

Figure 14 shows a flat V-groove weld with typical filler and cover passes.

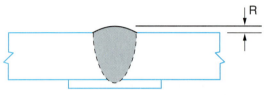

R = FACE REINFORCEMENT NOT TO EXCEED ⅛" OR AS SPECIFIED BY CODE

PROFILE OF ACCEPTABLE V-GROOVE WELD WITH BACKING

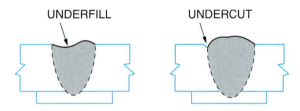

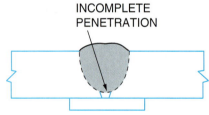

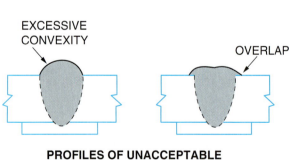

PROFILES OF UNACCEPTABLE V-GROOVE WELDS WITH BACKING

108F11.EPS

Figure 11 ◆ Acceptable and unacceptable profiles of V-groove welds with backing.

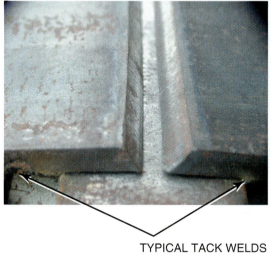

TYPICAL TACK WELDS

108F12.EPS

Figure 12 ◆ Tack-weld the practice coupon.

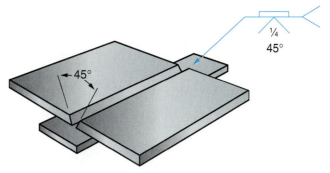

NOTE: THE ACTUAL NUMBER OF WELD BEADS WILL VARY DEPENDING ON THE PLATE THICKNESS

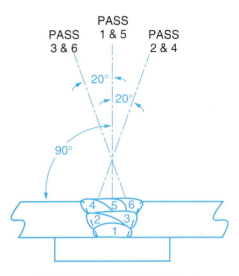

STRINGER BEAD SEQUENCE

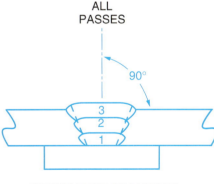

WEAVE BEAD SEQUENCE

108F13.EPS

Figure 13 ◆ Multiple-pass 1G weld sequences and work angles.

5.2.0 Horizontal Welds (2G Position)

Horizontal welds can be made with or without backing. This section explains welding horizontal beads and making horizontal V-groove welds with backing.

108F14.EPS

Figure 14 ◆ Typical V-groove weld filler and cover pass.

5.2.1 Practicing Horizontal Beads

Before welding a V-groove in the horizontal position, practice running horizontal stringer beads by building a pad in the horizontal position. Use ³⁄₃₂", ⅛", or ⁵⁄₃₂" E7018 electrodes. The electrode drag angle should be 10° to 15°, and the work angle should be 0°. To control the weld puddle the electrode work angle can be dropped slightly, but no more than 10°. Follow these steps to practice welding beads in the horizontal position:

Step 1 Tack-weld a flat plate welding coupon in the horizontal position.

Step 2 Run the first pass along the bottom of the coupon using ³⁄₃₂", ⅛", or ⁵⁄₃₂" E7018 electrodes.

Step 3 Clean the weld bead with a chipping hammer and wire brush.

Step 4 Weld the second bead just above the first bead.

Step 5 Continue running beads to complete the pad. *Figures 15* and *16* show the building of a pad in the horizontal position. Degrees shown are approximate.

5.2.2 Practicing Horizontal V-Groove Welds with Backing

Practice horizontal V-groove welds using ³⁄₃₂", ⅛", or ⁵⁄₃₂" E7018 electrodes and stringer beads. Use a 10° to 15° drag angle, but adjust the electrode work angle to tie in the weld as needed. Pay particular attention at the termination of the weld to fill the crater.

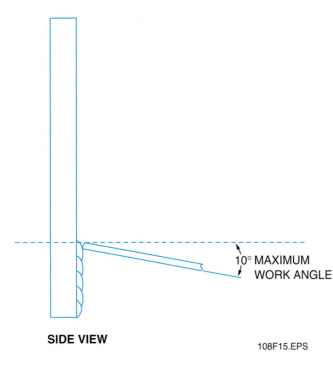

SIDE VIEW

10° MAXIMUM WORK ANGLE

108F15.EPS

Figure 15 ◆ Work angle in the horizontal position.

108F16.EPS

Figure 16 ◆ A pad built in the horizontal position.

Follow these steps to practice welding V-groove welds with metal backing in the horizontal position:

Step 1 Tack-weld the practice coupon together following the example given earlier. Use the standard or alternate horizontal weld coupon as directed by your instructor.

Step 2 Tack-weld the weld coupon in the horizontal position.

Step 3 Run the root pass.

Step 4 Clean the weld.

Step 5 Run the filler and cover passes, cleaning the weld between passes, to complete the weld.

Figure 17 shows a horizontal V-groove weld with metal backing. Degrees shown are approximate.

5.3.0 Vertical Welds (3G Position)

Vertical V-groove welds can be made with or without backing. This section explains welding vertical beads and making vertical V-groove welds with backing.

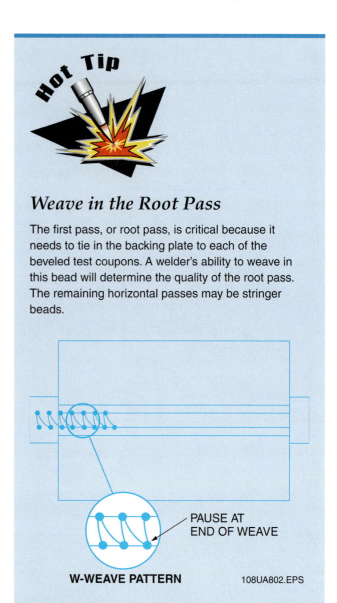

Weave in the Root Pass

The first pass, or root pass, is critical because it needs to tie in the backing plate to each of the beveled test coupons. A welder's ability to weave in this bead will determine the quality of the root pass. The remaining horizontal passes may be stringer beads.

PAUSE AT END OF WEAVE

W-WEAVE PATTERN   108UA802.EPS

5.3.1 Practicing Vertical Beads

Before welding a joint in the vertical position, practice running vertical stringer beads by build-

NOTE: THE ACTUAL NUMBER OF WELD BEADS WILL VARY DEPENDING ON THE PLATE THICKNESS

ing a pad in the vertical position. Use ³⁄₃₂", ¹⁄₈", or ⁵⁄₃₂" E7018 electrodes. Set the amperage in the lower part of the range for better puddle control. The electrode should be at a 0° to 10° push angle with a 0° work angle. Use a slight side-to-side motion to control the weld puddle. The width of the stringer beads should be no more than three times the electrode diameter (³⁄₈" for ¹⁄₈" electrodes).

Follow these steps to practice welding beads in the vertical position:

Step 1   Tack-weld a flat plate welding coupon in the vertical position.

Step 2   Run the first pass along the vertical edge of the coupon using ³⁄₃₂", ¹⁄₈", or ⁵⁄₃₂" E7018 electrodes.

Step 3   Clean the weld bead.

Step 4   Continue running beads to complete the pad. Clean the weld between each pass. *Figure 18* shows the travel angle in the vertical position. Degrees shown are approximate.

5.3.2 Practicing Vertical V-Groove Welds with Backing

Practice vertical V-groove welds using ³⁄₃₂", ¹⁄₈", or ⁵⁄₃₂" E7018 electrodes with stringer beads and/or weave beads. Set the amperage in the lower part of the range for better puddle control. Use a 0° to

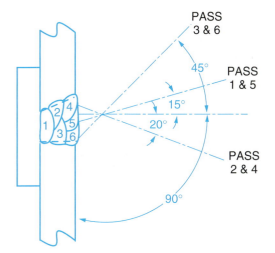

ALTERNATE JOINT REPRESENTATION

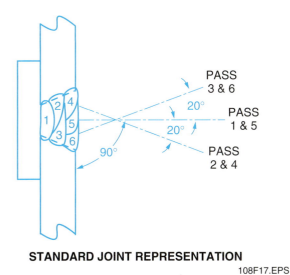

STANDARD JOINT REPRESENTATION

108F17.EPS

Figure 17 ◆ Multiple-pass 2G weld sequences and work angles.

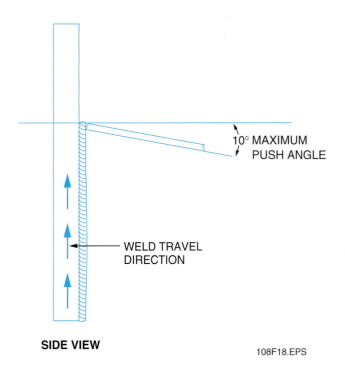

SIDE VIEW

108F18.EPS

Figure 18 ◆ Travel angle in the vertical position.

10° push angle, but adjust the electrode work angle to tie in the weld on one side or the other as needed. If using a weave bead, pause slightly at each toe to penetrate and fill the crater to prevent undercut. Pay particular attention at the termination of the weld to fill the crater.

Follow these steps to practice welding V-groove welds with metal backing in the vertical position:

Step 1 Tack-weld the practice coupon together following the example given earlier.

Step 2 Tack-weld the weld coupon in the vertical position.

Step 3 Run the root pass with a tight weave technique, pausing at each side.

Step 4 Clean the weld.

Step 5 Run the filler and cover passes, cleaning each pass, to complete the weld.

Figure 19 shows a vertical V-groove weld with metal backing. Degrees shown are approximate.

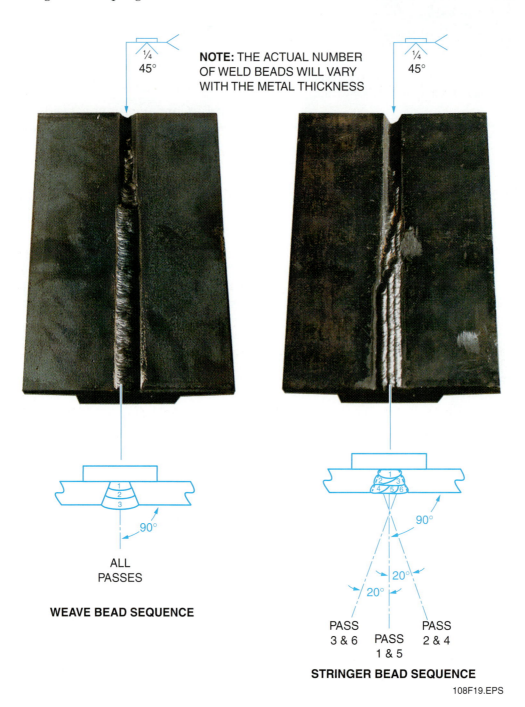

NOTE: THE ACTUAL NUMBER OF WELD BEADS WILL VARY WITH THE METAL THICKNESS

WEAVE BEAD SEQUENCE

STRINGER BEAD SEQUENCE

108F19.EPS

Figure 19 ◆ Multiple-pass 3G weld sequences and work angles.

AWS ENTRY LEVEL WELDER – TRAINEE MODULE 29108-03

5.4.0 Overhead Welds (4G Position)

Overhead V-groove welds can be made with or without backing. This section explains welding overhead beads and making overhead V-groove welds with backing.

5.4.1 Practicing Overhead Beads

Before welding a V-groove in the overhead position, practice running overhead stringer beads by building a pad in the overhead position. Use ³⁄₃₂", ⅛", or ⁵⁄₃₂" E7018 electrodes. The electrode should have a 0° work angle and a 10° to 15° drag angle.

Follow these steps to practice welding beads in the overhead position:

Step 1 Tack-weld the welding coupon in the overhead position.

Step 2 Run the first pass along the edge of the coupon using ³⁄₃₂", ⅛", or ⁵⁄₃₂" E7018 electrodes.

Step 3 Clean the weld bead with a chipping hammer and wire brush.

Step 4 Continue running and cleaning the beads to complete the pad. *Figure 20* shows building a pad in the overhead position.

5.4.2 Practicing Overhead V-Groove Welds with Backing

Practice making overhead V-groove welds using ³⁄₃₂", ⅛", or ⁵⁄₃₂" E7018 electrodes and stringer beads. Use a 10° to 15° drag angle, but adjust the electrode work angle to tie in the weld as needed. Pay particular attention at the termination of the weld to fill the crater.

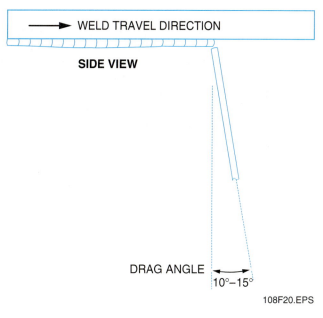

Figure 20 ◆ Building a pad in the overhead position.

Follow these steps to practice welding V-groove welds with metal backing in the overhead position:

Step 1 Tack-weld the practice coupon together following the example given earlier.

Step 2 Tack-weld the weld coupon in the overhead position.

Step 3 Run the root pass.

Step 4 Clean the weld with a chipping hammer and wire brush.

Step 5 Run the filler and cover passes, cleaning after each pass, to complete the weld.

Figure 21 shows an overhead V-groove weld with metal backing. Degrees shown are approximate.

Avoid Molten Showers

Try not to stand directly under the coupons while welding because hot slag and molten metal can cause very severe burns. Always wear the proper protective equipment and clothing for overhead welding.

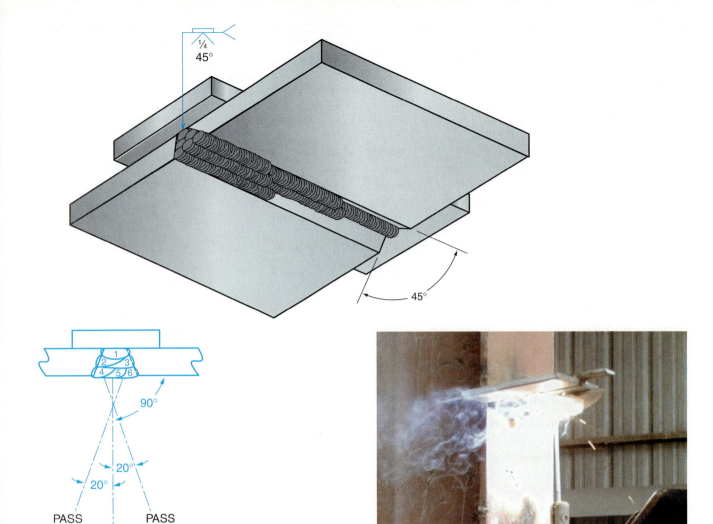

STRINGER BEAD SEQUENCE

OVERHEAD WELDING IN PROGRESS

108F21.EPS

Figure 21 ◆ Multipass 4G weld sequences and work angles.

Summary

The ability to make groove welds on plate with metal backing using low-hydrogen electrodes is an essential welding skill. These skills are necessary to progress to more difficult pipe welding procedures. Several groove weld styles and combination styles, including V-grooves with backing, were covered in this module. Groove welds are made in the flat, horizontal, vertical, and overhead positions. Proper preparation of the equipment, base metal, and backing is important for making quality groove welds. Welds in the vertical and overhead positions are typically required for welder certification on plate. Practice these welds until you can consistently produce acceptable welds.

Review Questions

1. A weld made in an opening in a part is called a _____ weld.
   a. fillet
   b. groove
   c. flange
   d. tack

2. The portion of a joint where the members approach closest to each other is called the _____.
   a. joint root
   b. depth of bevel
   c. groove radius
   d. root opening

3. When preparing welding coupons for V-groove welds with backing, cut the metal into 3" by 7" rectangles, with the 7" lengths beveled at approximately _____.
   a. 10°
   b. 14°
   c. 22½°
   d. 33⅓°

4. All of the following are true about running low-hydrogen electrodes *except* _____.
   a. maintain a short arc
   b. take only a few from the oven at a time
   c. always whip them
   d. do not use ones missing flux

5. The V-groove with backing is a common groove weld normally made with _____ electrodes.
   a. fast-fill
   b. fill-freeze
   c. fast-freeze
   d. low-hydrogen

6. A way to eliminate the need for back gouging, and to quickly and effectively prepare a joint for welding, is to use _____.
   a. tape backing
   b. steel backing
   c. a backing weld run with E7018 electrodes
   d. a backing weld run with fast-freeze electrodes

7. When using weave beads on flat V-groove welds, keep the electrode angle at a _____ work angle with a 10° to 15° drag angle.
   a. 0°
   b. 45°
   c. 60°
   d. 90°

8. Before welding a joint in the horizontal position, practice running horizontal stringer beads by building a _____ in the horizontal position.
   a. pad
   b. root
   c. crater
   d. puddle

9. When using stringer beads on V-groove welds in the vertical position, keep the electrode at a 0° to 10° push angle, with a _____ work angle.
   a. 0°
   b. 45°
   c. 60°
   d. 90°

10. Before welding a V-groove in the overhead position, you should _____.
    a. practice running overhead stringer beads
    b. tack-weld the coupon in the horizontal position
    c. build a pad in the vertical position
    d. clean the weld with a wire brush

Favio Kinsman

Welder
Miami, FL

Favio Kinsman is a 21-year-old welder who started his career by taking a welding course at an adult education facility. That training got him an entry-level job, but he didn't stop there. He continued his training in night school. Soon he was promoted to a better paying job as a welder. He has since had the opportunity to travel and work on large projects.

How did you get your start?
When I was 17 years old, I was interested in welding and took a welding course at an adult education center, where I learned gas welding and arc welding. With that training, I was able to get a job as a welder/helper at a local cement plant.

Did your education stop there?
No. I continued to go to school at night to learn more about welding. I was soon certified at my job and promoted to welder. The money was good, and I learned a lot about connecting joists and reading welding prints. A year later, I was offered a job as an ironworker/welder at $16.15 an hour. Things went well, and I continued to learn. I later had jobs in Boston and Puerto Rico. In Puerto Rico, I helped build a large coliseum.

What do you like about welding?
I enjoy the challenges that go with working on large structures, as well as climbing and working in difficult welding positions. I've made excellent money in my four years as a welder, and I expect to keep working in the trade. I would like to own my own home and raise a family.

Performance Accreditation Tasks

The Performance Accreditation Tasks (PATCs) correspond to and support the learning objectives in AWS EG2.0-95, Guide for the Training and Qualification of Welding Personnel: Entry-Level Welder.

Note that in order to satisfy all learning objectives in AWS EG2.0-95, the instructor must also use the PATCs contained in the second level in the Welding Contren™ Learning Series.

PATCs provide specific acceptable critera for performance and help to ensure a true competency-based welding program for students.

The following tasks are designed to evaluate your ability to run V-groove welds with backing with SMAW equipment in all positions using E7018 low-hydrogen electrodes. Perform each task when you are instructed to do so by your instructor. As you complete each task, take it to your instructor for evaluation. Do not proceed to the next task until instructed to do so by your instructor.

V-GROOVE WELDS WITH BACKING IN THE FLAT (1G) POSITION

Using ³⁄₃₂", ⅛", and ⁵⁄₃₂" E7018 electrodes, make a V-groove weld with steel backing on carbon steel plate in the flat position as indicated.

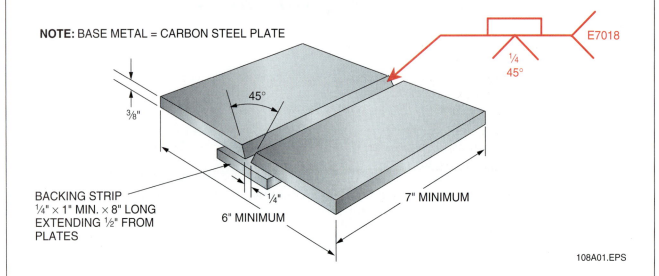

NOTE: BASE METAL = CARBON STEEL PLATE

45°

³⁄₈"

E7018

¼
45°

BACKING STRIP
¼" × 1" MIN. × 8" LONG
EXTENDING ½" FROM
PLATES

¼"

6" MINIMUM

7" MINIMUM

108A01.EPS

Criteria for Acceptance:

- Uniform rippled appearance on the bead face
- Craters and restarts filled to the full cross-section of the weld
- Uniform weld wize ±¹⁄₁₆"
- Acceptable weld profile in accordance with AWS D1.1
- Smooth transition with complete fusion at the toes of the weld
- No porosity
- No overlap
- No undercut
- No inclusions
- No cracks
- Acceptable guided bend test results

V-GROOVE WELDS WITH BACKING IN THE HORIZONTAL (2G) POSITION

Using ³⁄₃₂", ⅛", and ⁵⁄₃₂" E7018 electrodes, make a V-groove weld with steel backing on carbon steel plate in the horizontal position as indicated.

NOTE: BASE METAL = CARBON STEEL PLATE

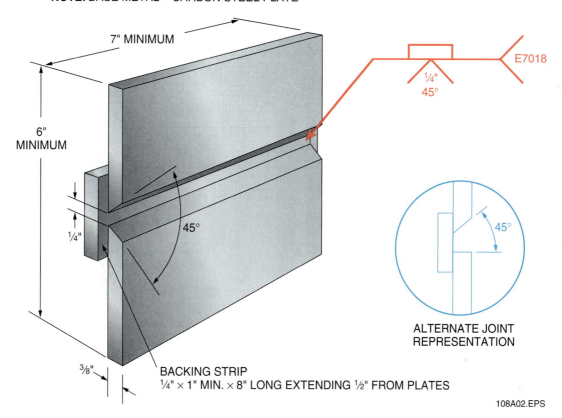

ALTERNATE JOINT REPRESENTATION

BACKING STRIP
¼" × 1" MIN. × 8" LONG EXTENDING ½" FROM PLATES

108A02.EPS

Criteria for Acceptance:

- Uniform rippled appearance on the bead face
- Craters and restarts filled to the full cross-section of the weld
- Uniform weld wize ±¹⁄₁₆"
- Acceptable weld profile in accordance with AWS D1.1
- Smooth transition with complete fusion at the toes of the weld
- No porosity
- No undercut
- No overlap
- No inclusions
- No cracks
- Acceptable guided bend test results

V-GROOVE WELD WITH BACKING IN THE VERTICAL (3G) POSITION

Using ³⁄₃₂", ⅛", and ⁵⁄₃₂" E7018 electrodes, make a V-groove weld with steel backing on carbon steel plate in the vertical position as indicated.

NOTE: BASE METAL = CARBON STEEL PLATE

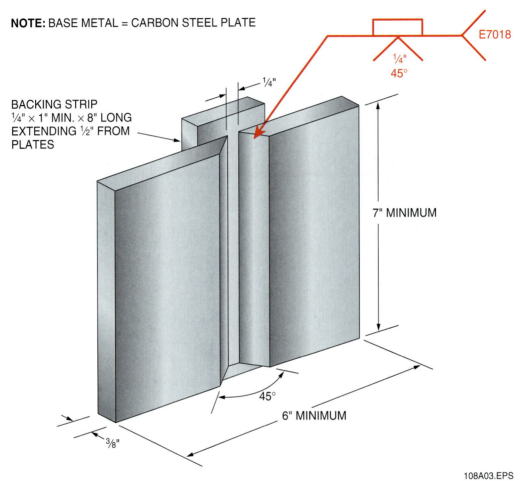

BACKING STRIP
¼" × 1" MIN. × 8" LONG
EXTENDING ½" FROM
PLATES

¼"

E7018

¼"
45°

7" MINIMUM

45°

6" MINIMUM

⅜"

108A03.EPS

Criteria for Acceptance:

- Uniform rippled appearance on the bead face
- Craters and restarts filled to the full cross-section of the weld
- Uniform weld wize ±¹⁄₁₆"
- Acceptable weld profile in accordance with AWS D1.1
- Smooth transition with complete fusion at the toes of the weld
- No porosity
- No overlap
- No undercut
- No inclusions
- No cracks
- Acceptable guided bend test results

_____
_____
_____
_____
_____
_____
_____
_____
_____
_____
_____

V-GROOVE WELD WITH BACKING IN THE OVERHEAD (4G) POSITION

Using ³⁄₃₂", ⅛", and ⁵⁄₃₂" E7018 electrodes, make a V-groove weld with steel backing on carbon steel plate in the overhead position as indicated.

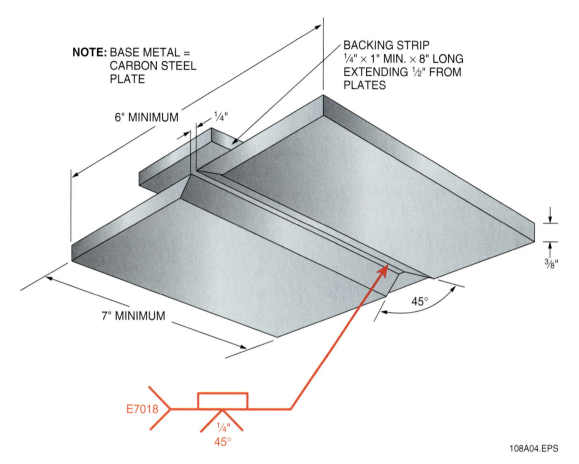

NOTE: BASE METAL = CARBON STEEL PLATE

BACKING STRIP ¼" × 1" MIN. × 8" LONG EXTENDING ½" FROM PLATES

6" MINIMUM   ¼"

7" MINIMUM

³⁄₈"

45°

E7018

¼"
45°

108A04.EPS

Criteria for Acceptance:

- Uniform rippled appearance on the bead face
- Craters and restarts filled to the full cross-section of the weld
- Uniform weld wize ±¹⁄₁₆"
- Acceptable weld profile in accordance with AWS D1.1
- Smooth transition with complete fusion at the toes of the weld
- No porosity
- No undercut
- No overlap
- No inclusions
- No cracks
- Acceptable guided bend test results

Additional Resources

This module is intended to present thorough resources for task training. The following reference works are suggested for further study. These are optional materials for continued education rather than for task training.

Welding Technology, 2001. J. W. Giachino, W. R. Weeks, G. S. Johnson. Homewood, IL: American Technical Publishers, Inc.

Modern Welding, 1997. A. D. Althouse, C. H. Turnquist, W. A. Bowditch, K. E. Bowditch. Tinley Park, IL: The Goodheart-Willcox Company, Inc.

CONTREN™ LEARNING SERIES—USER UPDATES

The NCCER makes every effort to keep these textbooks up-to-date and free of technical errors. We appreciate your help in this process. If you have an idea for improving this textbook, or if you find an error, a typographical mistake, or an inaccuracy in NCCER's Contren™ textbooks, please write us, using this form or a photocopy. Be sure to include the exact module number, page number, a detailed description, and the correction, if applicable. Your input will be brought to the attention of the Technical Review Committee. Thank you for your assistance.

Instructors—If you found that additional materials were necessary in order to teach this module effectively, please let us know so that we may include them in the Equipment/Materials list in the Instructor's Guide.

Write: Curriculum Revision and Development Department
National Center for Construction Education and Research
P.O. Box 141104, Gainesville, FL 32614-1104

Fax: 352-334-0932

E-mail: curriculum@nccer.org

Craft _____ Module Name _____

Copyright Date _____ Module Number _____ Page Number(s) _____

Description _____

_____

_____

_____

(Optional) Correction _____

_____

_____

(Optional) Your Name and Address _____

_____

_____

Joint Fit-Up
and Alignment

Course Map

This course map shows all of the modules in the AWS Entry Level Welder – Phase 1 curriculum. The suggested training order begins at the bottom and proceeds up. Skill levels increase as you advance on the course map. The local Training Program Sponsor may adjust the training order.

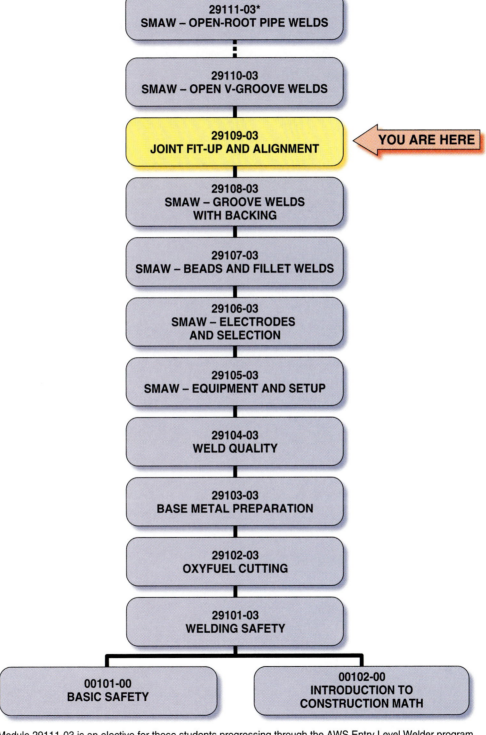

AWS ENTRY LEVEL WELDER—PHASE 1

29111-03*
SMAW – OPEN-ROOT PIPE WELDS

29110-03
SMAW – OPEN V-GROOVE WELDS

29109-03
JOINT FIT-UP AND ALIGNMENT ← YOU ARE HERE

29108-03
SMAW – GROOVE WELDS WITH BACKING

29107-03
SMAW – BEADS AND FILLET WELDS

29106-03
SMAW – ELECTRODES AND SELECTION

29105-03
SMAW – EQUIPMENT AND SETUP

29104-03
WELD QUALITY

29103-03
BASE METAL PREPARATION

29102-03
OXYFUEL CUTTING

29101-03
WELDING SAFETY

00101-00
BASIC SAFETY

00102-00
INTRODUCTION TO CONSTRUCTION MATH

*Module 29111-03 is an elective for those students progressing through the AWS Entry Level Welder program.

29109CMAP.EPS

Figures

Table

Joint Fit-Up and Alignment

Objectives

When you have completed this module, you will be able to do the following:

1. Identify and explain job code specifications.
2. Use fit-up gauges and measuring devices to check joint fit-up.
3. Identify and explain distortion and how it is controlled.
4. Fit up joints using plate and pipe fit-up tools.
5. Check for joint misalignment and poor fit-up before and after welding.

Prerequisites

Before you begin this module, it is recommended that you successfully complete Modules 00100-00 through 29108-03.

Required Trainee Materials

1. Pencil and paper
2. Appropriate personal protective equipment

1.0.0 ◆ INTRODUCTION

Joint design and setup affect the safety and quality of a completed weldment. Because joint design and setup are so important, they are covered by written codes and specifications that must be followed. Special tools to measure and aid fit-up are also available. This module will explain the codes and the special tools and measuring devices to aid setup. In addition, joint inspection techniques and procedures to ensure that proper joint setup is maintained during welding will also be explained.

2.0.0 ◆ JOB CODE SPECIFICATIONS

Whenever a bridge, building, ship, or pressure vessel is welded, the manufacturer and the buyer must reach agreement on how each weld will be made. To eliminate the need to write a new code for each job, government agencies, societies, and associations have developed codes. These codes are used universally to ensure safety and quality when welds are made.

2.1.0 Governing Codes and Standards

A welding code is a detailed listing of the rules and principles that apply to specific welded products. Codes ensure that safe and reliable welded products will be produced and that persons associated with the welding operation will be safe. All welding should be performed following the guidelines and specifications outlined in various codes, which are specified by clients when they place orders or let contracts. In addition, when codes are specified, the use of these codes is mandated with the force of law by one or more government jurisdictions. Always check the contract, order, or project specification for the specified code(s). Some of the more common codes are:

- *API 1104, Standards for Welding of Pipelines and Related Facilities* – American Petroleum Institute (API); used for pipelines.
- *ASME Boiler and Pressure Vessel Code* – American Society of Mechanical Engineers (ASME); of the 11 sections, Sections II, V, and IX pertain to materials, nondestructive testing, and welding and brazing qualifications used for pressure vessels and piping systems.
- *ASME B31.1, Power Piping* – American Society of Mechanical Engineers (ASME); used for pressure piping.

- *AWS D1.1, Structural Welding Code* – American Welding Society (AWS); used for bridges and buildings.

2.2.0 Code Changes

Periodically, codes are reviewed and changes made. Normally this occurs yearly, but some codes may go more than one year before they are updated. When a code is updated, the entire code is reissued with an updated year prefix, or addendum sheets (new pages) for the areas of the code affected by the changes are issued. Changes are typically noted to alert readers to what has changed from the prior edition. The AWS code indicates changes to the code with a double vertical line in the margin alongside the changed area. Editorial changes to the AWS code are indicated by a single vertical line.

The ASME updates the entire code every three years. An addendum is issued yearly. The yearly addendum is identified by placing the letter A in front of the year on the cover of the code. API updates the entire code every five years.

It is important to recognize code changes. When a client specifies a code year, the code for that year, and that year only, is to be used unless otherwise specified by the client. When referring to a code, be sure the year matches the specifications for the job. *Figure 1* shows various codes.

2.3.0 Welding Procedure Specifications

A welding procedure specification (WPS) is a written set of instructions for producing sound welds. Each WPS is written and tested in accordance with a particular welding code or specification and must be in accordance with industry practice. All welding requires that acceptable industry standards be followed, but not all welds require a WPS. If a weld does require a WPS, the WPS must be followed. The consequences of not following a required WPS are severe. Consequences include producing an unsafe weldment that could endanger life as well as the rejection of the weldment and lawsuits. When it is required, always follow the WPS. The requirement for the use of a WPS is often listed on job blueprints as a note or in the tail of the welding symbol. If you are unsure whether the welding being performed requires a WPS, do not proceed until you check with your supervisor.

Each WPS is written by an individual who knows welding codes, specifications, and acceptable industry practices. It then becomes the responsibility of each manufacturer or contractor to test and qualify the WPS before using it. The WPS is tested by welding test coupons. Then, the coupons are tested according to the code. For shielded metal arc welding (SMAW) of complete penetration groove welds, the testing required includes nondestructive testing (NDT), tensile strength tests, and root, face, or side bend tests. The results of the testing are recorded on a procedure qualification record (PQR). The WPS and PQR must be kept on file.

Many different formats are used for WPSs, but they all contain the same essential information. Information typically found on a WPS includes:

- *Scope* – The welding process and base metals to which the WPS applies as well as the governing code and/or specification to be used.
- *Base metal* – The chemical composition or specification of the applicable base metal using the industry-standard identification; for example, A36 for carbon steel.
- *Welding process* – The welding process to be used to make the weld. For example, SMAW, gas metal arc welding (GMAW), or gas tungsten arc welding (GTAW).
- *Filler metal* – The composition, identifying type, or classification of the filler metal to be used, as well as the electrode sizes for base metals of different thicknesses and various positions.

AWS Sheet Metal Welding Code

The AWS completed the revised and updated American National Standards Institute (ANSI)-approved edition of the *Sheet Metal Welding Code* in 2000. The number of the *Sheet Metal Welding Code* is *AWS D9.1M/D9.1-2000*.

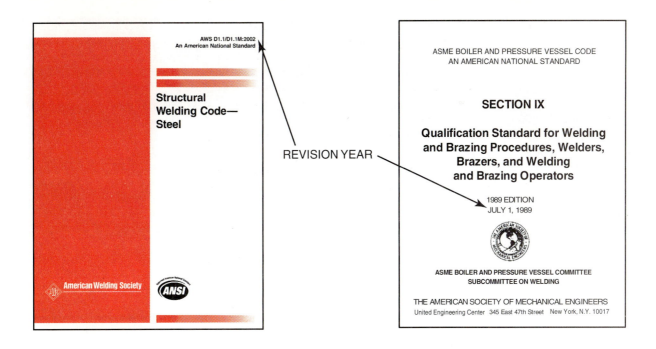

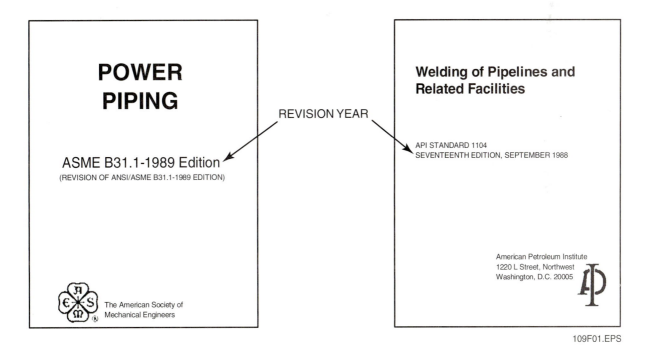

Figure 1 ◆ Codes.

- *Type of current* – Either alternating current (AC) or direct current (DC). If DC is specified, the polarity must also be given.
- *Arc voltage and travel speed* – An arc voltage range. Ranges for travel speed for automatic processes are mandatory and are recommended for semiautomatic processes.
- *Joint design and tolerances* – Joint design details, tolerances, and welding sequences that are

given as a cross-sectional drawing or as a reference to drawings or specifications.
- *Joint and surface preparation* – The methods that can be used to prepare joint faces and the degree of surface cleaning required.
- *Tack welding* – Details pertaining to tack welding. Tack welders must use the WPS.
- *Welding details* – The size of electrodes to use for different portions and positions of the joint, the

arrangement of welding passes to fill the joint, and the pass width and weave limitations.

- *Positions of welding* – The welding positions that can be used; for example, 1G, 3F, 5G.
- *Peening* – The details of peening and the type of tool to be used.
- *Heat input* – The details to control heat input.
- *Second-side preparation* – The method used to prepare the second side when joints are welded from two sides.
- *Postheat treatment* – The details about postheat treatment or a reference to a separate document.

Figure 2 shows a portion of a typical WPS.

3.0.0 ◆ FIT-UP GAUGES AND MEASURING DEVICES

Before making a weld, the joint must be fit up and checked to ensure it conforms to the WPS or site quality standards. The most common tools used to lay out and check joint fit-ups are straightedges, squares, levels, and Hi-Lo gauges.

3.1.0 Straightedges

Straightedges are used to scribe straight lines and check joint alignment. Many have calibrations along their length for measuring. Straightedges, particularly longer ones, are typically fabricated on the job from small channel or angle iron. When using a straightedge, be careful not to apply heat to it by placing it on hot metal or near the flame of a cutting torch or welding arc. Heat can cause the

straightedge to become permanently distorted. Before using a straightedge, check it for straightness by visually sighting down the edge. If it is distorted and cannot be straightened, destroy it.

3.2.0 Squares

Two types of squares are commonly used for layout: pipefitter's squares and combination try squares. Pipefitter's squares, which are used to measure angles and check squareness, are available in a variety of sizes.

Combination squares are smaller than pipefitter's squares, with blades typically 12" or 18" long. They have replaceable attachments that slide and lock along a groove on the blade. Attachments include a combination 90°/45° level head, a centering head, and a protractor head. The combination attachment is used to check and lay out 90° and 45° angles, to check level, and to measure depth. The centering head is used to measure round stock and to locate the center of shafts or other round objects. The protractor head is used to lay out and check angles. When using squares, avoid heat, which could cause them to distort. Also, protect squares from welding spatter, which could stick to measuring surfaces, giving a false reading. To protect their accuracy, squares should be stored in protective areas when not in use. *Figure 3* shows a pipefitter's square and combination square.

3.3.0 Levels

Levels come in a variety of sizes and shapes. Most levels designed for plate and piping work are

Austin Industrial

Austin

An Austin Industries Company

Specification:
Date:
Revision:
Page: Of

Welding Procedure Specification

TITLE: _____

PROCESS	APPROX. NUMBER OF PASSES	ROD OR ELECT. SIZE	CURRENT	VOLTAGE	FILLER METALS SFA SPEC. CLASS	TYPE	
						F. NO.	A. NO.

JOINTS (QW-402)
Groove Design _____
Backing: Yes _____ No. _____
Backing Material (Type) _____
Other _____
_____

BASE METALS (QW-403)
P. No. _____ Group _____ to P. No. _____ Group _____
Thickness Range _____
Pipe Dia. Range _____
Other _____

FILLER METALS (QW-404)
F. No. _____ Other _____
A. No. _____ Other _____
Spec. No. (SFA) _____
AWS No. (Class) _____
Size of Electrode _____
Size of Filler _____
Electrode-Flux (Class) _____
Consumable Insert _____
Other _____

POSITION (QW-405)
Position of Groove _____
Welding Progression _____
Other _____

PREHEAT (QW-406)
Preheat Temp. _____
Interpass Temp. _____
Preheat Maintenance _____
Other _____
_____

POSTWELD HEAT TREATMENT (QW-407)
Temperature _____
Time Range _____
Other _____

GAS (QW-408)
Shielding Gas(es) _____
Percent Composition (mixtures) _____
_____
Flow Rate _____
Gas Backing _____
Trailing Shielding Gas Composition _____
_____
Other _____

ELECTRICAL CHARACTERISTICS (QW-409)
Current AC or DC _____ Polarity _____
Amps (range) _____ Volts (range) _____
Other _____

TECHNIQUE (QW-410)
String or Weave bead _____
Orifice or Gas Cup Size _____
Initial & Interpass Cleaning _____
(Brushing, Grinding, etc.) _____
_____
Method of Back Gouging _____
Oscillation _____
Contact Tube to Work Distance _____
Multiple or Single Pass (per side) _____
Multiple or Single Electrodes _____
Travel Speed (Range) _____
Other _____

JOINT ___ AL AND FIT-UP

SUPPORTING PQR NO(S) _____
_____
_____
_____

Written By _____

Figure 2 ◆ Portion of a typical WPS.

109F02.EPS

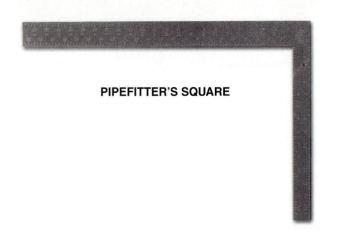

PIPEFITTER'S SQUARE

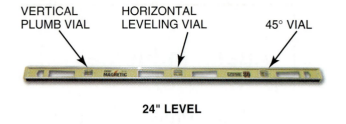

VERTICAL PLUMB VIAL

HORIZONTAL LEVELING VIAL

45° VIAL

MAGNETIC

24" LEVEL

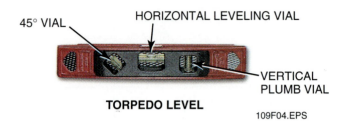

45° VIAL

HORIZONTAL LEVELING VIAL

VERTICAL PLUMB VIAL

TORPEDO LEVEL

109F04.EPS

Figure 4 ◆ Common levels.

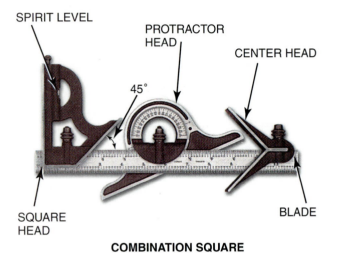

SPIRIT LEVEL

PROTRACTOR HEAD

CENTER HEAD

45°

SQUARE HEAD

BLADE

COMBINATION SQUARE

109F03.EPS

Figure 3 ◆ Pipefitter's square and combination square.

made from magnesium or aluminum. Some have magnetized bases. Levels are used to check that layouts are level (horizontal) and **plumb** (vertical). Levels use a bubble in a glass vial to check level and plumb. Centering the bubble between the lines marked on the vial indicates level when the level is in the horizontal position and plumb when the level is in the vertical position. Some levels also have a vial set at 45° to check 45° angles. There are also special levels that have an adjustable protractor scale that can be set to check any angle.

The torpedo level is another type of level. It is about 9" long and tapered on each end. The torpedo level has three vials: one to check level, one to check plumb, and one to check 45° layouts. Protect levels from heat and weld spatter, which can break the glass tube. Also, be aware that weld spatter can stick to the bottom of the level, giving a false reading. *Figure 4* shows common levels.

3.4.0 Hi-Lo Gauges

The primary purpose of a Hi-Lo gauge is to check for pipe joint misalignment, although plate joint misalignment can also be checked. The name of the gauge comes from the relationship between the alignment of one pipe to the other pipe, which is called **high-low.**

To check for internal misalignment, Hi-Lo gauges have two prongs, or alignment stops, that are pulled tightly against the inside diameter of the joint so that one stop is flush with each side of the joint. The variation between the two stops is read on a scale marked on the gauge. To measure misalignment with a Hi-Lo gauge, insert the prongs of the gauge into the joint gap. Pull up on the gauge until the prongs are snug against both inside surfaces, and read the misalignment indicated on the scale. *Figure 5* shows checking internal misalignment with a Hi-Lo gauge.

Many Hi-Lo gauges also have the capability to check:

- Pipe wall thickness
- Joint gap
- Weld reinforcement
- Scribe lines for socket welds

Many Hi-Lo gauges also have a scale that indicates the material thickness when the gauge is inserted to check for internal misalignment. *Figure 6* shows a typical material thickness scale on a Hi-Lo gauge.

The alignment stops are tapered for checking the joint gap. They are ³⁄₃₂" thick and taper to ¹⁄₁₆" thick at both ends. Check the gap by inserting the tapered head and visually checking the opening.

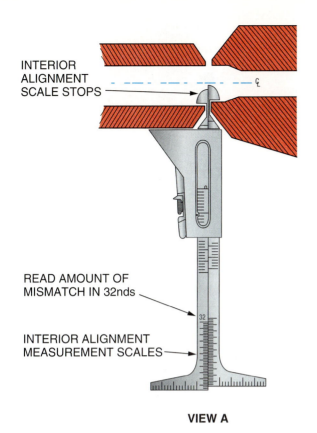

INTERIOR ALIGNMENT SCALE STOPS

℄

READ AMOUNT OF MISMATCH IN 32nds

32

INTERIOR ALIGNMENT MEASUREMENT SCALES

VIEW A

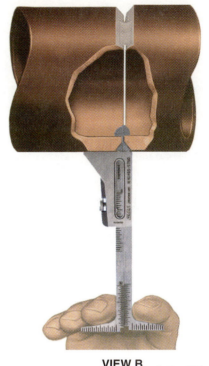

VIEW B

109F05.EPS

Figure 5 ◆ Checking internal misalignment.

Hot Tip

Hi-Lo Application

When checking for internal misalignment, take two gauge readings: one before tacking together the two pieces of pipe and one after tack welding. The Hi-Lo gauge is designed to fit into the gap between the two pipes, even after tacking. It typically comes with both metric and English scales.

Figure 7 shows the tapered alignment stops on a Hi-Lo gauge.

Weld reinforcement is measured by placing one foot of the Hi-Lo gauge on the pipe and the other on the weld face, making sure the foot on the pipe surface is flush. Weld reinforcement is read on the material thickness scale. *Figure 8* shows measuring weld reinforcement with a Hi-Lo gauge.

When fitting up socket welds, the pipe is scribed so that a measurement can be made to ensure that a gap is left between the end of the pipe and the socket. This gap prevents cracking when the pipe expands during the welding process. The scribe-line scale on the feet of the Hi-Lo gauge can be used to measure from the socket to the scribed line to ensure the gap is as specified on the WPS or site quality specifications. Most codes require a minimum ⅟₁₆" gap, but actual requirements for your job may vary due to pipe size or base metal type. *Figure 9* shows measuring scribe lines with a Hi-Lo gauge.

CAUTION

Actual requirements for socket weld gaps must be verified for your job by checking the WPS or site quality standards. Failure to follow the WPS can cause equipment damage or result in a flawed weld.

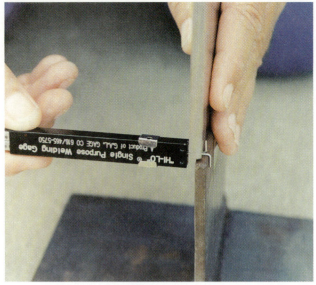

Figure 6 ◆ Material thickness scale.

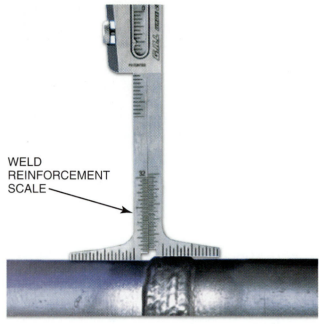

WELD
REINFORCEMENT
SCALE

109F08.EPS

Figure 8 ◆ Measuring weld reinforcement.

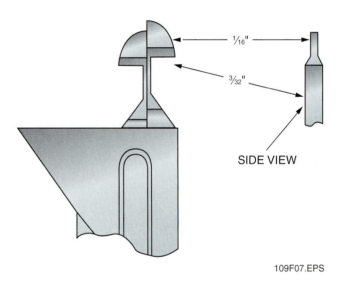

SIDE VIEW

109F07.EPS

Figure 7 ◆ Tapered alignment stops.

4.0.0 ◆ FIT-UP TOOLS

Many tools are available to aid plate and pipe fit-up. Some are special tools designed specifically for fit-up; others are common hand tools. Before using any tool, be sure to inspect it for damage or missing parts. Never use a tool for a job for which it was not intended. Be sure to clean and return all tools when you have finished with them.

4.1.0 Positioning Parts of a Weldment

Hydraulic jacks, chain falls, and come-alongs are used to position parts of a weldment. These types of positions are described in the following sections.

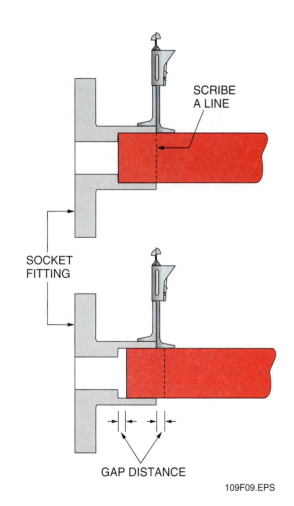

SCRIBE
A LINE

SOCKET
FITTING

GAP DISTANCE

109F09.EPS

Figure 9 ◆ Measuring the distance between scribe lines on a socket weld.

4.1.1 Hydraulic Jacks

When using hydraulic jacks, never weld directly on the jack base or ram. To secure the jack, a ring can be tack-welded to act as a socket for the base and/or ram. When welding near hydraulic jacks, protect the ram from weld spatter. If the ram is off the floor, be sure to secure it with rope or chain. Monitor the jack for oil leaks. Oil will create a fire hazard, and oil in the area to be welded will create weld defects (porosity). Oil leaks must be cleaned up before welding. *Figure 10* shows using a hydraulic jack to aid joint fit-up.

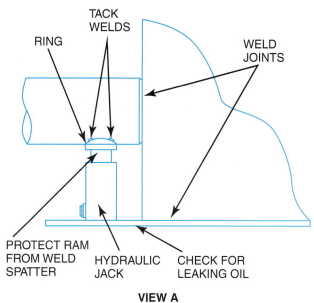

VIEW A

4.1.2 Chain Hoists (Chain Falls)

Chain hoists, also called falls, are used to lift or lower weldment parts. Always position the chain hoist directly over the center of gravity of the object being lifted. Secure the chain hoist over the weldment with a sling or chain wrapped around a structural member.

Never hang chain hoists from piping, ducts, conduit, or raceways. These items are not designed to carry external loads, and damage could result. The ground should be attached directly to the item to be welded. *Figure 11* shows using a chain hoist to aid joint fit-up.

VIEW B

109F10.EPS

Figure 10 ◆ Hydraulic jack.

4.1.3 Come-Alongs

Come-alongs can be used for vertical lifting and pulling at angles. Often, more than one come-along or a chainfall and a come-along are used to precisely position a weldment. Before welding, check grounding to be sure the welding current will not pass through the come-along. The ground should be attached directly to the item to be welded. *Figure 12* shows using a come-along to aid joint fit-up.

Chain Collectors and Retainers

Many chain hoists are equipped with buckets, canvas bags, or other devices designed to collect the chain loops to prevent them from interfering with the workpiece. If the hoist you are using includes this feature, make sure the collector or retainer is aligned correctly with the drop of the chain. Otherwise, the collector or retainer may tilt, dumping the heavy load of chain on top of you. Wear appropriate personal protective equipment, including a hard hat, when using chain hoists or falls.

SLING AROUND
STRUCTURAL MEMBER   CHAIN HOIST

WELD JOINT

109F11.EPS

Figure 11 ◆ Chain hoist.

4.2.0 Plate Fit-Up Tools

The most common method of holding a joint in place after it has been fitted up is to tack-weld it in place. This works very well for small weldments and joints that are straight. However, large weld-

COME-   SLING AROUND   CHAIN
ALONG   MEMBER   FALL

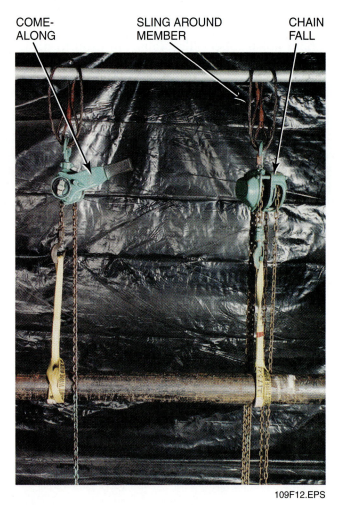

109F12.EPS

Figure 12 ◆ Come-along.

ments or long or thick joints often require some type of mechanical means, as well as tack welding, to set up and hold them in place for welding. The most common tools for plate fit-up are strongbacks, clips, yokes, and wedges. In addition, special plate alignment tools are available.

4.2.1 Strongbacks

Strongbacks are typically made on the job site from heavy bar stock. They are notched at the weld joint to allow access to the joint so that welds can be made without interference. The strongback can be placed on the face or root side of the weldment. The plates to be joined are clamped or tack-welded flush against the strongback, which holds the joint in alignment. If possible, when tack welding strongbacks, place the tack welds on only one side of the strongback. This will make the strongback easy to remove by striking it with a hammer on the side opposite the tack welds. *Figure 13* shows a welded strongback.

4.2.2 Clips, Yokes, and Wedges

Clips, yokes, and wedges are also made on the job site. They can be used to align joints and then hold them in place during welding. Clips are welded to the edge of one plate and then wedges are positioned on the other plate and driven under the clips to force the joint into alignment. Yokes work in a similar manner. A yoke is welded to one plate. A slotted plate is then placed over the yoke, and a wedge is driven under the yoke to force the joint into alignment. *Figure 14* shows yokes and wedges being used to align and hold a joint.

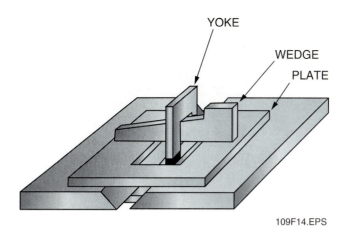

Figure 14 ◆ Yokes and wedges.

4.2.3 Strongbacks with Yokes and Wedges or Bolts

If the joint needs to be forced into alignment, yokes and wedges or bolts can be used with a strongback. When yokes and wedges are used, the strongback is positioned on the weldment straddling the joint. Yokes are positioned over the strongback and welded to the plates. Wedges are then driven under the yokes to force the joint into alignment. When bolts are used, the strongback must be made of channel stock or reinforced angle iron. Holes are cut in the channel or angle iron so that the bolts can pass through. The strongback is then positioned on the weldment straddling the joint. The position of the bolts is marked, and the strongback is removed. Bolts long enough to protrude through the strongback or threaded stock are then welded onto the plates.

The strongback is then positioned with the bolts protruding. Nuts are placed on the bolts and tightened with a ratchet or impact wrench, pulling the joint into position. *Figure 15* shows aligning a joint with a strongback.

4.2.4 Plate Alignment Tools

Special tools are manufactured for aligning plate. A typical configuration of an alignment tool consists of a yoke, threaded adjusting rod, gap plate, and root bar *(Figure 16)*. The alignment tool is used by straddling the joint opening with the yoke. The gap plate can be changed to match the specified root opening. The threaded adjusting rod is lowered until the opening in the gap plate is below the joint and the root bar can be inserted. The threaded adjusting rod is then tightened, bringing the joint into alignment. If necessary, the root bar can be adjusted to compensate for uneven joint thicknesses.

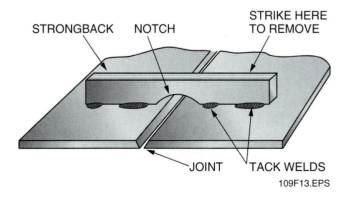

Figure 13 ◆ Strongback.

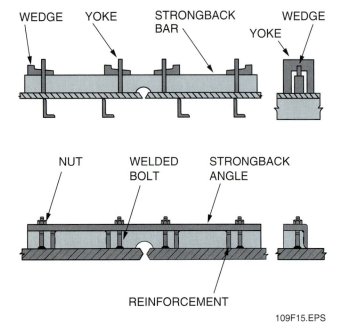

Figure 15 ◆ Aligning a joint with a strongback.

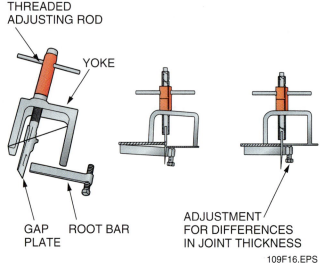

Figure 16 ◆ Plate alignment tool.

4.3.0 Pipe Fit-Up Tools

There are many different manufacturers of pipe fit-up tools. The tools they produce are similar, but before using a tool, be sure to review the guidelines provided by its manufacturer.

4.3.1 Pipe Jacks and Rollers

Pipe jacks and rollers (Figure 17) are used to support pipe for fit-up and welding. Pipe jacks typically have either a V-head or roller head and a height adjustment. Rollers, which can be floor-stand or table models, can be adjusted horizontally for various pipe diameters.

Adjustable jack stands and rollers are necessary tools in the pipeline welding industry. They allow the welder to position various sizes of pipe for alignment and to manually turn the pipe as

Nonferrous Alignment Tools

In some industries, fabricated alignment tools (wedges, strongbacks, and yokes) must be constructed of nonferrous materials, such as brass, to prevent sparking. This is common in some segments of the petrochemical industry.

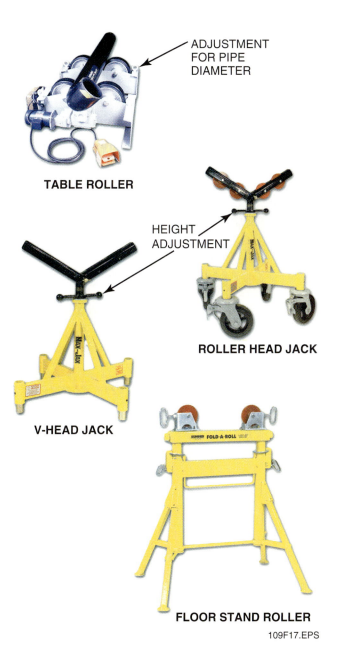

TABLE ROLLER

ADJUSTMENT FOR PIPE DIAMETER

HEIGHT ADJUSTMENT

ROLLER HEAD JACK

V-HEAD JACK

FLOOR STAND ROLLER

109F17.EPS

Figure 17 ◆ Pipe jacks and rollers.

needed so the weld can be performed from the top, which is the preferred position. When fabricating piping systems in place, however, jack stands and rollers cannot be used because of positioning limitations. One way to deal with this is to weld straight sections of piping systems using jack stands and rollers. The straight sections can then be welded in place to the rest of the piping system.

> **WARNING!**
>
> Jack stands and roller assemblies should not be field-fabricated. Load capacities and other safety considerations are built into these devices by their manufacturers, so only commercial jack stands and rollers should be used. Field-fabricated jack stands and rollers can fail and cause injury or even death.

> **CAUTION**
>
> When using pipe rollers to tack or weld pipe, be sure to ground directly to the pipe to prevent arcing damage to the rollers and roller bearings.

4.3.2 Chain Clamps

Chain clamps are used to align and hold pipe for fit-up and tacking. Some manufacturers use link chain, others use chain similar to bicycle chain. Regardless of the type of chain used, the basic procedure for using chain clamps is the same. The chain, which is anchored to one side of the clamp, is passed around the pipe and secured. The slack in the chain is then removed using a screw jack to pull the pipe tightly against the clamp. *Figure 18* shows a chain hold-down clamp mounted on a jack stand.

The chain clamp shown here acts more as a chain vise than a clamp. It does not provide precise alignment. Clamps of this type usually are

Field-Fabricated Jack Stands

A mechanic was crushed to death when a bus on which he was working fell off a set of jack stands. The stands were fabricated of plate steel by a local welding shop. The top plate of each was completely flat; they had no lips, which commercial stands always have. The front tires of the bus were not chocked, and there was nothing to prevent it from falling off the stands.

The investigation of the accident revealed the jack stands had not been tested or certified for their rated capacity, nor were they marked with such a capacity. They were not fabricated in accordance with commercial jack stand construction. The lips on commercial jack stands cradle the area being supported. Also, commercial jack stands normally have three legs that help them compensate for irregular surfaces.

Sections of large-diameter pipe supported by jack stands may not be as large as buses, but they can cause serious or fatal injuries if they fall from their supports. Always use commercial jack stands that have been tested and certified.

109F18.EPS

Figure 18 ◆ Chain hold-down clamp.

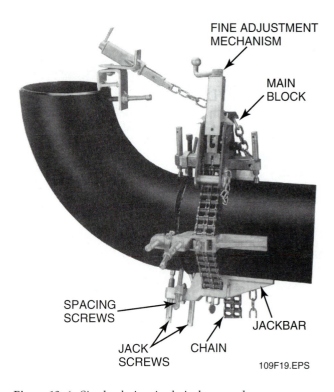

FINE ADJUSTMENT MECHANISM

MAIN BLOCK

SPACING SCREWS

JACKBAR

JACK SCREWS

CHAIN

109F19.EPS

Figure 19 ◆ Single-chain, single-jackscrew clamp.

used to secure the free end of the pipe on the pipe stand rather than to align and secure the ends to be welded. As the screw jack is tightened, it tends to pull the ends of the pipe out of alignment with each other.

Several types of chain clamps can be used to provide precise fit and alignment:

* Single-chain, single-jackscrew clamps
* Single-chain, double-jackscrew clamps
* Double-chain, double-jackscrew clamps
* Rim clamps

The single-chain, single-jackscrew clamp shown in *Figure 19* typically will align and re-form out-of-round pipe up to Schedule 40 wall thickness. It also can be used on pipe over Schedule 40 as an alignment tool as long as no re-forming is required. The single-chain clamp has a single jackscrew on each jack bar with optional spacing

screws that allow the gap to be set quickly and precisely.

In addition to fitting up straight pipe, attachments for the precision fit-up clamps allow them to be used to fit up elbows, Ts, and flanges. The attachments connect by chain to the main block and to the elbow, T, or flange with a chain or clamp. The attachment has a fine adjustment crank to remove slack from the chain to support the elbow, T, or flange.

4.3.3 Other Pipe Alignment and Clamping Tools

Cage clamps are designed to align and clamp pipe of one specific size. As pipe size changes, so must the size of the cage clamp being used. A hammer is usually required to position and remove a cage clamp, but the hammer must be used carefully. Hammering on the clamp often can deform it. This deformity will be carried on to the next fit performed with that clamp. A cage clamp is shown in *Figure 20*.

The rim clamp is a more versatile, nonchain design. It is recommended for applications where heavy-duty re-forming is required. The jack-screws on a rim clamp exert pressure on specified high points for precise alignment. These clamps are ideal for tasks in which 100% weld and grind is required before the clamp can be released.

4.3.4 Clamping Devices for Small-Diameter Pipe

Typical chain clamps are too bulky to use on small-diameter pipe ranging in size from ⅜" to 2¾". A simple shop-built clamp for small-diameter pipe can be made from a piece of angle iron. The pipe is laid in the angle, and C-clamps or wires are used to secure the pipe. Small blocks are placed on both sides of the angle to hold it in position. When extreme accuracy is required, special clamps for small-diameter pipe can be used. These clamps have multiple jaws that automatically clamp and align the pipe when the clamping screws are turned. Special stainless steel clamps are available for working with stainless steel piping. *Figure 21* shows clamps for small-diameter pipe fit-up.

109F21.EPS

Figure 21 ◆ Shop-built pipe clamps.

109F20.EPS

Figure 20 ◆ Cage clamp.

4.3.5 Pipe Pullers

Pipe pullers are clamps, often chain-type, that are used to pull together two pipe workpieces so that they can be joined by welding. These types of pullers are often applied in field fabrication or addition projects.

In the petrochemical industry, for example, a new unit may be added to an existing facility. The piping must be joined through the use of manifolds and other piping designs. In order to make the final tie-ins from existing piping to the new piping, pipe pullers are often used to align the joining parts precisely.

CAUTION

Do not force joints into alignment and then hold them in place with welds. Stress will be exerted at the welded joint in such an alignment, and cracking could result. Precise alignment and fitting should be accomplished before finalizing the installation with the weld.

Pipe pullers are frequently used to pull together sections of large-diameter pipe for welding. Fillers or **consumable inserts** may be installed between the faces of the joints to achieve precisely the necessary gap between the parts. These inserts or fillers may become a permanent part of the finished weld. *Figure 22* shows a pipe puller.

4.3.6 Flange Alignment Tools

Various methods and tools can be used to make sure a flange facing aligns precisely with the pipe before it is permanently welded in place. Flange pins can be installed in the flange boltholes to provide a reference point from which leveling can be

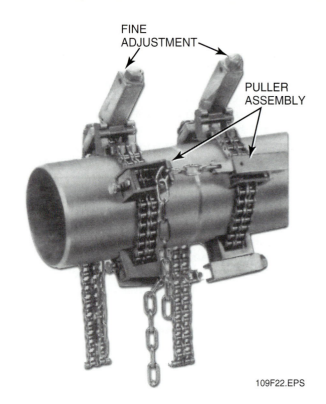

FINE ADJUSTMENT

PULLER ASSEMBLY

109F22.EPS

Figure 22 ◆ Pipe puller.

measured for tack welds or complete welding. Usually a minimum of two flange pins are installed in two of the boltholes to provide a precise reference point from the center of one bolthole to another. From this flat, level benchmark, other tools can be used to adjust the flange to the pipe for true and precise alignment. Multi-squares are often used with various adjustable flange levelers, such as the tool shown in *Figure 23*.

Gaps in Aligning and Fitting

Why is it so necessary to maintain a gap between two pieces, especially pipe pieces, when aligning and fitting prior to welding?

Types of Welded Flanges

Some of the more common flanges generally installed using welded connections are the welding neck flange, the lap joint flange, the socket weld flange, and the slip-on flange. Threaded flanges may also be back-welded, as determined by the process. Welding neck flanges are preferred for use in severe service applications, such as those involving high pressure, subzero temperatures, or elevated temperatures.

AWS ENTRY LEVEL WELDER – TRAINEE MODULE 29109-03

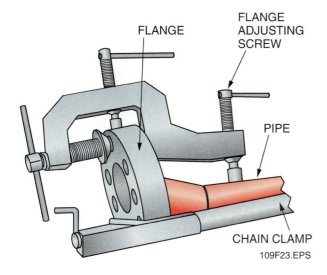

Figure 23 ◆ Flange alignment tool.

5.0.0 ◆ WELDMENT DISTORTION

Besides the obvious cosmetic problems resulting from poor alignment and fit, proper application of alignment and fit-up methods must be applied to deal with distortion. Distortion is the expansion and contraction of metal as it responds to changes in temperature. Tools such as chain clamps, rim clamps, C-clamps, strongbacks, and other devices can be used to redirect this expansion or contraction. This is referred to as controlling the distortion rather than preventing it. Distortion cannot be prevented as long as the metal experiences changes in temperatures created by the welding process. It can, however, be controlled, resulting in a weldment that is relatively free of damaging stress.

5.1.0 Causes of Distortion

Distortion is caused by the nonuniform expansion and contraction of the weld metal and adjacent base metal during the heating and cooling cycles of welding. When a metal block is heated, it expands uniformly in all directions. As the metal block cools, it contracts uniformly in all directions. If the metal block is restricted on two sides as it is heated, expansion cannot take place in the direction of restriction. Because the metal block must expand the same amount, all expansion occurs in the unrestricted directions. As the metal block cools, it contracts uniformly in all directions, leaving it narrower in the restricted direction and longer in the unrestricted direction. *Figure 24* illustrates the expansion, contraction, and distortion control of a block of steel.

During welding, these same forces act on the weld metal and base metal. As the weld metal is

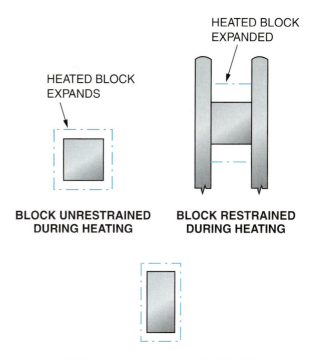

RESTRAINED BLOCK IS DISTORTED
AFTER COOLING

Figure 24 ◆ Controlling distortion.

deposited and fuses with the base metal, it is at its maximum expanded state. Upon cooling, it attempts to contract but is restricted by the surrounding base metal. Stresses develop within the weld until the stress reaches the yield point of the base metal, causing the base metal and weld metal to stretch. These same stresses also cause the base metal to move. When the base metal is heated during the welding process, it will attempt to expand but will be restricted by the surrounding cooler base metal. Some movement will occur, but as the weld cools, the base metal that was welded will contract more due to the stress caused by the restriction of the surrounding base metal. This causes distortion. Even when the weld is at room temperature, stress equal to the strength of the base metal will be locked in the weldment. This stress is called **residual stress.** *Figure 25* shows how stress and distortion affect a weld.

5.2.0 Correlation of Metal
Properties and Distortion

The degree of distortion is directly related to the stresses generated during welding. Two of the metal properties that figure prominently in these stresses are the **coefficient of thermal expansion (linear)** and the **specific heat per unit volume.** Metals with higher values for these two properties

Latent Distortion

If a weldment is clamped firmly during welding to prevent any movement, it will stay straight as long as the clamps are in place. Even after cooling, when the clamps are removed, residual stress will cause some distortion. When weldments that have residual stress are machined, they distort as the machining process removes metal, allowing the residual stress to overcome the strength of the base metal.

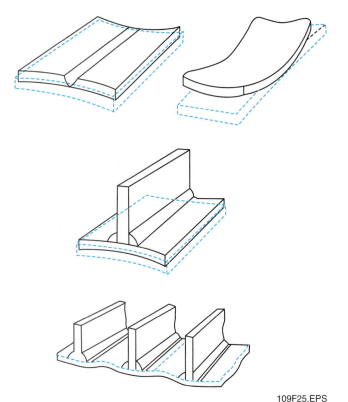

109F25.EPS

Figure 25 ◆ Stress and distortion in welds.

will experience a higher degree of distortion than those with lower values. Stainless steel has a much higher coefficient of thermal expansion and specific heat per unit volume than carbon steel. It therefore is more likely to be affected by distor-

tion. *Table 1* compares the coefficients of thermal expansion and the specific heats per unit volume for some common metals.

5.3.0 Controlling Distortion

Distortion can be controlled using a variety of techniques and tools. The following sections explain how to control distortion to minimize its effect on weldments.

5.3.1 Clamping and Bracing

Chain clamps, rim clamps, strongbacks, and other field-fabricated support devices are often used to hold weldments firmly in place during the welding process in order to prevent warping caused by distortion. These devices must be designed, constructed, and installed such that their overall support can withstand the stresses of welding. They must also be left in place long enough to allow the weldment to cool to **ambient temperature.** Some distortion may still take place once the devices are removed. This is caused by residual stress present in the weldment.

5.3.2 Tack Welding

A tack weld holds parts of a weldment in proper alignment until the finish welds are made. In order for them to be effective pay attention to the number of tack welds, their length, and the distance between them. If too few tacks are made, the

Table 1 Thermal Expansion and Specific Heat per Unit Volume for Common Metals

Metal	Coefficient of Thermal Expansion @ 20°C	Specific Heat Per Unit Volume J/g°C
Aluminum	$22–24 \times 10^{-6}/°C$	0.900
Stainless steel (316)	$14–17 \times 10^{-6}/°C$	0.500
Steel	$12–13 \times 10^{-6}/°C$	0.448
Copper	$16–17 \times 10^{-6}/°C$	0.390

chance of the joint closing up as the welding proceeds is much greater. Clamps and other fit-up and aligning devices must be used to maintain proper alignment and the proper gap width during the process.

If tack welds are to be included with the main weld, they must be applied by qualified welders. They must also follow the qualifying WPS for the project, if one applies. Likewise, the consumables used as fillers must meet the same filler requirements as the material for the finished welds. For those tack welds that require removal prior to the finish weld, use extreme care to avoid causing defects in the material.

Note
The qualifying WPS that specifies the tack welding procedures, as well as the consumables to be used, is typically designed as part of the engineering plans. It is not normally a function or responsibility of the welder. However, the welder must be qualified to perform the tasks and follow the procedures specified in the WPS.

5.3.3 *Amount of Weld Material*

The more weld metal placed in a joint, the greater the forces of shrinkage. Properly sizing a weld not only minimizes distortion, it also saves weld metal and time. Excess reinforcement on the face of a weld increases the forces of distortion and adds nothing to the strength of the weld. Excess face reinforcement actually reduces the strength of a weld and is therefore prohibited by welding codes. The face of fillet welds may be slightly convex, flat, or slightly concave. Groove welds should have slight reinforcement of no more than ⅛" for butt or corner welds. *Figure 26* shows proper weld reinforcement.

Note
Refer to your site's WPS for specific requirements on weld reinforcement. The information in this module is provided as a general guideline only. The site WPS must be followed for all welds covered by a procedure. Check with your supervisor if you are unsure of the WPS for your application.

Proper fit-up and edge preparation also reduce the amount of weld required. Open-root joints should have a root opening from ¼₆" to ⅛". To con-

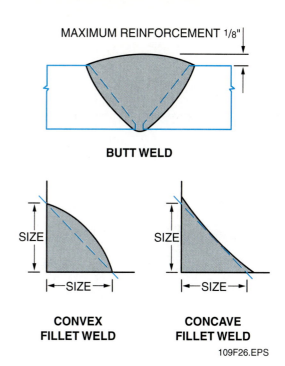

MAXIMUM REINFORCEMENT 1/8"

BUTT WELD

SIZE | SIZE

CONVEX FILLET WELD

SIZE | SIZE

CONCAVE FILLET WELD

109F26.EPS

Figure 26 ◆ Weld reinforcement.

trol burn-through, a root land of ¼₆" to ⅛" is used. Each side of the open-root joint is beveled from 30° to 37½° (included angle of 60° to 75°). The greater the angle of bevel, the more weld will be required to fill the joint. The bevel angle must be sufficient to allow access to the root but not so large that it requires excess weld metal to fill. Welds made with GMAW or GTAW generally require larger openings than welds made with SMAW because the nozzles used with GMAW and GTAW are larger in diameter than the electrodes used with SMAW. These nozzles require more room in the joint to reach the root. *Figure 27* shows an open V-groove joint preparation.

CAUTION
When preparing a joint, refer to your site's WPS for specific requirements. The information in this manual is provided as a general guideline only. The site WPS must be followed for all welds covered by a procedure. Check with your supervisor if you are unsure of the WPS for your application.

When possible, a double V-groove should be used in place of a single V-groove (*Figure 28*). The double V-groove requires half the weld metal as compared to the single V-groove. Also, welding from both sides reduces distortion since the forces of distortion will be working against each other.

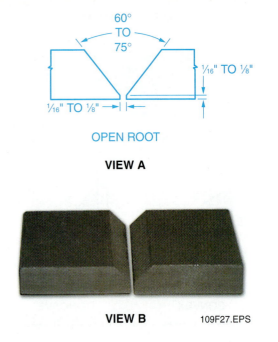

VIEW A

VIEW B 109F27.EPS

Figure 27 ◆ Open V-groove preparation.

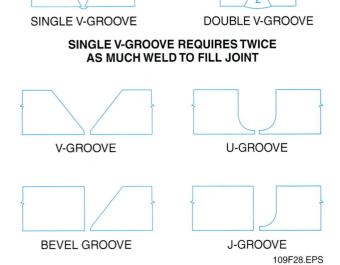

SINGLE V-GROOVE REQUIRES TWICE AS MUCH WELD TO FILL JOINT

109F28.EPS

Figure 28 ◆ Types of groove styles.

On thick joints, U- and J-grooves also require less weld metal than V-grooves or bevel grooves, although U- and J-grooves require more preparation time (*Figure 28*).

5.3.4 Backing Strips on Groove Welds

Molten metal that extends beyond the back or opposite side of the groove is called penetration. Controlling the amount and shape of this penetra-

tion is important to the function of the weldment. It is also important because controlling the penetration helps reduce distortion by controlling the weld material uniformly. Various forms of backing materials or strips can be applied to the back side of plate weldments. The strips typically have a groove machined directly into them that controls the shape and size of the penetration material once it cools. Backing strips can be purchased in rolls from which the desired length can be cut and then taped directly to the back of the prepared joint. If the backing strip is designed to become part of the permanent weldment, it must be made of material similar to the alloy or metal being welded. If the backing strip is to be removed, it normally is made of dissimilar metal to prevent it from being welded into the joint. The backing strip can then be peeled from the back side of the finished weldment. Backing strips of copper, for example, are often used on weldments of stainless steel.

When thick metal backing strips are used on groove welds, the root opening normally opens to ¼", and the bevel is reduced to 22½° (a total angle of 45°). A root land is not required because the backing will prevent burn-through. Distortion can be reduced considerably through this control of the size and shape of the penetration. *Figure 29* shows different backing methods applied to groove welds.

5.3.5 Gas Backing on Pipe Welds

When welding pipe, the depth of penetration to the inside of the pipe must be controlled. Penetrating too deeply into the interior of the pipe

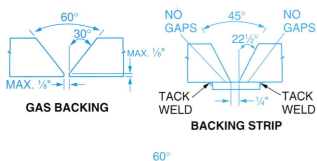

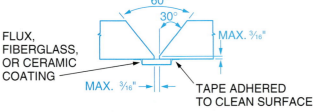

109F29.EPS

Figure 29 ◆ Groove welds with backing strip.

interferes with the flow within the pipe, causing turbulence and other problems. Controlling this penetration is also necessary to reduce stress on joints. However, it is difficult, if not impossible, to gain access to the interior of the pipe. Various methods are used to provide backing and to control penetration in pipe welding.

Gas backing is one method used when welding pipe joints. When gas backing is used, some method of containing the gas inside the pipe must be used *(Figure 30)*. For a short section of pipe, the ends of the pipe can be capped by taping or clamping a metal disk over the ends. For larger or longer pipe sections, however, this method may not be feasible due to the amount of gas that would be required to fill the pipe. For these cases, inflatable bladders, water-soluble plugs, or soft plastic bags can be inserted into the pipe near the weld joint. The inflatable bladders can be fished out of the pipe after the weld cools. The water-soluble plugs or soft plastic bags can either be blown or washed out of the pipe after the weld cools.

The type of gas used will depend on the base metal of the pipe being welded. Generally, nitrogen or carbon dioxide is used for carbon steel piping; argon is used for stainless steel, aluminum, and alloy steel piping.

CAUTION

The pipe end opposite the end into which gas enters must have a vent hole to purge the gas and prevent pressure from building.

5.3.6 Backing Rings

Backing rings are flat metal strips that have been rolled to fit inside a pipe. They can be ordered in a variety of base metal types to match the base metal being welded. Some of these rings are split, making them easier to insert and adjust to piping that may be slightly out of round. Backing rings are available that have three or more nubs around the outside of the ring. The diameter of the nub is the root opening required for the pipe diameter being welded. Nubs can be pegs, buttons, or indents punched into the ring. Pegs must be removed after the joint is tack-welded. They are removed by striking them with a chipping hammer. Buttons can be removed or left in position to be melted during the root pass. *Figure 31* shows backing rings.

The joint preparation for backing rings *(Figure 32)* is similar to the preparation for open-root joints. The joint angle should be 60° or 75° with a root opening and land up to ⅛" (or as specified on the WPS). If backing rings with nubs are used, the root opening will be set automatically by making sure the beveled ends of the pipe are positioned firmly against the nubs.

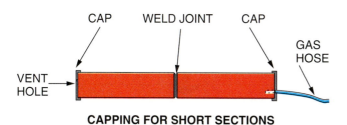

CAPPING FOR SHORT SECTIONS

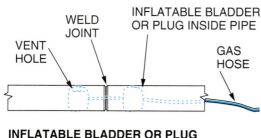

INFLATABLE BLADDER OR PLUG FOR LONG SECTION OF PIPE

109F30.EPS

Figure 30 ◆ Gas backing.

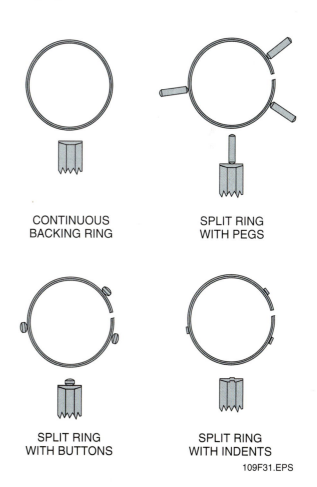

109F31.EPS

Figure 31 ◆ Backing rings.

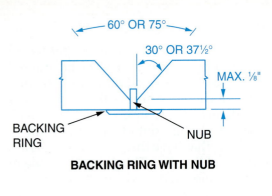

BACKING RING WITH NUB

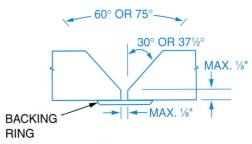

BACKING RING WITHOUT NUB

109F32.EPS

Figure 32 ◆ Pipe joint with backing ring.

5.3.7 Consumable Inserts on Pipe Welds

Consumable inserts are similar to backing rings. As the name implies, they are completely consumed during welding and become part of the weld metal. Because they are part of the finished weld, they must match the filler metal requirements for the weld being made. Gas backing often is required when using consumable inserts. Consumable inserts come in a variety of shapes and are identified by class numbers, as shown in *Figure 33*.

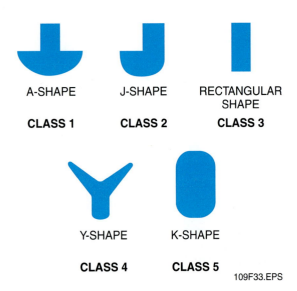

Figure 33 ◆ Consumable inserts.

109F33.EPS

Consumable inserts can be purchased in 50' coils and in preformed split rings. The rectangular Class 3 insert also is available as a solid ring.

Joint preparation for consumable inserts *(Figure 34)* can consist of a V-groove or a J-groove, with the V-groove having a joint angle of 75° and a root land of ⅟₁₆". The J-groove joint angle is 20° with a ⅟₁₆" root land. The consumable insert usually is tacked firmly against the prepared ends of the pipe.

5.3.8 Inserts on Socket Joints

Socket joints generally are used on pipe that is 5" or smaller in diameter. A socket joint uses a prefabricated fitting containing sockets on the ends; the pipe slips into these sockets. The fitting and pipe are joined using a fillet weld. All the common fittings, such as elbows, flanges, couplings, reducers, and even valves, are available as socket fittings. Using socket joints is a quick and easy method for joining pipe evenly.

In order to eliminate stress, distortion, and possible cracking by expansion during heating, the end of the pipe must not touch the bottom of the socket. Codes covering socket welds require a ⅟₁₆" to ³⁄₃₂" gap between the end of the pipe and the bottom of the socket fitting.

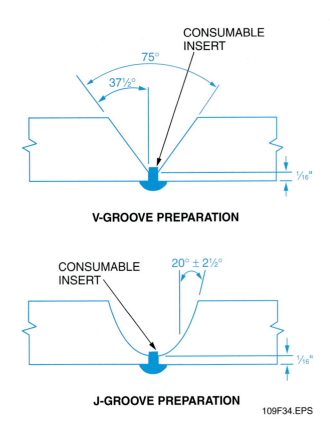

Figure 34 ◆ Joint preparation for consumable inserts.

109F34.EPS

When preparing the joint, there are two ways to ensure that the gap between the end of the pipe and the socket fitting is maintained. One way is to scribe the pipe a preset distance from the end. The gap can be checked by measuring from the scribed line to the socket and then adding the socket depth. The scribed lines also can be used to check setup after the joint is tacked up or welded.

The second method of ensuring the correct gap is to use a gap ring. A gap ring is a split ring that is formed to the gap required for the pipe diameter being welded. The gap ring is placed in the bottom of the socket end and becomes a permanent part of the joint. Gap rings (gap-o-lets) are the fastest way to fit up socket welds and are available for various pipe sizes in packages of 20 each or more. *Figure 35* shows two methods of preparing pipe joints.

5.3.9 Intermittent Welding

Many joints do not require welds along their full length. Using short intermittent welds instead of continuous welds will reduce distortion. Often, stiffeners, brackets, and braces can be intermittently welded rather than continuously welded.

5.3.10 Backstep Welding

Backstep welding *(Figure 36)* is a welding technique in which the general progression of welding is from left to right, but the weld beads are deposited in short increments from right to left. This technique reduces distortion by minimizing and interrupting the heat input.

5.3.11 Welding Sequence

A welding sequence involves placing welds at different points on a weldment so that shrinkage forces in one location are counteracted by shrinkage forces in another location. Welding sequences can be quite complicated for large complex weldments or quite simple for less complicated weldments. When possible, use a welding sequence because it is a very effective means of controlling distortion. A simple welding sequence is to make short welds on alternating sides of the joint. This can be used for fillet or groove, intermittent, or continuous welding. *Figure 37* shows examples of intermittent welding sequences.

Welding sequences are often performed by two individuals welding on opposite sides of a joint at the same time. This type of welding is often called buddy welding. It is a very effective way to control distortion.

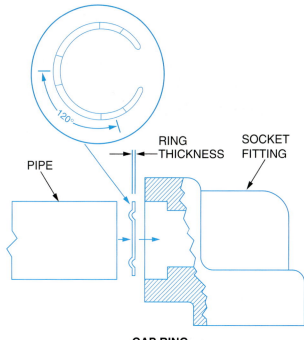

GAP RING

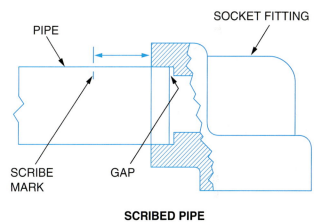

SCRIBED PIPE

109F35.EPS

Figure 35 ◆ Pipe joint preparation.

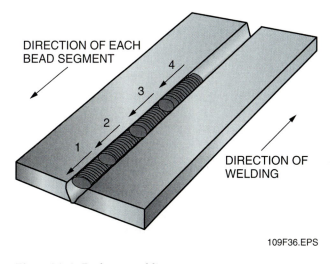

109F36.EPS

Figure 36 ◆ Backstep welding.

Alternating Welds

When possible, an alternating weld should be placed directly adjacent to or across from its matching weld on the opposite side. On lengthy welds, however, this sequence may not be advisable because it could leave long runs unwelded in between these welds. In these cases, stagger the welds from side to side.

If welding alternately on either side of the joint is not possible, or if one side must be completed first, a joint preparation may be used that deposits more weld metal on the second side. The greater contraction resulting from depositing more weld metal on the second side will help counteract the distortion on the first side.

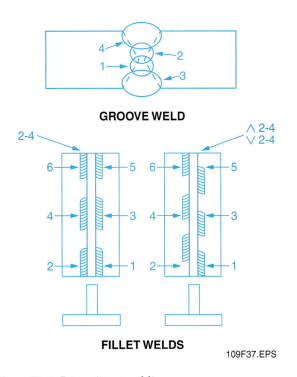

Figure 37 ◆ Intermittent welding sequences.

5.3.12 *Heat Treatments*

Distortion can be controlled by preheating and postheating. Applying heat before welding is called preheating. Applying heat immediately after welding is called postheating. Preheating and postheating will be covered in a later module.

Welding must be stopped periodically to allow inspection of the weld in progress. These hold points should be scheduled strategically. Any work that will be obstructed after welding is complete, and therefore will not be easily inspected, must be inspected before the weld covers it. The following are among the items to be checked:

- Preheat temperature
- Postheat temperature

- Compliance with WPS requirements
- Weld root pass
- Weld layers

6.0.0 ◆ CHECKING JOINT MISALIGNMENT AND POOR FIT-UP

The quality of joint preparation and fit-up directly affects the quality of the completed weld. After a weld is made, it is very costly and time-consuming to repair a defect. By thoroughly checking the joint fit-up, potential problems can be avoided. Follow these steps to check a fit-up:

Step 1 Determine if the weld is covered by a WPS. If the weld has a WPS, obtain a copy and follow it to make the weld.

Step 2 Check that the base metal type and grade are as specified. If either the base metal type or grade is wrong, contact your supervisor.

Step 3 Check that the joint surfaces are free of contamination such as grease, oil, moisture, and rust. Also, check the surfaces parallel to the root and face of the weld. If there is contamination, clean the joint before continuing.

Step 4 Check the joint surface for cracks or laminations. If you find either, contact your supervisor.

Step 5 Check that the edge preparation is as specified for the joint *(Figure 38)*. Check the:
- Groove type
- Root opening
- Root face
- Included angle
- Bevel angle
- Base metal thickness

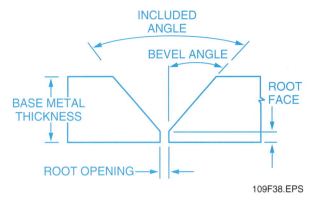

Figure 38 ◆ Edge preparation.

Step 6 Check the fit-up of backing strips, backing rings, and consumable inserts if required. Backing strips and backing rings must be tight against the joint along or around its entire length with even root spacing. Consumable inserts must fit tightly at the root of the weld. Correct any fit-up problems before proceeding.

Step 7 Check that the parts of the weldment are square, level, and plumb, and check that any angles are as specified. Correct any problems before proceeding.

Step 8 Check that the tack welds have been cleaned and feathered. Correct any problems before proceeding.

Step 9 Check that the welding process to be used is as specified on the WPS.

Step 10 Check that the consumables (type and size) are as specified on the WPS. Also check that arrangements have been made for the heated storage of consumables if required on the WPS.

Step 11 Check that the welding sequence is available if required.

Step 12 Check that the welding machine settings are as specified on the WPS.

Step 13 Check that arrangements have been made for preheating, interpass temperature control, and postheating if required on the WPS.

Step 14 Check, if required, that the welder who will perform the welding is certified to the WPS.

Summary

It is very important to perform proper joint fit-up and alignment to ensure an acceptable weld. The proper fit-up measuring devices and tools must be used to accomplish this. Welders must know how to compensate for welding distortion as well as how to check for poor fit-up after the weld has been completed.

Government agencies, professional societies, and associations have written guidelines for joint fit-up and alignment. These guidelines not only help ensure quality welds, but also help ensure safe welds and welding environments. All fit-up procedures should follow these guidelines, which have been proven over the years by use and testing. When specified, these codes must be used under penalty of law.

Review Questions

1. A detailed listing of the rules and principles that apply to specific welded products is called a _____.
   a. standard
   b. procedure
   c. punch list
   d. welding code

2. The written set of instructions for producing sound welds is called a _____.
   a. contract
   b. welding code
   c. WPS
   d. PQR

3. Welding test coupons prepared for a job are tested according to the _____.
   a. manufacturer's guidelines
   b. WPS guidelines
   c. applicable code
   d. PQR

4. Which tool is often fabricated on the job from small channel or angle iron?
   a. Straightedge
   b. Plumb bob
   c. Square
   d. Level

5. Before welding a 45° joint, check the angle of the joint with a _____.
   a. straightedge
   b. Hi-Lo gauge
   c. framing square
   d. combination square

6. The primary use of a Hi-Lo gauge is to check for _____.
   a. pipe joint misalignment
   b. I-beam joint alignment
   c. 90° joints
   d. 45° joints

7. Secure a chain hoist over a weldment with a sling or chain wrapped around _____.
   a. a structural member
   b. piping
   c. conduit
   d. raceways

8. When using a strongback to hold two weldment parts that have been aligned and fitted up, the tack welds used to hold the strongback to the weldment parts should be placed only on _____ of the strongback.
   a. one side
   b. the notched side
   c. vertical sides
   d. opposite sides

9. The stress locked in a weldment at room temperature, and equal to the strength of the base metal, is called _____.
   a. base stress
   b. residual stress
   c. thermal conductivity
   d. coefficient of expansion

10. The more weld metal placed into a joint, the greater the forces of _____.
    a. stress
    b. magnetism
    c. shrinkage
    d. thermal conductivity

Val Baumann

Welding Student
Northern Arizona Vocational Institute of Technology
Heber, Arizona

Val Baumann is a good example of why you should never give up searching for your place in life. His story:

Val dropped out of high school and was later kicked out of a charter school. Many people figured he had no future, but his parents, along with one of his former school principals, never lost faith. As a last resort, they decided to try him in welding school. There Val found his niche. He did so well that after the first semester he became a student aide. Soon he was helping to teach other kids how to weld. He became a class leader and a role model for other students. Val capped his comeback by winning the welding gold medal at the state competition of the Vocational Industrial Clubs of America (VICA). As a result of that accomplishment, he was invited to compete at the national level, where he did very well.

Success is never accidental. You get out of life what you put into it. In this case, some caring people kept believing when it would have been easy to give up. Val found what he was looking for and turned what could have been a wasted life into a success story.

Trade Terms Introduced in This Module

Addendum: Supplementary information typically used to describe corrections or revisions to documents.

Ambient temperature: The room temperature of the atmosphere completely surrounding an object on all sides.

Coefficient of thermal expansion (linear): The change in length per unit length of material for a 1°C change in temperature.

Consumable insert: Preplaced filler metal that is completely fused into the root of the joint during welding, becoming part of the weld.

High-low: The discrepancy in the internal alignment of two sections of pipe.

Level: A line on a horizontal axis parallel with the earth's surface (horizon).

Peening: The mechanical working of metal using hammer blows.

Plumb: A line on a vertical axis perpendicular to the earth's surface (horizon).

Residual stress: Stress remaining in a weldment as a result of heat.

Specific heat per unit volume: The quantity of heat required to raise one unit mass of a substance by one unit degree.

Tack-weld: A weld made to hold parts of a weldment in proper alignment until the final weld is made.

Additional Resources

This module is intended to present thorough resources for task training. The following reference works are suggested for further study. These are optional materials for continued education rather than for task training.

The Procedure Handbook of Arc Welding, 1994. Cleveland, OH: The Lincoln Electric Company.
Welding Handbook, 2001. Miami, FL: The American Welding Society.

CONTREN™ LEARNING SERIES—USER UPDATES

The NCCER makes every effort to keep these textbooks up-to-date and free of technical errors. We appreciate your help in this process. If you have an idea for improving this textbook, or if you find an error, a typographical mistake, or an inaccuracy in NCCER's Contren™ textbooks, please write us, using this form or a photocopy. Be sure to include the exact module number, page number, a detailed description, and the correction, if applicable. Your input will be brought to the attention of the Technical Review Committee. Thank you for your assistance.

Instructors—If you found that additional materials were necessary in order to teach this module effectively, please let us know so that we may include them in the Equipment/Materials list in the Instructor's Guide.

Write: Curriculum Revision and Development Department
National Center for Construction Education and Research
P.O. Box 141104, Gainesville, FL 32614-1104

Fax: 352-334-0932

E-mail: curriculum@nccer.org

Craft _____ Module Name _____

Copyright Date _____ Module Number _____ Page Number(s) _____

Description _____

_____

_____

_____

(Optional) Correction _____

_____

_____

(Optional) Your Name and Address _____

_____

_____

Shielded Metal Arc Welding – Open V-Groove Welds

Course Map

This course map shows all of the modules in the AWS Entry Level Welder – Phase 1 curriculum. The suggested training order begins at the bottom and proceeds up. Skill levels increase as you advance on the course map. The local Training Program Sponsor may adjust the training order.

AWS ENTRY LEVEL WELDER—PHASE 1

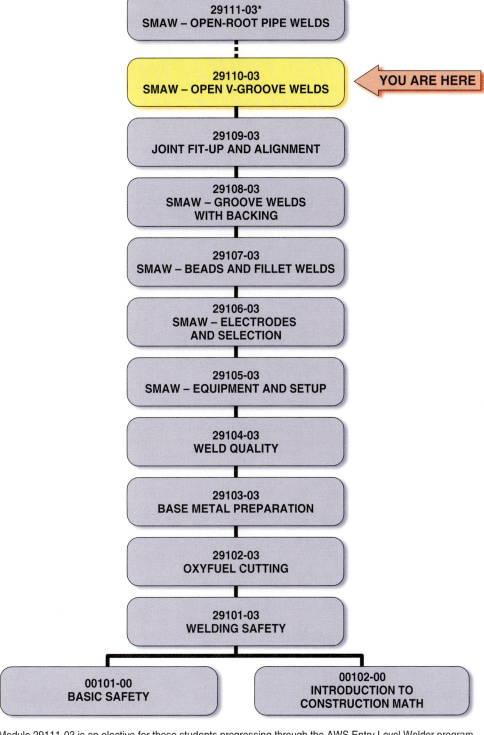

29111-03*
SMAW – OPEN-ROOT PIPE WELDS

29110-03
SMAW – OPEN V-GROOVE WELDS

YOU ARE HERE

29109-03
JOINT FIT-UP AND ALIGNMENT

29108-03
SMAW – GROOVE WELDS
WITH BACKING

29107-03
SMAW – BEADS AND FILLET WELDS

29106-03
SMAW – ELECTRODES
AND SELECTION

29105-03
SMAW – EQUIPMENT AND SETUP

29104-03
WELD QUALITY

29103-03
BASE METAL PREPARATION

29102-03
OXYFUEL CUTTING

29101-03
WELDING SAFETY

00101-00
BASIC SAFETY

00102-00
INTRODUCTION TO
CONSTRUCTION MATH

*Module 29111-03 is an elective for those students progressing through the AWS Entry Level Welder program.

29110CMAP.EPS

Figures

Shielded Metal Arc Welding – Open V-Groove Welds

Objectives

When you have completed this module, you will be able to do the following:

1. Prepare shielded metal arc welding (SMAW) equipment for open-root V-groove welds.
2. Perform open-root V-groove welds in the:
   - Flat (1G) position
   - Horizontal (2G) position
   - Vertical (3G) position
   - Overhead (4G) position

Prerequisites

Before you begin this module, it is recommended that you successfully complete Modules 00100-00 through 29109-03.

Required Trainee Materials

1. Pencil and paper
2. Appropriate personal protective equipment

1.0.0 ◆ INTRODUCTION

An open V-groove weld is a nonstandard term for one type of open-root groove weld without backing or an insert. In this module, it is an open-root V-groove weld without backing. As explained in a previous module, V-groove welds are commonly used for welding thick plate. The open-root V-groove weld is easy to prepare because it only requires single-face beveling of the matching members. However, without backing, welding the root pass requires practice to perfect.

This module will explain the joint preparation and performance of open-root V-groove welds on plate using E6010 and E7018 electrodes in all posi-tions. It will also prepare the welder to advance to the more difficult pipe welding training modules. Performance demonstrations will require that the welds meet the qualification standards in the *ASME Boiler and Pressure Vessel Code, Section XI— Welding and Brazing Qualifications* for visual and destructive testing.

2.0.0 ◆ WELDING EQUIPMENT SETUP

Before welding can take place, safety procedures and practices must be reviewed, the area has to be made ready, the welding equipment must be set up, and the metal to be welded must be prepared. The following sections explain how to set up arc welding equipment for welding.

2.1.0 Safety Practices

The following is a summary of safety procedures and practices that must be observed while cutting or welding. Keep in mind that this is just a sum-mary. Complete safety coverage is provided in the Level One module, *Welding Safety.* If you have not completed that module, do so before continu-ing. Above all, be sure to wear appropriate pro-tective clothing and equipment when welding or cutting.

2.1.1 Protective Clothing and Equipment

- Always use safety goggles with a full face shield or a helmet. The goggles, face shield, or helmet lens must have the proper light-reducing tint for the type of welding or cutting to be performed. Never directly or indirectly view an electric arc without using a properly tinted lens.
- Wear proper protective leather and/or flame-retardant clothing along with welding gloves

that will protect you from flying sparks and molten metal as well as heat.

- Wear 8" or taller high-top safety shoes or boots. Make sure that the tongue and lace area of the footwear will be covered by a pant leg. If the tongue and lace area is exposed or the footwear must be protected from burn marks, wear leather spats under the pants or chaps and over the front top of the footwear.
- Wear a solid material (non-mesh) hat with a bill pointing to the rear or, if much overhead cutting or welding is required, a full leather hood with a welding face plate and the correct tinted lens. If a hard hat is required, use a hard hat that allows the attachment of rear deflector material and a face shield.
- If a full leather hood is not worn, wear a face shield and snug-fitting welding goggles over safety glasses for gas welding or cutting. Either the face shield or the lenses of the welding goggles must be an approved shade 5 or 6 filter. For electric arc welding, wear safety goggles and a welding hood with the correct tinted lens (shade 9 to 14).
- If a full leather hood is not worn, wear earmuffs or at least earplugs to protect ear canals from sparks.

2.1.2 Fire/Explosion Prevention

- Never carry matches or gas-filled lighters in your pockets. Sparks can cause the matches to ignite or the lighter to explode, causing serious injury.
- Never perform any type of heating, cutting, or welding until a hot work permit is obtained and an approved fire watch is established. Most work site fires caused by these types of operations are started by cutting torches.
- Never use oxygen to blow off clothing. The oxygen can remain trapped in the fabric for a time. If a spark hits the clothing during this time, the clothing can burn rapidly and violently out of control.
- Make sure that any flammable material in the work area is moved or shielded by a fire-resistant covering. Approved fire extinguishers must be available before any heating, welding, or cutting operations are attempted.
- Never release a large amount of oxygen or use oxygen as compressed air. Its presence around flammable materials or sparks can cause rapid and uncontrolled combustion. Keep oxygen away from oil, grease, and other petroleum products.
- Never release a large amount of fuel gas, especially acetylene. Methane and propane tend to concentrate in and along low areas and can be

ignited a considerable distance from the release point. Acetylene is lighter than air but is even more dangerous. When mixed with air or oxygen, it will explode at much lower concentrations than any other fuel.

- To prevent fires, maintain a neat and clean work area and make sure that any metal scrap or slag is cold before disposal.
- Before cutting or welding containers such as tanks or barrels, check to see if they have contained any explosive, hazardous, or flammable materials including petroleum products, citrus products, or chemicals that decompose into toxic fumes when heated. As a standard practice, always clean and then fill any tanks or barrels with water or purge them with a flow of inert gas to displace any oxygen.

2.1.3 Work Area Ventilation

- Make sure confined space procedures are followed before conducting any welding or cutting in the confined space.
- Never use oxygen in confined spaces for ventilation purposes.
- Always perform cutting or welding operations in a well-ventilated area. Cutting or welding operations involving zinc or cadmium materials or coatings result in toxic fumes. For long-term cutting or welding of such materials, always wear an approved full face, supplied-air respirator (SAR) that uses breathing air supplied externally of the work area. For occasional, very short-term exposure, a HEPA-rated or metal-fume filter may be used on a standard respirator.
- Make sure confined spaces are ventilated properly for cutting or welding purposes.

2.2.0 Preparing the Welding Area

To practice welding, a welding table, bench, or stand is needed. The welding surface must be steel, and provisions must be made for placing welding coupons out of position. *Figure 1* shows a typical welding station.

To set up the area for welding, follow these steps:

Step 1   Check to be sure the area is properly ventilated. Use doors, windows, and fans.

Step 2   Check the area for fire hazards. Remove any flammable materials before proceeding.

Step 3   Check the location of the nearest fire extinguisher. Do not proceed unless the extinguisher is charged and you know how to use it.

110F01.EPS

Figure 1 ◆ Welding station.

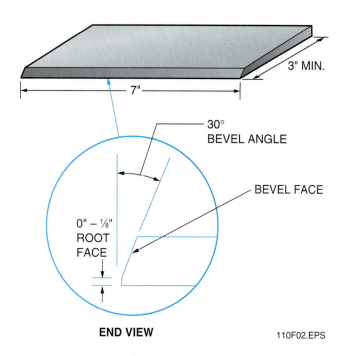

END VIEW

110F02.EPS

Figure 2 ◆ Metal cut for open-root V-groove weld coupons.

Step 4 Position a welding table near the welding machine.

Step 5 Set up flash shields around the welding area.

2.3.0 Preparing the Weld Coupons

If possible, the weld coupons should be ⅜"-thick carbon steel. If this size is not readily available, ¼"- to ¾"-thick steel can be used for practice welds. Open-root welds generally have a bevel angle of 30°.

 Note

Carbon steel, ⅜" thick will be required for the performance qualification tests.

Clean the steel plate prior to welding by using a wire brush or grinder to remove heavy mill scale or corrosion. Prepare weld coupons to practice the following welds:

- *Stringer bead welds* – The coupons can be any size or shape that is easily handled.
- *Open-root V-groove welds* – As shown in *Figure 2*, cut the metal into 3" × 7" rectangles with one of the 7" lengths beveled at 30°. Grind a 0" to ⅛" root face on the bevel.

 Note

The welding codes allow the root face and opening on open-root welds to be 0" to ⅛". Adjust the root face and root opening as needed when you start your welding practices.

Follow these steps to prepare each V-groove weld coupon:

Step 1 Check the bevel face. There should be no slag and a 0" to ⅛" root face, and the bevel angle should be 30°. An example of this is shown in *Figure 3*.

Step 2 Center the beveled strips with a 0" – ⅛" root opening and tack-weld them in place. Place the tack welds on the back side of the joint. Use three to four ¼" tack welds.

2.4.0 Electrodes

Obtain a small quantity of the electrodes to be used. For the welding exercises in this module, ⅛" and 5⁄32" E6010 and 3⁄32" and ⅛" E7018 electrodes will be used.

 Note

The electrode sizes are recommendations only. Other electrode sizes may be substituted depending on site conditions. Check with your instructor for the electrode sizes to use.

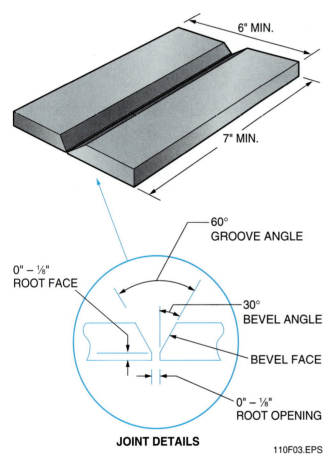

6" MIN.

7" MIN.

60°
GROOVE ANGLE

0" – ⅛"
ROOT FACE

30°
BEVEL ANGLE

BEVEL FACE

0" – ⅛"
ROOT OPENING

JOINT DETAILS

110F03.EPS

Figure 3 ◆ Open-root V-groove weld coupon.

For E6010 electrodes, use a whipping motion when depositing the stringer bead to control the weld puddle. Then remove all slag between passes.

For E7018 electrodes, keep the following in mind:

- Do not whip low-hydrogen electrodes.
- Maintain a short arc because a low-hydrogen weld relies on its slag to protect the molten metal. It does not have a heavy gaseous shield.
- Remove all slag between passes.
- Remove only a small number of electrodes at a time from the oven because low-hydrogen electrodes pick up moisture quickly.
- Do not use low-hydrogen electrodes with chipped or missing flux.

For either E6010 or E7108 electrodes, observe the following:

- When running stringer beads, use a slight side-to-side motion and a slight pause at the weld toe to tie in and flatten the bead.
- When running weave beads, use a slight pause at the weld toe to tie in and fill the bead.
- Set the amperage in the lower portion of the suggested range for better puddle control when welding vertically.

Obtain only the electrodes to be used for a particular welding exercise at one time. Have some

LASH

To ensure good welding results, remember LASH.

Length of arc (L) – The distance between the electrode and the base metal (usually one times the electrode diameter).

Angle (A) – Two angles are critical:

- Travel angle – The longitudinal angle of the electrode in relation to the axis of the weld joint
- Work angle – The traverse angle of the electrode in relation to the axis of the weld joint

Speed (S) – Travel speed is measured in inches per minute (IPM). The width of the weld will determine if the travel speed is correct.

Heat (H) – Controlled by the amperage setting and dependent upon the electrode diameter, base metal type, base metal thickness, and the welding position.

---

type of pouch or rod holder to store the electrodes in to prevent them from becoming damaged. Never store electrodes loose on a table. They may end up on the floor where they are a tripping hazard and can become damaged. Some type of metal container such as a pail must also be available to discard hot electrode stubs.

WARNING!

Do not throw electrode stubs on the floor. They roll easily, and someone could step on one, slip, and fall.

2.5.0 Preparing the Welding Machine

Identify a welding machine to use. *Figure 4* shows typical welding machines. Proceed with the following welding set-up steps:

Step 1 Verify that the welding machine can be used for DC (direct current) welding.

Step 2 Check to be sure that the welding machine is properly grounded through the current receptacle.

Step 3 Check the area for proper ventilation.

Step 4 Verify the location of the current disconnect.

Step 5 Set the polarity to direct current electrode positive (DCEP).

Step 6 Connect the clamp of the workpiece lead to the workpiece.

Step 7 Set the amperage for the electrode type and size to be used. Typical settings are:

Electrode	Size	Amperage
E6010	$\frac{1}{8}$"	65A to 130A
E6010	$\frac{5}{32}$"	90A to 175A
E7018	$\frac{3}{32}$"	50A to 110A
E7018	$\frac{1}{8}$"	90A to 150A

Note

Amperage recommendations vary by manufacturer, position, current type, and electrode brand. For specific recommendations, refer to the manufacturer's recommendations for the electrode being used.

Step 8 Check to be sure the electrode holder is not grounded.

Step 9 Turn on the welding machine.

3.0.0 ◆ OPEN-ROOT V-GROOVE WELDS

The open-root V-groove weld is a common groove weld normally made on plate and pipe. Practicing the open-root V-groove weld on plate will prepare the welder to make the more difficult pipe welds covered in the next module.

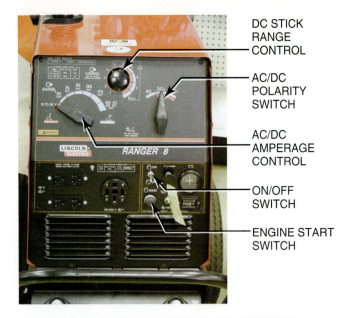

DC STICK
RANGE
CONTROL

AC/DC
POLARITY
SWITCH

AC/DC
AMPERAGE
CONTROL

ON/OFF
SWITCH

ENGINE START
SWITCH

PORTABLE 200A ENGINE-DRIVEN AC/DC WELDER

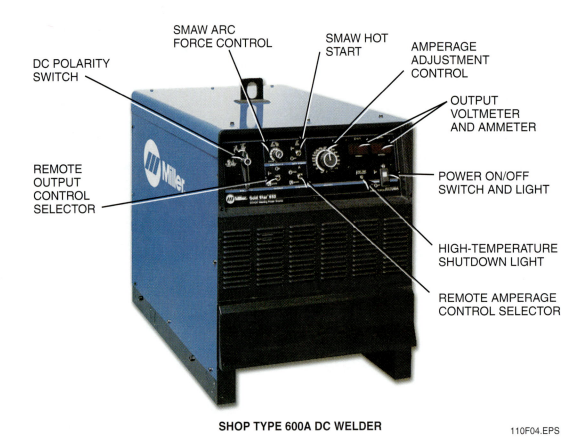

SMAW ARC
FORCE CONTROL

SMAW HOT
START

AMPERAGE
ADJUSTMENT
CONTROL

DC POLARITY
SWITCH

OUTPUT
VOLTMETER
AND AMMETER

REMOTE
OUTPUT
CONTROL
SELECTOR

POWER ON/OFF
SWITCH AND LIGHT

HIGH-TEMPERATURE
SHUTDOWN LIGHT

REMOTE AMPERAGE
CONTROL SELECTOR

SHOP TYPE 600A DC WELDER

110F04.EPS

Figure 4 ◆ Typical DC and AC/DC welding machines.

The performance qualification requirements for this module for visual and destructive testing are based on the *ASME Boiler and Pressure Vessel Code, Section IX, welder certification test.*

3.1.0 Root Pass

The most difficult part of making an open-root V-groove weld is the root pass. The root pass is made from the V-groove side of the joint and must have complete penetration, but not an excessive amount of burn-through root reinforcement. The penetration is controlled with a technique called *running a keyhole.* A keyhole, as shown in *Figure 5,* is a hole made when the root faces of two plates are melted away by the arc.

The molten metal flows to the back side of the keyhole, forming the weld. The keyhole should be about ³⁄₁₆" to ¼" in diameter. If the keyhole is larger, there will be excess burn-through and slag inclusions. If the keyhole closes up, there will be a lack of penetration. Root reinforcement should be flush to ⅛".

The keyhole is controlled by using fast-freeze E6010 electrodes and by whipping the electrode. Move the electrode forward about ³⁄₁₆" to ¼" and then back about ⅛" and pause. Use care not to increase the arc length when moving the electrode. If the keyhole starts to grow, decrease the pause and increase the forward length of the whip. If the keyhole starts to close up, pause slightly longer and decrease the length of the whip. Other factors that affect the keyhole size are the amperage setting, land size, and root opening. All these items can be adjusted within the param-

eters allowed by the welding codes or your site quality procedures.

After the root pass is run, it should be cleaned and inspected. Inspect the root face for excess buildup or undercut, as shown in *Figure 6,* which could trap slag when the filler pass is run. Remove excess buildup or undercut with a hand grinder by grinding the face of the root pass with the edge of a grinding disk. Use care not to grind through the root pass or widen the groove.

3.2.0 Groove Weld Positions

As illustrated in *Figure 7,* groove welds can be made in all positions. The weld position for plate is determined by the axis of the weld and the orientation of the workpiece. Groove weld positions for plate are flat, or 1G (groove); horizontal, or 2G; vertical, or 3G; and overhead, or 4G.

In the 1G and 2G positions, the weld axis can be inclined up to 15°. Any weld axis inclination for the other positions varies with the rotational position of the weld face as specified in AWS standards.

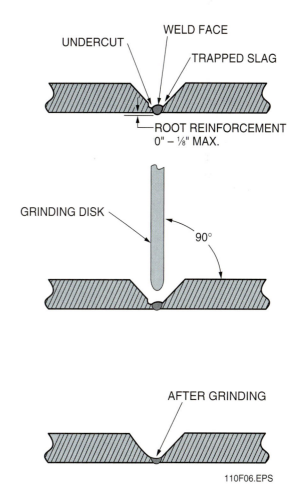

110F06.EPS

Figure 6 ◆ Grinding a root pass.

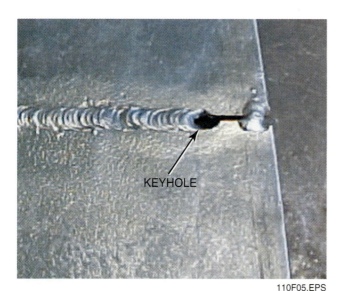

110F05.EPS

Figure 5 ◆ Root pass and keyhole.

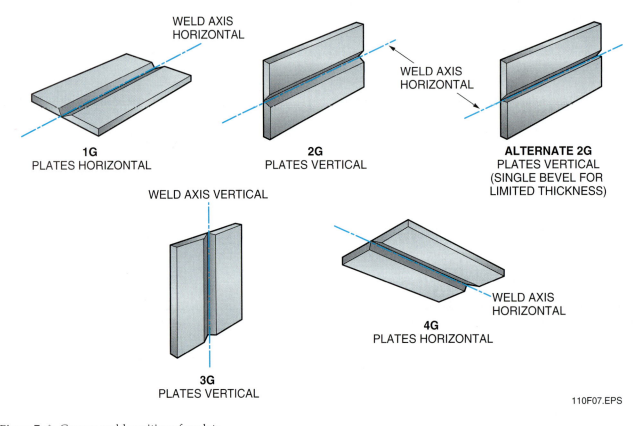

Figure 7 ◆ Groove weld positions for plate.

110F07.EPS

3.3.0 Acceptable and Unacceptable Groove Weld Profiles

Groove welds should be made with slight reinforcement and a gradual transition to the base metal at each toe. As shown in *Figure 8,* groove welds must not have excess reinforcement, underfill, excessive undercut, or overlap. If a groove weld has any of these defects, it must be repaired.

4.0.0 ◆ SMAW OF OPEN-ROOT V-GROOVE WELDS

This section of the module explains how to perform the following open-root V-groove welds:

- Flat (1G) welds
- Horizontal (2G) welds
- Vertical (3G) welds
- Overhead (4G) welds

4.1.0 Practicing Flat Open-Root V-Groove Welds (1G Position)

Practice flat open-root V-groove welds, as depicted in *Figure 9,* using ⅛" E6010 and ³⁄₃₂" E7018 electrodes. Use weave beads and keep the electrode angle 90° to the plate surface (0° work angle) with a 10° to 15° drag angle. When using stringer beads, use a 10° to 15° drag angle, but adjust the electrode work angle to tie in and fill the bead. Pay particular attention at the termination of the weld to fill the crater. If the practice coupon gets too hot between passes, you may have to cool it in water.

 WARNING!

Use pliers to handle the hot practice coupons. Wear gloves when placing the practice coupon in water. Steam will rise off the coupon and can burn or scald unprotected hands.

R = FACE AND ROOT REINFORCEMENT PER CODE NOT TO EXCEED ⅛" MAX.

ACCEPTABLE WELD PROFILE

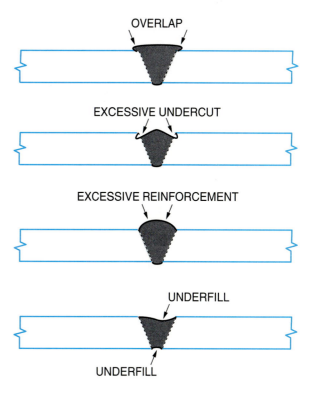

UNACCEPTABLE WELD PROFILES

110F08.EPS

Figure 8 ◆ Acceptable and unacceptable groove weld profiles.

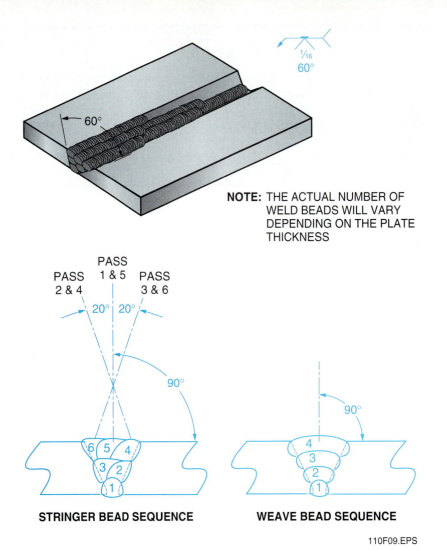

NOTE: THE ACTUAL NUMBER OF
WELD BEADS WILL VARY
DEPENDING ON THE PLATE
THICKNESS

PASS 1 & 5
PASS 2 & 4
PASS 3 & 6

STRINGER BEAD SEQUENCE

WEAVE BEAD SEQUENCE

110F09.EPS

Figure 9 ◆ Multipass 1G weld sequences and work angles.

CAUTION

Cooling with water is only done on practice coupons. Never cool test coupons or on-the-job welds with water. Cooling with water can cause weld cracks and affect the mechanical properties of the base metal.

Follow these steps to practice open-root V-groove welds in the flat position:

Step 1 Tack-weld the practice coupon together as explained earlier.

Step 2 Place the weld coupon in a flat position.

Step 3 Run the root pass using ⅛" E6010 electrodes.

Step 4 Clean the root pass. Grind if required.

Step 5 Run the filler and cover passes, cleaning the weld after each pass, to complete the weld using ³⁄₃₂" E7018 electrodes.

4.2.0 Horizontal Welds (2G Position)

Horizontal welds can be made on a vertically positioned flat surface or on a horizontal open-root V-groove. This section explains welding horizontal beads and making horizontal open-root V-groove welds.

4.2.1 Practicing Horizontal Beads

Before welding an open-root V-groove in the horizontal position, practice running horizontal stringer beads by building a pad in the horizontal position (*Figure 10*). Use ⅛" E6010 electrodes. The

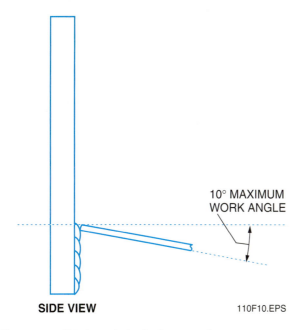

SIDE VIEW

110F10.EPS

Figure 10 ◆ Work angle in the horizontal position.

NOTE: THE ACTUAL NUMBER OF WELD BEADS WILL VARY DEPENDING ON THE PLATE THICKNESS

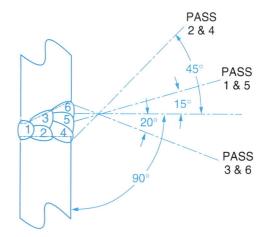

ALTERNATE JOINT REPRESENTATION

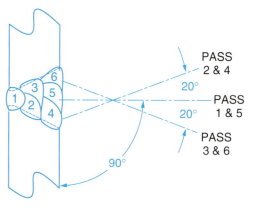

STANDARD JOINT REPRESENTATION

110F11.EPS

Figure 11 ◆ Multipass 2G weld sequences and work angles.

electrode angle should be a 10° to 15° drag angle and perpendicular to the plate. To control the weld puddle, the electrode can be dropped slightly but no more than 10°. Follow these steps to practice welding beads in the horizontal position:

Step 1 Tack-weld the flat plate welding coupon into the horizontal position.

Step 2 Run the first pass along the bottom of the coupon using ⅛" E6010 electrodes.

Step 3 Clean the weld bead.

Step 4 Continue running beads to complete the pad. Clean the weld beads after each pass.

4.2.2 Practicing Horizontal Open-Root V-Groove Welds

A horizontal open-root V-groove weld is shown in *Figure 11*. Practice horizontal open-root V-groove welds using ⅛" E6010 and E7018 electrodes and stringer beads. For the electrode angles, use a 10° to 15° drag angle and adjust the work angle as required. Pay particular attention at the termination of the weld to fill the crater.

Follow these steps to practice welding open-root V-groove welds in the horizontal position:

Step 1 Tack-weld the practice coupon together as explained earlier. Use the standard or alternate horizontal weld coupon as directed by your instructor.

Step 2　Tack-weld the flat plate weld coupon into the horizontal position.

Step 3　Run the root pass using E6010 electrodes.

Step 4　Clean the weld.

Step 5　Run the filler and cover passes, cleaning the weld after each pass, to complete the weld. Use E7018 electrodes and stringer beads.

4.3.0 Vertical Welds (3G Position)

Vertical welds can be made on a vertically positioned flat surface or on a vertical open-root V-groove, as shown in *Figure 12*. This section explains welding vertical beads and making vertical open-root V-groove welds.

4.3.1　Practicing Vertical Beads

Note

Perform this practice if needed. Otherwise, move on to the next section.

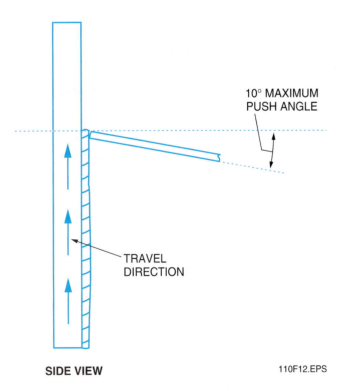

10° MAXIMUM PUSH ANGLE

TRAVEL DIRECTION

SIDE VIEW　　　　　110F12.EPS

Figure 12 ◆ Travel angle in the vertical position.

Before welding a joint in the vertical position, practice running vertical weave beads by building a pad in the vertical position (see *Figure 12*). Use ⅛" E6010 electrodes. Set the amperage in the lower part of the range for better puddle control. The electrode travel should be upward with a 0° to 10° push angle and perpendicular to the plate (0° work angle) for the first pass. A 0° to 10° work angle toward the previous bead should be used for subsequent beads. Use a side-to-side motion with a slight pause at each weld toe to tie in and to control the weld puddle. The width of the weave beads should be no more than three times the electrode diameter. Follow these steps to practice welding beads in the vertical position:

Step 1　Tack-weld the flat plate welding coupon into the vertical position.

Step 2　Run the first pass along the vertical edge of the coupon. Use a whipping motion.

Step 3　Clean the weld bead.

Step 4　Continue running beads and cleaning each pass to complete the pad.

4.3.2　Practicing Vertical Open-Root V-Groove Welds

Figure 13 is an illustration of a vertical open-root V-groove weld. Practice vertical open-root V-groove welds using ⅛" E6010 for the root weld pass and ³⁄₃₂" or ⅛" E7018 electrodes and weave beads for the remaining passes. Set the amperage in the lower part of the range for better puddle control. Use an upward travel direction with a 0° to 10° push angle, but adjust the electrode work angle to tie in the weld on one side or the other as needed. Pause slightly at each weld toe to penetrate and fill the crater to prevent undercut. Pay particular attention at the termination of the weld to fill the crater. Follow these steps to practice open-root V-groove welds in the vertical position:

Step 1　Tack-weld the practice coupon together as explained earlier.

Step 2　Tack-weld the weld coupon into the vertical position.

Step 3　Run the root pass using E6010 electrodes.

Step 4　Clean the weld.

Step 5　Run the filler and cover passes with E7018 ⅛" electrodes to complete the weld. Clean the weld between each pass.

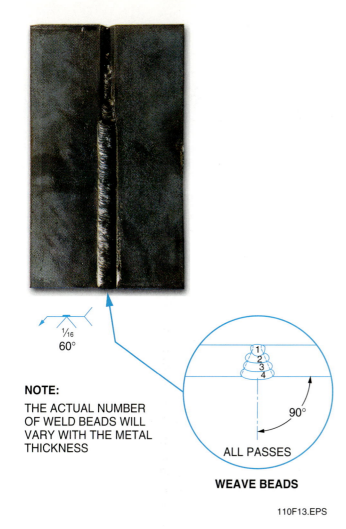

NOTE:

THE ACTUAL NUMBER
OF WELD BEADS WILL
VARY WITH THE METAL
THICKNESS

Figure 13 ◆ Multipass 3G weld sequences and work angles.

4.4.0 Overhead Welds (4G Position)

Overhead welds, as shown in *Figure 14*, can be made on an overhead flat surface or on an overhead open-root V-groove.

4.4.1 Practicing Overhead Beads

Before welding a joint in the overhead position, practice running overhead stringer beads by building a pad in the overhead position. Use ⅛" E6010 electrodes. The electrode work angle should be 90° with the plate with a 10° to 15° drag angle for the first pass. Then the work angle should be 10° toward the previous beads for the following beads. Follow these steps to practice welding beads in the overhead position:

Step 1 Tack-weld the flat plate welding coupon into the overhead position.

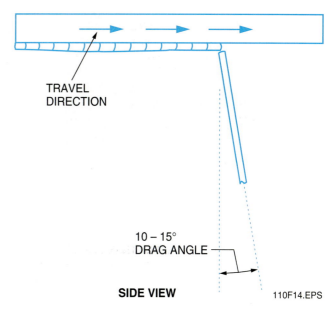

Figure 14 ◆ Building a pad in the overhead position.

Step 2 Run the first pass along the edge of the coupon.

Step 3 Clean the weld bead.

Step 4 Continue running beads, cleaning the weld between passes, to complete the pad.

4.4.2 Practicing Overhead Open-Root V-Groove Welds

An overhead open-root V-groove weld is shown in *Figure 15*. Practice running overhead open-root V-groove welds using ⅛" E6010 and ³⁄₃₂" and ⅛" E7018 electrodes and stringer beads. Use a 10° to 15° drag angle, but adjust the electrode work angle to tie in the weld as needed. Pay particular attention at the termination of the weld to fill the crater. Follow these steps to practice welding open-root V-groove welds in the overhead position:

Step 1 Tack-weld the practice coupon together as explained earlier.

Step 2 Tack-weld the weld coupon into the overhead position.

Step 3 Run the root pass using E6010 electrodes.

Step 4 Clean the weld.

Step 5 Run the filler and cover passes with E7018 ⅛" electrodes to complete the weld. Clean the weld after each pass.

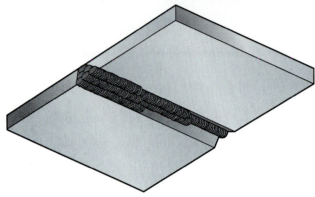

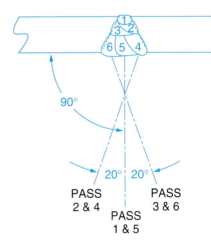

STRINGER BEAD SEQUENCE

110F15.EPS

Figure 15 ◆ Multipass 4G weld sequences and work angles.

Summary

The ability to make open-root V-groove welds on plate is an essential welding skill. A welder must possess this skill to perform welding jobs and to progress to more difficult pipe welding procedures. Making open-root V-groove welds requires the welder to set up and prepare the welding equipment and the welding coupons. V-groove welds can be made in the flat, horizontal, vertical, and overhead positions. Practice these welds until you can consistently produce acceptable welds.

Review Questions

1. Open-root welds generally have a bevel angle of _____.
   a. 10°
   b. 14°
   c. 22½°
   d. 30°

2. Welding codes allow the root face on open-root welds to be within a range of _____.
   a. $\frac{1}{32}$" to $\frac{3}{32}$"
   b. $\frac{1}{16}$" to $\frac{3}{32}$"
   c. $\frac{1}{32}$" to $\frac{1}{8}$"
   d. 0" to $\frac{1}{8}$"

3. When preparing an open-root V-groove weld coupon, center the beveled strips with a _____ root opening and then tack-weld them in place.
   a. 0" to $\frac{1}{8}$"
   b. $\frac{1}{16}$" to $\frac{1}{8}$"
   c. $\frac{1}{32}$" to $\frac{1}{16}$"
   d. $\frac{3}{32}$" to $\frac{5}{32}$"

4. Practicing open-root V-groove welds on plate will prepare the welder to make the more difficult _____ welds.
   a. flat
   b. pipe
   c. vertical
   d. horizontal

5. When making an open-root V-groove weld, the hole made when the two root faces of two plate edges are melted away by the arc is called a(n) _____.
   a. bead
   b. overlap
   c. keyhole
   d. root pass

6. When the plates are vertical and the axis of the weld is vertical, the code for this position of groove weld is _____.
   a. 1G
   b. 2G
   c. 3G
   d. 4G

7. When making stringer beads on flat open-root V-groove welds, keep the electrode angle at 90° to the plate surface with a _____ drag angle.
   a. 5° to 10°
   b. 10° to 15°
   c. 15° to 20°
   d. 20° to 22°

8. To practice welding open-root V-groove welds in the horizontal position, run the first pass along the bottom of the coupon using _____ E6010 electrodes.
   a. 1⁄32"
   b. 3⁄32"
   c. 1⁄8"
   d. 3⁄8"

9. When practicing horizontal open-root V-groove welds, pay particular attention at the termination of the weld to fill the _____.
   a. pad
   b. root
   c. crater
   d. puddle

10. Before welding a joint in the vertical position, practice running vertical weave beads by building a _____ in the vertical position.
   a. pad
   b. root
   c. crater
   d. puddle

Jerry Trainor

Vice President
Education Employment and Training
Kawerak, Inc.
Nome, Alaska

Jerry was born in Pablo, Montana, and grew up in Yakima Valley after his family moved to Washington state. He learned on the job to become a journeyman welder/fabricator. He even spent some time as a fourth-grade teacher, where he learned he had a natural talent for teaching. In the early 1980s, he became Assistant Professor for welding technology at Lewis-Clark State College in Idaho. Over the years and at several institutions, he has developed an impressive list of credentials as a vocational trainer, teacher, and administrator.

How did you become interested in the welding industry?
I got interested in welding as a result of taking a wood and metal class during my senior year in high school. Then I got my first welding job in Davis, California, building a food processing plant for Hunt's Foods. From that time on until 1983, I welded on many construction and production shop projects.

What do you think it takes to be a success in your trade?
To be successful in welding, a person has to take pride in doing the best job possible, understand the field of metallurgy, and be able to fabricate simple and complex projects. I enjoy fabricating and designing different implements, working on complex drawings of steel structures, and working with other professionals.

What are some of the things you do in your job?
At present, my position has very little to do with welding. I am in charge of developing training programs in various trade areas and delivering training programs to the various villages in the Bering Strait region of Alaska. Kawerak, Inc. is the nonprofit arm of the Bering Strait Regional Corporation. We provide a variety of services to Alaska natives who reside within this region. My department has delivered individual training programs in carpentry and heavy equipment operations to rural villages, and more than 129 students have completed the training courses. This has been very successful, because the village residents do not have to leave their villages to attend training programs. They also receive college credits from the University of Alaska Fairbanks for their efforts.

What would you say to someone entering the trade today?
My advice would be to become as proficient as possible in all phases of welding, from using a cutting torch to welding in shielded metal arc welding (SMAW) through gas tungsten arc welding (GTAW). Take an interest in the metallurgy and production of different metal materials, understand complex blueprints, and be open to taking advanced theory courses through a community college or other providers. Always be ready to go the extra mile and put in a full day's work for a full day's pay. Be proud of your skills and abilities and take pride in always doing the best job possible. Be able to walk away from a job knowing that your part of the construction project was done with your best expertise and efforts.

Performance Accreditation Tasks

The Performance Accreditation Tasks (PATCs) correspond to and support the learning objectives in AWS EG2.0-95, Guide for the Training and Qualification of Welding Personnel: Entry-Level Welder.

Note that in order to satisfy all learning objectives in AWS EG2.0-95, the instructor must also use the PATCs contained in the second level in the Welding Contren™ Learning Series.

PATCs provide specific acceptable criteria for performance and help to ensure a true competency-based welding program for students.

The following tasks are designed to evaluate your ability to run open-root V-groove welds with SMAW equipment in all positions using E6010 and E7018 electrodes. Perform each task when you are instructed to do so by your instructor. As you complete each task, take it to your instructor for evaluation. Do not proceed to the next task until instructed to do so by your instructor.

OPEN-ROOT V-GROOVE WITH E6010 AND E7018 ELECTRODES IN THE FLAT POSITION

Using ⅛" and ⁵⁄₃₂" E6010 and ³⁄₃₂" E7018 electrodes, make an open-root V-groove weld on carbon steel plate in the flat position as indicated.

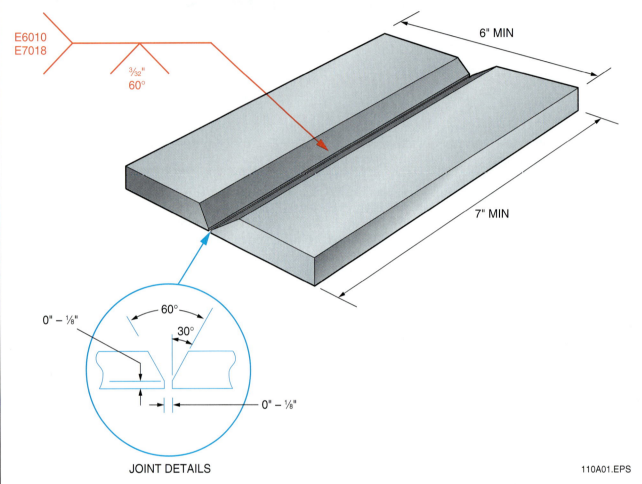

JOINT DETAILS

110A01.EPS

Criteria for Acceptance:

- Uniform rippled appearance on the bead face　　　———————
- Craters and restarts filled to the full cross-section of the weld　　———————
- Uniform weld wize　　———————
- Acceptable weld profile in accordance with the ASME ASME *Boiler and Pressure Vessel Code*　　———————
- Smooth transition with complete fusion at the toes of the weld　　———————
- Complete uniform root penetration at least flush with the base metal to a maximum buildup of ⅛"　　———————
- No porosity　　———————
- No excessive undercut　　———————
- No inclusions　　———————
- No cracks　　———————
- Acceptable guided bend test results　　———————

OPEN-ROOT V-GROOVE WITH E6010 AND E7018 ELECTRODES IN THE HORIZONTAL POSITION

Using ⅛" E6010 and ³⁄₃₂" E7018 electrodes, make an open-root V-groove weld on carbon steel plate in the horizontal position as shown.

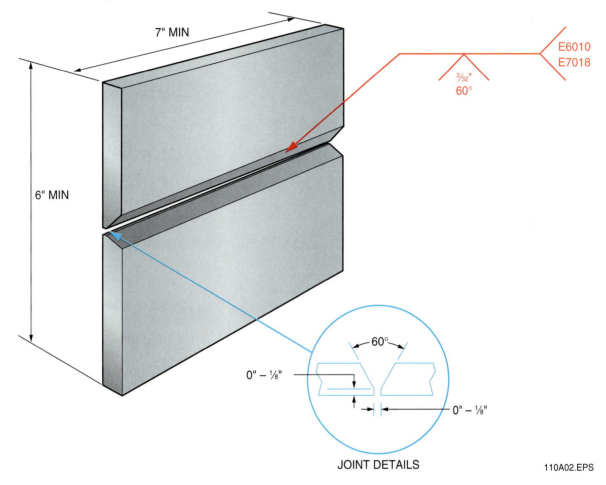

JOINT DETAILS

110A02.EPS

Criteria for Acceptance:

- Uniform rippled appearance on the bead face
- Craters and restarts filled to the full cross-section of the weld
- Uniform weld wize
- Acceptable weld profile in accordance with the ASME *Boiler and Pressure Vessel Code*
- Complete uniform root penetration at least flush with the base metal to a maximum buildup of ⅛"
- Smooth transition with complete fusion at the toes of the weld
- No porosity
- No excessive undercut
- No inclusions
- No cracks
- Acceptable guided bend test results

OPEN-ROOT V-GROOVE WITH E6010 AND E7018 ELECTRODES IN THE VERTICAL POSITION

Using ⅛" E6010 and ³⁄₃₂" E7018 electrodes, make an open-root V-groove weld on carbon steel plate in the vertical position as shown.

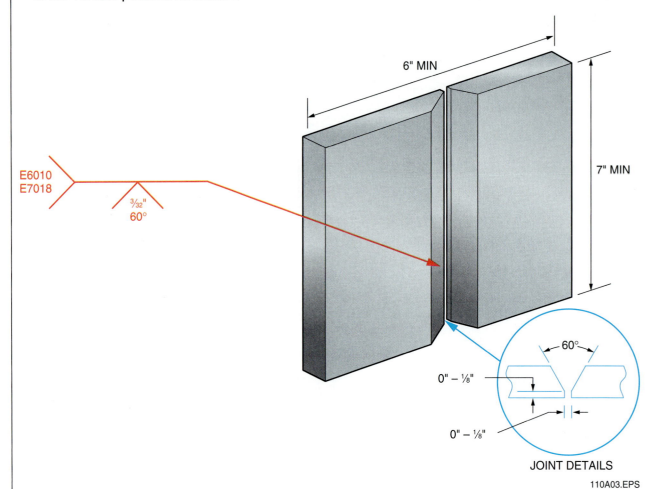

E6010
E7018

³⁄₃₂"
60°

6" MIN

7" MIN

60°

0" – ⅛"

0" – ⅛"

JOINT DETAILS

110A03.EPS

Criteria for Acceptance:

- Uniform rippled appearance on the bead face
- Craters and restarts filled to the full cross-section of the weld
- Uniform weld wize
- Acceptable weld profile in accordance with the ASME *Boiler and Pressure Vessel Code*
- Complete uniform root penetration at least flush with the base metal to a maximum buildup of ⅛"
- Smooth transition with complete fusion at the toes of the weld
- No porosity
- No excessive undercut
- No inclusions
- No cracks
- Acceptable guided bend test results

OPEN-ROOT V-GROOVE WITH E6010 ELECTRODES IN THE OVERHEAD POSITION

Using ⅛" E6010 and ³⁄₃₂" E7018 electrodes, make an open-root V-groove weld on carbon steel plate in the overhead position as indicated.

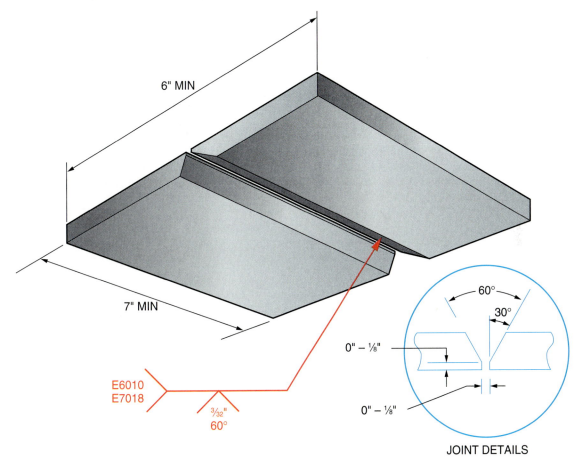

6" MIN

7" MIN

E6010
E7018

³⁄₃₂"
60°

60°

30°

0" – ⅛"

0" – ⅛"

JOINT DETAILS

110A04.EPS

Criteria for Acceptance:

- Uniform rippled appearance on the bead face   _____
- Craters and restarts filled to the full cross-section of the weld   _____
- Uniform weld wize   _____
- Acceptable weld profile in accordance with the ASME *Boiler and Pressure Vessel Code*   _____
- Complete uniform root penetration at least flush with the base metal to a maximum buildup of ⅛"   _____
- Smooth transition with complete fusion at the toes of the weld   _____
- No porosity   _____
- No excessive undercut   _____
- No inclusions   _____
- No cracks   _____
- Acceptable guided bend test results   _____

Additional Resources

This module is intended to present thorough resources for task training. The following reference works are suggested for further study. These are optional materials for continued education rather than for task training.

Welding Technology, 2001. J. W. Giachino, W. R. Weeks, G. S. Johnson. Homewood, IL: American Technical Publishers, Inc.

Modern Welding, 2000. A. D. Althouse, C. H. Turnquist, W. A. Bowditch, and K. E. Bowditch. Tinley Park, IL: The Goodheart-Willcox Company, Inc.

Shielded Metal Arc Welding – Open-Root Pipe Welds

Course Map

This course map shows all of the modules in the AWS Entry Level Welder – Phase 1 curriculum. The suggested training order begins at the bottom and proceeds up. Skill levels increase as you advance on the course map. The local Training Program Sponsor may adjust the training order.

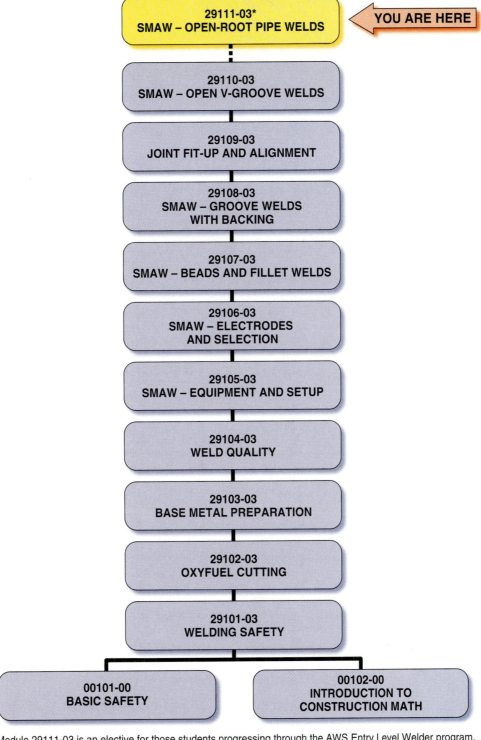

AWS ENTRY LEVEL WELDER—PHASE 1

29111-03*
SMAW – OPEN-ROOT PIPE WELDS

YOU ARE HERE

29110-03
SMAW – OPEN V-GROOVE WELDS

29109-03
JOINT FIT-UP AND ALIGNMENT

29108-03
SMAW – GROOVE WELDS WITH BACKING

29107-03
SMAW – BEADS AND FILLET WELDS

29106-03
SMAW – ELECTRODES AND SELECTION

29105-03
SMAW – EQUIPMENT AND SETUP

29104-03
WELD QUALITY

29103-03
BASE METAL PREPARATION

29102-03
OXYFUEL CUTTING

29101-03
WELDING SAFETY

00101-00
BASIC SAFETY

00102-00
INTRODUCTION TO CONSTRUCTION MATH

*Module 29111-03 is an elective for those students progressing through the AWS Entry Level Welder program.

29111CMAP.EPS

Figures

Table

Shielded Metal Arc Welding – Open-Root Pipe Welds

Objectives

When you have completed this module, you will be able to do the following:

1. Prepare shielded metal arc welding (SMAW) equipment for open-root V-groove pipe welds.
2. Identify and explain open-root V-groove pipe welds.
3. Perform SMAW for open-root pipe welds in the:
   - Flat (1G-ROTATED) position
   - Horizontal (2G) position
   - Multiple (5G) position
   - Multiple inclined (6G) position

Prerequisites

Before you begin this module, it is recommended that you successfully complete Modules 00100-00 through 29110-03.

Required Trainee Materials

1. Pencil and paper
2. Appropriate personal protective equipment

1.0.0 ◆ INTRODUCTION

The open-root V-groove weld is the most common weld used for joining medium- and thick-walled pipe. It is used for both critical and noncritical piping systems. The welds for critical piping must be extremely strong, at least as strong as the pipe itself. Critical piping is used in high-pressure steam systems, oil pipelines, nuclear power plants, race car roll cages, and other critical applications where the strength of the weld is essential to the safety of people and the environment. Such welds must be expertly made and thoroughly inspected.

Except for the root pass, weld passes are made with an E7018 or stronger electrode and are usually tested. Any defects in the weld are repaired. The joint is easy to prepare because it only requires the beveling of the matching members. The E6010 electrode is used for the root pass because a backing ring is not used. This electrode is a fast-freeze electrode that solidifies quickly and gives good penetration into the root of the joint, even with a small root gap. However, welding the root pass does take practice to perfect.

This module explains the preparation and welding of open-root V-groove welds on pipe in all positions, using E6010 electrodes for the root pass and E7018 electrodes for the remaining passes. Performance demonstrations will require that the welds meet the qualification standards in the *ASME Boiler and Pressure Vessel Code, Section IX, Welding and Brazing Qualifications*. For the purposes of this training module, the dimensions are representative of codes and may not be specific to one exact code. Always refer to the applicable code or site-specific specifications.

2.0.0 ◆ ARC WELDING EQUIPMENT SETUP

Before welding can take place, safety procedures and practices must be reviewed, the area has to be made ready, the welding equipment must be set up, and the metal to be welded must be prepared. The following sections explain how to set up arc welding equipment for welding.

2.1.0 Safety Practices

The following is a summary of safety procedures and practices that must be observed while cutting or welding. Keep in mind that this is just a summary. Complete safety coverage is provided in the Level One module, *Welding Safety.* If you have not completed that module, do so before continuing. Above all, be sure to wear appropriate protective clothing and equipment when welding or cutting.

2.1.1 Protective Clothing and Equipment

- Always use safety goggles with a full face shield or a helmet. The goggles, face shield, or helmet lens must have the proper light-reducing tint for the type of welding or cutting to be performed. Never directly or indirectly view an electric arc without using a properly tinted lens.
- Wear proper protective leather and/or flame-retardant clothing along with welding gloves that will protect you from flying sparks and molten metal as well as heat.
- Wear 8" or taller high-top safety shoes or boots. Make sure that the tongue and lace area of the footwear will be covered by a pant leg. If the tongue and lace area is exposed or the footwear must be protected from burn marks, wear leather spats under the pants or chaps and over the front top of the footwear.
- Wear a solid material (non-mesh) hat with a bill pointing to the rear or, if much overhead cutting or welding is required, a full leather hood with a welding face plate and the correct tinted lens. If a hard hat is required, use a hard hat that allows the attachment of rear deflector material and a face shield.
- If a full leather hood is not worn, wear a face shield and snug-fitting welding goggles over safety glasses for gas welding or cutting. Either the face shield or the lenses of the welding goggles must be an approved shade 5 or 6 filter. For electric arc welding, wear safety goggles and a welding hood with the correct tinted lens (shade 9 to 14).
- If a full leather hood is not worn, wear earmuffs or at least earplugs to protect ear canals from sparks.

2.1.2 Fire/Explosion Prevention

- Never carry matches or gas-filled lighters in your pockets. Sparks can cause the matches to ignite or the lighter to explode, causing serious injury.
- Never perform any type of heating, cutting, or welding until a hot work permit is obtained and an approved fire watch is established. Most work site fires caused by these types of operations are started by cutting torches.
- Never use oxygen to blow off clothing. The oxygen can remain trapped in the fabric for a time. If a spark hits the clothing during this time, the clothing can burn rapidly and violently out of control.
- Make sure that any flammable material in the work area is moved or shielded by a fire-resistant covering. Approved fire extinguishers must be available before any heating, welding, or cutting operations are attempted.
- Never release a large amount of oxygen or use oxygen as compressed air. Its presence around flammable materials or sparks can cause rapid and uncontrolled combustion. Keep oxygen away from oil, grease, and other petroleum products.
- Never release a large amount of fuel gas, especially acetylene. Methane and propane tend to concentrate in and along low areas and can be ignited a considerable distance from the release point. Acetylene is lighter than air but is even more dangerous. When mixed with air or oxygen, it will explode at much lower concentrations than any other fuel.
- To prevent fires, maintain a neat and clean work area and make sure that any metal scrap or slag is cold before disposal.
- Before cutting or welding containers such as tanks or barrels, check to see if they have contained any explosive, hazardous, or flammable materials including petroleum products, citrus products, or chemicals that decompose into toxic fumes when heated. As a standard practice, always clean and then fill any tanks or barrels with water or purge them with a flow of inert gas to displace any oxygen.

2.1.3 Work Area Ventilation

- Make sure confined space procedures are followed before conducting any welding or cutting in the confined space.
- Never use oxygen in confined spaces for ventilation purposes.
- Always perform cutting or welding operations in a well-ventilated area. Cutting or welding operations involving zinc or cadmium materials or coatings result in toxic fumes. For long-term cutting or welding of such materials, always wear an approved full face, supplied-air respirator (SAR) that uses breathing air supplied externally of the work area. For occasional, very short-term exposure, a HEPA-rated

or metal-fume filter may be used on a standard respirator.

- Make sure confined spaces are ventilated properly for cutting or welding purposes.

2.1.4 Electrical Shock Prevention

- Floors must be kept dry. Use wooden platforms, rubberized carpet/floor coverings, or other insulated material.
- Only experienced electricians may work on electric arc welding machine power connections.
- Ensure that work benches are grounded.
- With the power off, check cable connectors. Look for cracked insulators, loose contacts, and worn or cut hoses.

2.2.0 Preparing the Welding Area

To practice welding, a welding table, bench, or stand is needed. The welding surface must be made of steel, and provisions must be made for supporting the pipe welding coupons for out-of-position welding. *Figure 1* shows a typical welding station.

To set up the area for welding, follow these steps:

Step 1 Check to be sure the area is properly ventilated. Use doors, windows, and fans as necessary.

Step 2 Check the area for fire hazards. Remove any flammable materials before proceeding.

Step 3 Check the location of the nearest fire extinguisher. Do not proceed unless the extinguisher is charged and you know how to use it.

Step 4 Position a welding table near the welding machine.

Step 5 Set up flash shields around the welding area.

111F01.EPS

Figure 1 ◆ Welding station.

2.3.0 Preparing Pipe Weld Coupons

Pipe weld coupons should be cut from 3" to 12" diameter Schedule 40 or Schedule 80 carbon steel pipe. Each welded joint will require two coupons of the same size and schedule pipe.

Figure 2 shows typical American Society of Mechanical Engineers (ASME) and American Petroleum Institute (API) pipe bevel specifications for bevel angles, root lands, and root openings.

To prepare weld coupons for open-root V-groove weld joints, follow these steps:

Step 1 Clean heavy rust or mill scale from the pipe with a grinder or wire brush.

Step 2 Bevel the end of the pipe to 30° or 37½° by any acceptable beveling method such as flame cutting or grinding.

Conserve Materials

Pipe for practice welding is expensive and difficult to obtain. Make every effort to conserve and not waste the material available. Reuse weld coupons until all surfaces have been welded on by cutting the coupon apart and reusing the pieces. Check with the instructor for the appropriate size coupon.

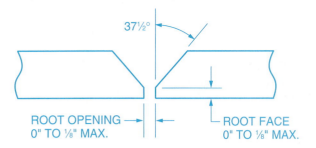

37½°

ROOT OPENING
0" TO ⅛" MAX.

ROOT FACE
0" TO ⅛" MAX.

TYPICAL ASME SPECIFICATION

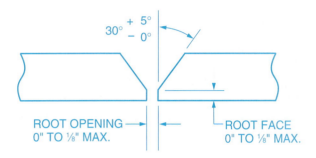

30° + 5° − 0°

ROOT OPENING
0" TO ⅛" MAX.

ROOT FACE
0" TO ⅛" MAX.

TYPICAL API AND AWS SPECIFICATION

111F02.EPS

Figure 2 ◆ ASME and API pipe bevel specifications.

Step 3   Cut off a section of beveled pipe end (2" minimum).

Note
For 1G welding, the coupons may have to be cut longer to fit the rollers being used.

Follow these steps to prepare the open V-groove weld pipe coupon:

Step 1   Check the bevel. There should be no slag, and the bevel angle should be 30° or 37½°.

Step 2   Grind or file a 0" to ⅛" root face on the bevel.

Note
Welding codes allow the root face on open-root welds to be from 0" to ⅛" wide and the root opening to be from 0" to ⅛". Select and adjust the root face and opening as needed when you start the welding practices.

Step 3   Align the two pipe sections so that the ID (inside diameter) surfaces are even all around. Small-diameter pipe can be aligned by clamping both pieces to a piece of angle iron. Larger-diameter pipe can be aligned with the aid of a pipe alignment jig or by holding a straight-edge across the joint, parallel to the pipe axis. The straightedge must be used all around the pipe in case one or both sections are distorted.

Step 4   Gap the root opening with pieces of filler wire, metal shims, or pieces of bare welding electrode of the correct diameter.

Step 5   When the root opening is correct and the pipe ends are aligned, make the first of four tack welds of no greater than 1".

Note
Longer tacks or more tacks may be required on heavy-wall or large-diameter pipe.

Step 6   After the first tack weld, check the root opening opposite the first tack weld. Adjust the gap if necessary and make the second tack weld on the opposite side from the first tack.

Step 7   Check the root opening again and weld the third tack midway between the first two tacks.

Step 8   Weld the fourth tack opposite the third tack and midway between the first and second tacks. There should now be four tack welds evenly spaced (90°) around the pipe coupon. This is illustrated by *Figure 3*.

Step 9   Clean and **feather** the tack welds with the edge of a grinding wheel. Feathering the ends of the tack welds with a grinder will help to fuse the tack welds into the root pass.

Beveling

The end of the joint must meet squarely with the mating pipe. To accomplish that, the sharp, inside edge (root face) of the bevel must be ground flat and square to the length of pipe. For the root gap to be uniform, final shaping should be done with a grinder.

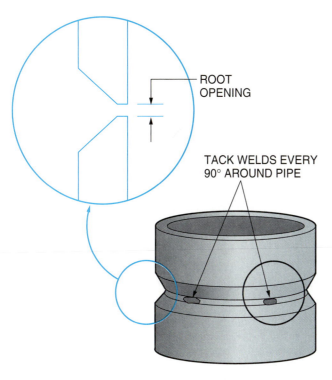

ROOT OPENING

TACK WELDS EVERY 90° AROUND PIPE

FEATHERED ENDS OF TACK WELD

111F03.EPS

Figure 3 ◆ Tacked open-root V-groove weld coupon.

2.4.0 Electrodes

Obtain a small quantity of the electrodes to be used. For the welding exercises in this module, ⅛" E6010 and ⅛" and ³⁄₃₂" E7018 electrodes will be used. Keep these general rules in mind when running E6010 and E7018 electrodes on pipe open V-joints.

- When making the root pass with E6010 electrodes, use the running keyhole technique to achieve proper root penetration.
- Use a whipping motion to control the puddle size with the E6010.
- Do not whip low-hydrogen E7018 electrodes.
- Maintain a short arc when running low-hydrogen electrodes.
- Set the amperage in the lower portion of the suggested range for better puddle control when welding vertically.
- Remove all slag between passes.

Do not take more E7018 electrodes from the oven than will be needed for one weld joint. Store the electrodes in some type of pouch or rod holder to prevent them from becoming damaged. Never store electrodes loose on a table. They may end up on the floor where they are a tripping hazard and can become damaged. Some type of metal container such as a pail must also be available to discard hot electrode stubs.

 WARNING!

Do not throw electrode stubs on the floor. They roll easily, and someone could step on one, slip, and fall.

2.5.0 Preparing the Welding Machine

Figure 4 shows typical AC/DC and DC welding machines. Identify a welding machine to use and then follow these steps to set it up for welding:

Step 1   Verify that the welding machine can be used for DC (direct current) welding.

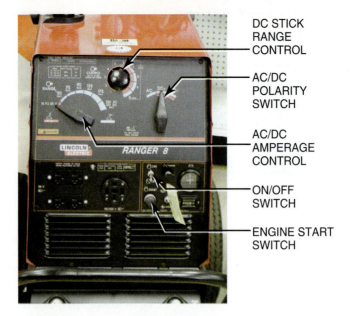

DC STICK
RANGE
CONTROL

AC/DC
POLARITY
SWITCH

AC/DC
AMPERAGE
CONTROL

ON/OFF
SWITCH

ENGINE START
SWITCH

PORTABLE 200A ENGINE-DRIVEN AC/DC WELDER

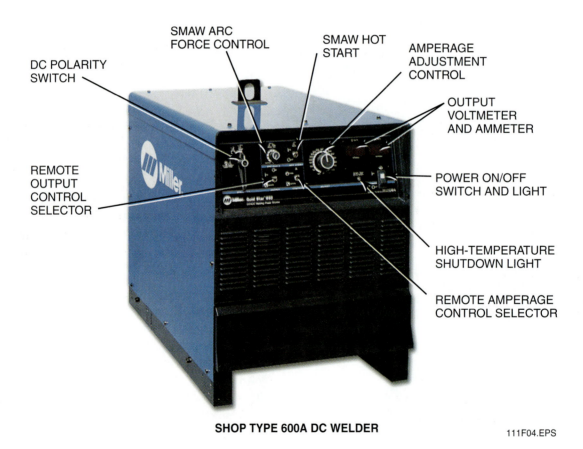

SMAW ARC
FORCE CONTROL

SMAW HOT
START

AMPERAGE
ADJUSTMENT
CONTROL

DC POLARITY
SWITCH

OUTPUT
VOLTMETER
AND AMMETER

REMOTE
OUTPUT
CONTROL
SELECTOR

POWER ON/OFF
SWITCH AND LIGHT

HIGH-TEMPERATURE
SHUTDOWN LIGHT

REMOTE AMPERAGE
CONTROL SELECTOR

SHOP TYPE 600A DC WELDER

111F04.EPS

Figure 4 ◆ Typical AC/DC and DC welding machines

LASH

To ensure good welding results, remember LASH.

Length of arc (L) – The distance between the electrode and the base metal (usually one times the electrode diameter).

Angle (A) – Two angles are critical:

* Travel angle – The longitudinal angle of the electrode in relation to the axis of the weld joint

* Work angle – The traverse angle of the electrode in relation to the axis of the weld joint

Speed (S) – Travel speed is measured in inches per minute (IPM). The width of the weld will determine if the travel speed is correct.

Heat (H) – Controlled by the amperage setting and dependent upon the electrode diameter, base metal type, base metal thickness, and the welding position.

Step 2   Check to be sure that the welding machine is properly grounded through the primary current receptacle.

Step 3   Check the area for proper ventilation.

Step 4   Verify the location of the primary current disconnect.

Step 5   Set the polarity to direct current electrode positive (DCEP).

Step 6   Set the amperage for the electrode type and size to be used. Typical settings are:

Electrode	Size	Amperage (Approximate Range)
E6010	$\frac{1}{8}$"	65A to 130A
E7018	$\frac{3}{32}$"	70A to 120A
E7018	$\frac{1}{8}$"	110A to 150A

Note

Amperage recommendations vary by manufacturer, position, current type, and electrode brand. For specific recommendations, refer to the manufacturer's recommendations for the electrode being used.

Step 7   Connect the clamp of the workpiece lead to the workpiece.

Step 8   Check to be sure the electrode holder is not grounded.

Step 9   Turn on the welding machine.

3.0.0 ◆ OPEN-ROOT V-GROOVE PIPE WELDS

The open-root V-groove weld is the most common groove weld normally made on pipe.

3.1.0 Open-Root Pass

The most difficult part of making an open-root V-groove weld is the root pass. The root pass must have complete penetration, but not an excessive amount of burn-through or root reinforcement. The extent of penetration is significantly important. If penetration is too little, it will create a weak joint. However, if there is too much penetration and excessive root reinforcement is created, the inside diameter of the pipe will be reduced and the flow will be restricted.

The penetration is controlled with a technique called *running a keyhole*. A keyhole-shaped hole is formed when the root faces on the two pipe edges are melted away by the arc (*Figure 5*). The molten metal flows to the back side of the keyhole, forming the weld. The keyhole should be about $\frac{3}{16}$" to $\frac{1}{4}$" in diameter. If the keyhole is larger, there will be excess burn-through with slag inclusions. If the keyhole closes up, a lack of penetration will result. Root reinforcement should be flush to $\frac{1}{8}$".

The keyhole is controlled by using fast-freeze E6010 electrodes that solidify quickly and by slightly whipping the electrode forward and back. Move the electrode forward about $\frac{3}{16}$" to $\frac{1}{4}$" and then back about $\frac{1}{8}$" and pause. Use care not to increase the arc length when moving the electrode. If the keyhole starts to grow, decrease the pause and increase the forward length of the whip. If the keyhole starts to close up, pause

---

Arc Strikes

When welding pipe, do not make arc strikes outside the weld groove on the surface of the pipe. An arc strike can cause a hardened spot that can crack as the pipe expands and contracts. An arc strike on the pipe surface is considered a defect and will require repair or rework.

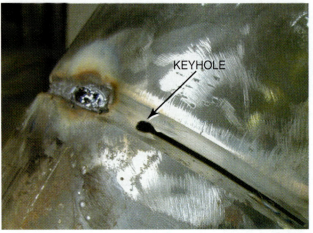

111F05.EPS

Figure 5 ◆ Root bead keyhole.

slightly longer and decrease the length of the whip. Other factors that affect the keyhole size are the amperage setting, land size, and root opening. All these items can be adjusted within the parameters allowed by the welding codes or your site quality procedures.

After the root pass is run, it should be cleaned and inspected. Check the root face for excess buildup or undercut that could trap slag when the filler pass is run. Remove excess buildup or undercut with a hand grinder by grinding the face of the root pass with the edge of a grinding disk. Most of the slag along the sides of the weld bead is removed by this grinding action, making it easier to add the next pass, called a *hot pass*. For high-pressure piping, it is essential that the hot pass is fused well with the root pass and the pipe walls. The hot pass must melt out any traces of slag from the root pass weld to eliminate any inclusions.

The hot pass is accomplished by running stringer bead(s) that cover both sides of the root pass using a short arc and high current. The hot pass should be accomplished within five minutes of the root pass cleaning and can be made using E6010 electrodes or low-hydrogen electrodes,

such as E7018 electrodes. After the hot pass bead or beads are cleaned, the filler and cover passes are applied. Use care not to grind through the root pass or widen the groove. *Figure 6* shows the effect of grinding a root pass. A narrow grinding wheel (not a cutoff wheel) is effective for this purpose.

Note that root passes are not always ground for pipe to be used in low- and medium-pressure systems. Grinding each root pass takes a long time. In these cases, the root pass slag is removed only by chipping and brushing before the hot pass bead or beads are applied.

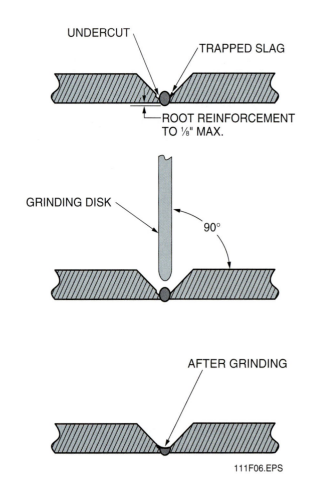

111F06.EPS

Figure 6 ◆ Grinding a root pass.

3.2.0 Pipe Groove Weld Test Positions

Groove welds may be made in all positions on pipe. The weld position is determined by the axis of the pipe. Four standard weld test positions, shown in *Figure 7*, are used with pipe:

- *Flat (1G-ROTATED) welding position* – The pipe axis is horizontal, and the pipe is slowly rotated while being welded on the top; weld beads are flat.
- *Horizontal (2G) welding position* – The pipe axis is vertical, and the pipe rotation is fixed; weld beads are horizontal.
- *Multiple (5G) welding position* – The pipe axis is horizontal, and the pipe rotation is fixed; weld beads are flat, vertical, and overhead.
- *Multiple inclined (6G) welding position* – The pipe axis is inclined, nominally at 45° ±5°, and pipe rotation is fixed; weld beads are horizontal, vertical, and overhead.

The pipe axis positions for 1G-ROTATED, 2G, and 5G can vary ±15° from the basic position. The pipe axis position for 6G can vary ±5° from the nominal 45° angle. See *Table 1*.

3.2.1 Tack Welds and Restarts on 5G and 6G Coupons

When 5G or 6G welds are to be destructively tested, the test specimens must be cut from the four regions illustrated in *Figure 8*, midway between the 12 o'clock and 3 o'clock, 3 o'clock and 6 o'clock, 6 o'clock and 9 o'clock, and 9 o'clock and 12 o'clock positions. (The 12 o'clock position is the top of the coupon groove; 6 o'clock is the bottom.) Tack welds and bead restarts should be avoided within the test coupon regions because they may cause the test specimens to fail. If necessary, stop a bead outside a specimen region and use a full electrode to get through the region.

3.3.0 Acceptable and Unacceptable Pipe Weld Profiles

Pipe groove welds should be made with slight reinforcement (not exceeding ⅛") and a gradual transition to the base metal at each toe. The root pass should have complete penetration. The root reinforcement on the inside of the pipe ranges from being flush to a maximum of ⅛". Pipe groove welds must not have excess reinforcement or underfill at the face or root, excessive undercut, or overlap. Excessively large cover passes will reduce the pipe's strength. They cause the stresses

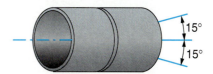

ROTATE PIPE AND DEPOSIT WELD AT OR NEAR THE TOP

PIPE HORIZONTAL (±15°) AND ROLLED TO KEEP WELD FLAT

1G-ROTATED POSITION

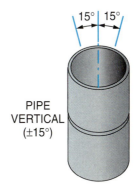

PIPE VERTICAL (±15°)

PIPE NOT ROTATED DURING WELDING

2G POSITION

PIPE NOT ROTATED DURING WELDING

PIPE HORIZONTAL (±15°)

5G POSITION

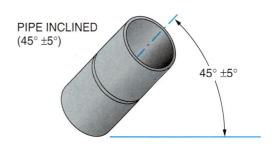

PIPE INCLINED (45° ±5°)

45° ±5°

PIPE NOT ROTATED DURING WELDING

6G POSITION

111F07.EPS

Figure 7 ◆ Four basic pipe groove weld test positions.

Table 1 Pipe Groove Weld Test Positions

Weld Test Position	Pipe Axis	Weld Beads	Pipe Position
1G	Horizontal	Flat	Slowly rotated
2G	Vertical	Horizontal	Fixed
5G	Horizontal	Flat, vertical, overhead	Fixed
6G	Inclined 45°	Horizontal, vertical, overhead	Fixed

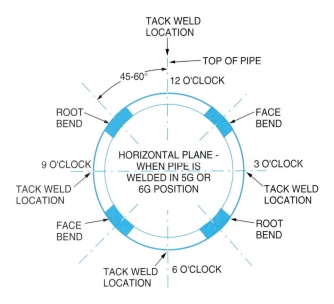

PIPES - ¹⁄₁₆" THROUGH ³⁄₈" THICK

BENT TEST SPECIMENS

111F08.EPS

Figure 8 ◆ Test specimen regions of 5G or 6G position pipe.

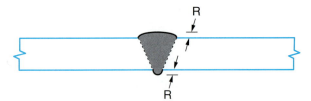

R = FACE AND ROOT REINFORCEMENT PER CODE
NOT TO EXCEED ¹⁄₈" MAX.

ACCEPTABLE WELD PROFILE

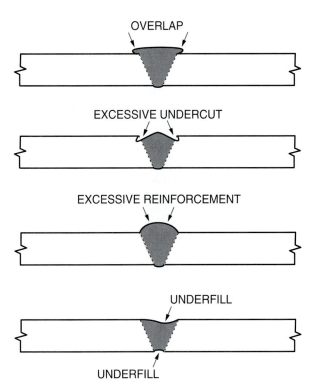

UNACCEPTABLE WELD PROFILES

111F09.EPS

Figure 9 ◆ Acceptable and unacceptable pipe groove weld profiles.

in the pipe to be concentrated along the sides of the weld and will not permit the pipe to expand and contract in a more uniform manner along its length. If a weld has any of these defects, as shown in *Figure 9*, it must be repaired.

Test Position

Welding pipe in the 6G position is a common welding test. A ring may be added to test for restricted accessibility (6GR).

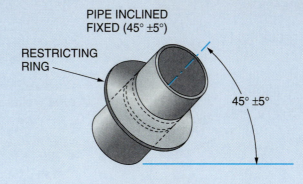

PIPE INCLINED
FIXED (45° ±5°)

RESTRICTING RING

45° ±5°

PIPE NOT ROTATED DURING WELDING

6GR POSITION

111UA1101.EPS

4.0.0 ◆ SMAW OF OPEN-ROOT V-GROOVE WELDS

This section of the module explains how to perform the following open-root V-groove welds:

- Flat (1G-ROTATED) position welds
- Horizontal (2G) position welds
- Multiple (5G) position welds
- Multiple inclined (6G) position welds

4.1.0 Flat (1G-ROTATED) Position Open-Root V-Groove Pipe Welds

Practice the 1G-ROTATED position open-root V-groove pipe welds using:

- ⅛" E6010 electrodes for the root pass
- ³⁄₃₂" E7018 electrodes for the hot pass (passes on top of the root)
- Either ³⁄₃₂" or ⅛" E7018 electrodes for the filler and cover passes

Keep the electrode angle 90° (0° work angle) to the pipe axis with a 10° to 15° drag angle. Use a whipping motion to run a keyhole for the root pass, paying particular attention to tie into the tack welds and restarts. Before running the hot pass (beads 2 and 3), clean the face of the root pass with a file or grinder.

When running the filler and cover passes, use stringer beads with a slight circular or side-to-side motion to ensure tie-in at the toes of the weld bead. Take particular care at the termination of the

weld to fill the crater. Run all passes at or near the top of the pipe as the pipe is rotated.

Follow these steps to practice open-root V-groove pipe welds in the 1G-ROTATED position:

Step 1 Tack-weld the practice pipe weld coupon together as explained earlier.

Step 2 Position the pipe weld coupon horizontally on two sets of rollers at a comfortable welding height.

Figure 10 shows roller supports commonly found in pipe welding shops.

Step 3 Make sure the workpiece clamp is attached directly onto the pipe coupon. This will prevent the welding current from passing through the roller bearings or arcing between the rollers and the pipe coupon.

CAUTION

Failure to attach the workpiece clamp to the pipe coupon can result in variations in welding current, damage to the roller bearings, and arcing on the rollers and pipe coupon.

Step 4 Run the root pass using ⅛" E6010 electrodes and a whipping motion. Position a tack weld at the 11 o'clock position, start the weld bead on the tack weld, and advance toward the 12 o'clock position.

TABLE ROLLER

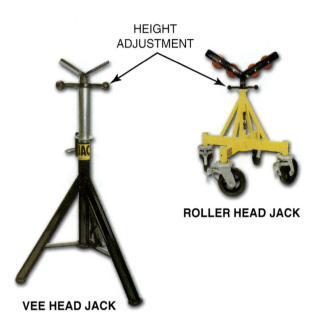

HEIGHT
ADJUSTMENT

ROLLER HEAD JACK

VEE HEAD JACK

FLOOR STAND ROLLER

111F10.EPS

Figure 10 ◆ Pipe roller supports.

Step 5 Roll the pipe as necessary to keep the weld in the flat position. Chip the restart between beads.

Step 6 Grind the root pass with a disk grinder to remove trapped slag.

Step 7 Use the same rolling procedure to make the hot pass (beads 2 and 3). Use ³⁄₃₂" E7018 electrodes.

Step 8 Use the same rolling procedure to make the filler/cover passes. Use ³⁄₃₂" or ¹⁄₈" E7018 electrodes and stringer beads. Pay particular attention at the tie-ins to prevent slag inclusions or excess buildup.

Figure 11 shows the bead sequence.

4.2.0 Vertical (2G) Position Open-Root V-Groove Pipe Welds

Practice 2G (vertical) position open-root V-groove pipe welds using:

- ¹⁄₈" E6010 electrodes for the root pass
- ³⁄₃₂" E7018 electrodes for the hot pass
- Either ³⁄₃₂" or ¹⁄₈" E7018 electrodes for the filler and cover passes

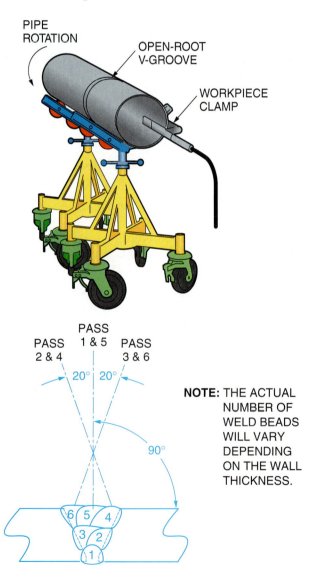

STRINGER BEAD SEQUENCE

111F11.EPS

Figure 11 ◆ Multiple-pass 1G weld sequences and work angles.

AWS ENTRY LEVEL WELDER – TRAINEE MODULE 29111-03

Filler Pass

Follow these guidelines when running the filler pass:

- Fill the crater to preclude cracking.
- Clean the crater of slag before restarting the arc; failure to do so will result in inclusions.
- Ensure high-strength, high-pressure pipe cleanliness by slightly grinding the crater.
- The location of the start and stop spots for each weld must be staggered.
- The weld groove should be filled level with spot beads prior to the cover pass.
- Avoid starting and stopping all weld passes in the same area.

Homemade Jack Stands

OSHA prohibits the use of homemade jack stands. Only stands that are manufactured to industry standards are acceptable.

Powered Pipe Roller

A pipe roller that is operated by an electric motor activated by a foot switch is shown here. These devices are primarily used for large, heavy pipe sections.

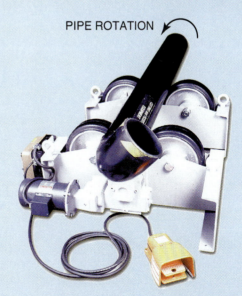

PIPE ROTATION

111P1101.EPS

The electrode should be at a 10° to 15° drag angle and at 90° (0° work angle) to the surface of the pipe for the root pass. To prevent the weld puddle from sagging, the electrode work angle can be dropped slightly but no more than 10°. Use a whipping motion to run a keyhole for the root pass, paying particular attention to tie into the tack welds and restarts.

Before running the hot pass (beads 2 and 3), clean the face of the root pass with a file or grinder. When running the hot and filler/cover passes, use stringer beads and a slight circular or side-to-side motion to ensure tie-in at the toes of the weld bead. Take particular care at the termination of the weld to fill the crater.

Figure 12 shows the 2G pipe weld coupon position and bead sequence.

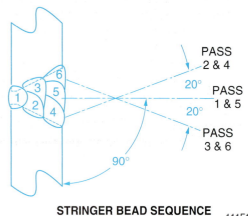

STRINGER BEAD SEQUENCE   111F12.EPS

Figure 12 ◆ Multiple-pass 2G weld sequences and work angles.

Cover Pass

What effect will excessively large cover passes have on the strength of a pipe?

Follow these steps to practice open-root V-groove pipe welds in the 2G position:

Step 1   Tack-weld the practice pipe weld coupon together as explained earlier.

Step 2   Tack-weld or fasten the pipe coupon to the positioning arm on the welding table with the axis of the pipe vertical.

Step 3   Position the arm to allow welding access all around the coupon.

Step 4   Run the root pass using ⅛" E6010 electrodes and a whipping motion. Start the root pass on a tack weld. Clean and chip the crater between restarts. Continue this procedure until the root pass is completed.

Step 5   Grind the root pass with a disk grinder to remove trapped slag.

Step 6   Run the hot pass (beads 2 and 3) using 3/32" E7018 electrodes.

Step 7   Run the filler/cover passes with either 3/32" or ⅛" E7018 electrodes and stringer beads.

4.3.0 Multiple (5G) Position Open-Root V-Groove Pipe Welds

Practice the 5G multiple position open-root V-groove pipe welds using:

- ⅛" E6010 electrodes for the root pass
- 3/32" E7018 electrodes for the hot pass
- Either 3/32" or ⅛" E7018 electrodes and stringer beads for the filler and cover passes

Weld uphill by starting at the bottom of the pipe and welding toward the top. Use a whipping motion to run a keyhole for the root pass, paying particular attention to tie into the tack welds and restarts. Before running the hot pass (beads 2 and 3), clean the face of the root pass with a file or grinder.

When running the filler and cover passes, use a slight circular or side-to-side motion to ensure tie-in at the toes of the weld bead. Pay particular attention at the termination of the weld to fill the crater. *Figure 13* shows the 5G pipe weld coupon and bead sequence.

Follow these steps to practice open-root V-groove pipe welds in the 5G position:

Step 1 Tack-weld the practice pipe weld coupon together as explained earlier.

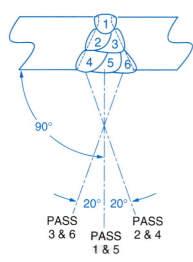

STRINGER BEAD SEQUENCE

111F13.EPS

Figure 13 ◆ Multiple-pass 5G bead sequences and work angles.

Step 2 Tack-weld or fasten the pipe weld coupon to the positioning arm with the pipe axis horizontal. Be sure to position the tack welds so that they are not in the areas from which the test coupons will be cut.

Step 3 Position the arm to allow welding access all around the coupon.

Step 4 Run the root pass using ⅛" E6010 electrodes. Start the root pass on the tack weld at the bottom of the pipe and weld uphill. When welding at the top and bottom of the coupon (flat and overhead positions), the electrode should usually point toward the pipe axis. When welding up the coupon sides (vertical welding), a slight electrode push angle (maximum of 10°) can be used to control the weld puddle.

Step 5 Grind the root pass with a disk grinder to remove trapped slag.

Step 6 Run the hot pass (beads 2 and 3) using ³⁄₃₂" E7018 electrodes.

Step 7 Use stringer beads to run the filler/cover passes with either ³⁄₃₂" or ⅛" E7018 electrodes.

4.4.0 Multiple Inclined (6G) Position Open-Root V-Groove Pipe Welds

Practice the 6G multiple inclined (45°) position open-root V-groove pipe welds using:

- ⅛" E6010 electrodes for the root pass
- ³⁄₃₂" E7018 electrodes for the hot pass
- Either ³⁄₃₂" or ⅛" E7018 electrodes and stringer beads for the filler and cover passes

Weld uphill by starting at the bottom of the pipe and welding toward the top. Use a whipping motion to run a keyhole for the root pass, paying particular attention to tie into the tack welds and restarts. Before running the hot pass (beads 2 and 3), clean the face of the root pass with a file or grinder.

When running the filler and cover passes, use a slight circular or side-to-side motion to ensure tie-in at the toes of the weld bead. Pay particular attention at the termination of the weld to fill the crater. *Figure 14* shows the 6G pipe weld coupon and bead sequence.

Follow these steps to practice open-root V-groove pipe welds in the 6G position:

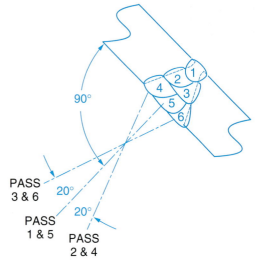

90°

PASS 3 & 6   20°

20°

PASS 1 & 5   PASS 2 & 4

STRINGER BEAD SEQUENCE

111F14.EPS

Figure 14 ◆ Multiple-pass 6G bead sequence and work angles.

Step 1   Tack-weld the practice pipe weld coupon together as explained earlier.

Step 2   Tack-weld or fasten the pipe weld coupon to the positioning arm with the pipe axis inclined 45° to the horizontal plane. Be sure to position the tack welds so that they are not in the areas from which the test specimen will be cut.

Step 3   Position the arm to allow welding access all around the coupon.

Step 4   Run the root pass using ⅛" E6010 electrodes. Start the root pass on the tack weld at the bottom of the pipe and weld uphill. When welding at the top and bottom of the coupon (flat and overhead welds), the electrode should point toward the pipe axis. When welding up the coupon sides (vertical welds), a slight electrode push angle (maximum of 10°) can be used to control the weld puddle.

Step 5   Grind the root pass with a disk grinder to remove trapped slag.

Step 6   Run the hot pass (beads 2 and 3) using ³⁄₃₂" E7018 electrodes.

Step 7   Use ³⁄₃₂" or ⅛" E7018 electrodes and stringer beads to run the filler/cover passes.

Summary

The ability to make open-root V-groove welds on pipe in all positions is one of the more difficult skills a welder must develop. The open-root V-groove weld is the most common weld joint used for joining medium- and thick-walled pipe. A welder must be able to set up the equipment, prepare the coupons for welding, and recognize unacceptable welds. Open-root V-groove welds can be made in the 1G-ROTATED, 2G, 5G, and 6G positions. Practice these welds until you can consistently produce acceptable welds.

Review Questions

1. When preparing to practice open-root pipe welds, each welded joint requires _____.
   a. a 6" section of pipe
   b. a 12" section of pipe
   c. two 12" sections of pipe
   d. two identical (size and schedule) coupons

2. To prepare an open V-groove weld coupon, the bevel angle should be _____.
   a. 10° or 15°
   b. 20° or 30°
   c. 30° or 37½°
   d. 60° or 75°

3. When preparing open V-groove weld pipe coupons, make _____ tack welds.
   a. two
   b. three
   c. four
   d. five

4. The most difficult part of making an open-root V-groove weld is the _____.
   a. bead
   b. overlap
   c. keyhole
   d. root pass

5. A keyhole-shaped hole is formed when the root faces on the two pipe edges are _____.
   a. melted away by the arc
   b. pulled together with a clamp
   c. removed with a grinder
   d. spread apart by a wedge

6. When the axis of the pipe being welded is horizontal and the pipe is slowly rotated while being welded on the top, the weld position is called _____.
   a. 1G
   b. 2G
   c. 5G
   d. 6G

7. When the axis of the pipe being welded is horizontal, and the pipe rotation is fixed, the weld position is called _____.
   a. 1G
   b. 2G
   c. 5G
   d. 6G

8. Pipe axis positions can vary ±15° from the basic position for the _____ pipe positions.
   a. 1G, 2G, and 6G
   b. 1G, 2G, and 5G
   c. 5G and 6G
   d. 1G and 2G

9. When practicing open-root V-groove pipe welds in the 1G-ROTATED position, use _____ electrodes for the hot pass (passes on top of the root).
   a. ⅛" E6010
   b. ⅛" E7018
   c. ³⁄₃₂" E6010
   d. ³⁄₃₂" E7018

10. When practicing open-root V-groove pipe welds in the 5G position, use _____ electrodes for the root pass.
    a. ³⁄₃₂" E7018
    b. ⅛" E7018
    c. ⅛" E6010
    d. ³⁄₃₂" E6010

Tom Atkinson

Execution Coordinator
Contract Management Group
Koch Petroleum Group LP
Corpus Christi, TX

Tom was born and raised in Robstown, Texas. He started his welding career in an agriculture class in high school. He then attended Del Mar College in Corpus Christi, Texas, graduating with an industrial education/welding combination degree. He went to work for the next five years as a gas metal arc welding (GMAW) and gas tungsten arc welding (GTAW) welder, gaining a lot of alloy-stainless experience. Later he went into business as a contract rig welder, spending four years in the oil fields and ten more years in the petrochemical and fuels industry.

During these years, contractors sent Tom to Illinois, Louisiana, Missouri, Tennessee, Arkansas, and New Mexico. He worked in at least 30 different facilities on and off as a supervisor. Then he was offered a job with Koch Refinery as an A-class welder. He took the job and took advantage of the education Koch offered. His inspection classes made it possible for him to pass the certified welding inspector (CWI) exam. He took craft instructor classes and became an instructor in welding and pipe fitting with the Construction Education Foundation.

Tom was often sent to Houston to help revise books on pipe fitting and welding with the NCCER. He took metallurgy, computer, and management classes, subjects he still takes when they are offered.

In 1996, Tom's welding and metallurgy knowledge earned him the first and second pages of the National Koch Discovery quarterly magazine. He was then promoted to execution coordinator with the Contract Management Group at Koch Petroleum Group in Corpus Christi. There, he manages multimillion-dollar jobs.

How did you become interested in the welding industry?

I went with my dad to a Tenneco facility where he worked, and I noticed the welds on the pipe in the facility. I told my dad that I sure would like to be able to make pretty welds like those. He said he had seen many men try and fail to make it in the welding field. He said I should think about something else besides welding, but I never did like the words *fail* or *can't*.

What do you think it takes to be a success in your trade?

First, you need education in the craft—the more, the better. Also, take all the keyboard and computer classes you can in high school and college. The typing I took in high school helped my success, and I wouldn't have made it without it. When you go to work in a trade, keep going to school to keep up with change, because change occurs constantly, dynamically, and—most of the time—without warning.

What are some of the things you do in your job?

I participate in contractor selection, monitor performance of contractor personnel, maintain project schedules, and ensure that procedures are being followed. I'm also responsible for improving productivity, enhancing the project schedule, and lowering costs by removing barriers in current work practices that inhibit the success of the project.

What do you like most about your job?

I like it when there is a challenge that requires thinking outside the box. Sometimes you can brainstorm and find a way to get things done. It may not be the same old way, and it may not be in our procedures, but when you get the buy-in of management and it works, that is when the procedure is changed—until someone comes up with a better idea. It is a good feeling coming up with new ideas that work.

What would you say to someone entering the trade today?

What really led to my success was that I was one of the best, if not the best, alloy welder in south Texas for welding in difficult, hard-to-get-at positions. What made me the best was my ability to change. I experimented with new ways of welding that sometimes worked. When I found new ways that worked, I changed the process—and changed it for everyone working around me when they saw that the new process beat the old way.

Constantly strive to find a better way. My dad once told me that everything was invented already. This was back in the late sixties when I was a kid. He would be the first to admit that change was just beginning, and the same is true today. I am constantly looking for better ways and new ideas in my job. My education and my adaptability to change enable me to succeed in this job, as they did when I was a welder.

Trade Term Introduced in This Module

Feather: Grind the tack weld to a tapered edge.

Performance Accreditation Tasks

The Performance Accreditation Tasks (PATCs) correspond to and support the learning objectives in AWS EG2.0-95, Guide for the Training and Qualification of Welding Personnel: Entry-Level Welder.

Note that in order to satisfy all learning objectives in AWS EG2.0-95, the instructor must also use the PATCs contained in the second level in the Welding Contren™ Learning Series.

PATCs provide specific acceptable criteria for performance and help to ensure a true competency-based welding program for students.

The following tasks are designed to evaluate your ability to run open-root V-groove pipe welds with SMAW equipment in the four standard test positions using E6010 and E7018 electrodes. Perform each task when you are instructed to do so by your instructor. As you complete each task, take it to your instructor for evaluation. Do not proceed to the next task until instructed to do so by your instructor.

OPEN-ROOT V-GROOVE PIPE WELD IN THE 1G-ROTATED POSITION

Using ⅛" E6010 electrodes for the root pass and ⅛" E7018 electrodes for the filler and cover passes, make an open-root V-groove weld on pipe in the 1G-ROTATED position as shown.

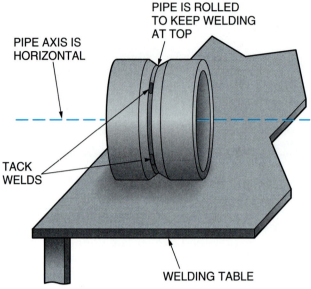

PIPE IS ROLLED TO KEEP WELDING AT TOP

PIPE AXIS IS HORIZONTAL

TACK WELDS

WELDING TABLE

111A01.EPS

Criteria for Acceptance:

- Uniform rippled appearance on the weld face   _____
- Craters and restarts filled to the full cross-section of the weld   _____
- Uniform weld size   _____
- Acceptable weld profile in accordance with the
  ASME *Boiler and Pressure Vessel Code, Section IX*   _____
- Smooth transition with complete fusion at the toes of the weld   _____
- Complete uniform root reinforcement at least flush with the
  inside of the pipe to a maximum buildup of ⅛"   _____
- No porosity   _____
- No excessive undercut   _____
- No inclusions   _____
- No cracks   _____
- Acceptable guided bend test results   _____

OPEN-ROOT V-GROOVE PIPE WELD IN THE 2G POSITION (PAGE 1 OF 2)

Using ⅛" E6010 electrodes for the root pass and ⅛" E7018 electrodes for the filler and cover passes, make an open-root V-groove weld on pipe in the 2G position as shown.

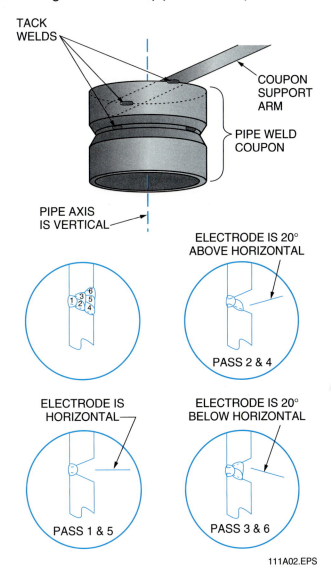

TACK WELDS

COUPON SUPPORT ARM

PIPE WELD COUPON

PIPE AXIS IS VERTICAL

ELECTRODE IS 20° ABOVE HORIZONTAL

PASS 2 & 4

ELECTRODE IS HORIZONTAL

PASS 1 & 5

ELECTRODE IS 20° BELOW HORIZONTAL

PASS 3 & 6

111A02.EPS

(continued)

OPEN-ROOT V-GROOVE PIPE WELD IN THE 2G POSITION (PAGE 2 OF 2)

Criteria for Acceptance:

- Uniform rippled appearance on the weld face   _____
- Craters and restarts filled to the full cross-section of the weld   _____
- Uniform weld size   _____
- Acceptable weld profile in accordance with the
  ASME *Boiler and Pressure Vessel Code, Section IX*   _____
- Smooth transition with complete fusion at the toes of the weld   _____
- Complete uniform root reinforcement at least flush with the
  inside of the pipe to a maximum buildup of ⅛"   _____
- No porosity   _____
- No excessive undercut   _____
- No inclusions   _____
- No cracks   _____
- Acceptable guided bend test results   _____

OPEN-ROOT V-GROOVE PIPE WELD IN THE 5G POSITION

Using ⅛" E6010 electrodes for the root pass and ⅛" E7018 electrodes for the filler and cover passes, make an open-root V-groove weld on pipe in the 5G position as shown.

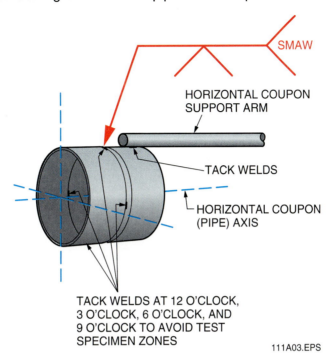

SMAW

HORIZONTAL COUPON SUPPORT ARM

TACK WELDS

HORIZONTAL COUPON (PIPE) AXIS

TACK WELDS AT 12 O'CLOCK, 3 O'CLOCK, 6 O'CLOCK, AND 9 O'CLOCK TO AVOID TEST SPECIMEN ZONES

111A03.EPS

Criteria for Acceptance:

- Uniform rippled appearance on the weld face _____
- Craters and restarts filled to the full cross-section of the weld _____
- Uniform weld size _____
- Acceptable weld profile in accordance with the ASME *Boiler and Pressure Vessel Code, Section IX* _____
- Smooth transition with complete fusion at the toes of the weld _____
- Complete uniform root reinforcement at least flush with the inside of the pipe to a maximum buildup of ⅛" _____
- No porosity _____
- No excessive undercut _____
- No inclusions _____
- No cracks _____
- Acceptable guided bend test results _____

OPEN-ROOT V-GROOVE PIPE WELD IN THE 6G POSITION

Using ⅛" E6010 electrodes for the root pass and ⅛" E7018 electrodes for the filler and cover passes, make an open-root V-groove weld on pipe in the 6G position as shown. If required for qualification test purposes, a restricting ring may be added to the 6G position coupon to form a 6GR position coupon.

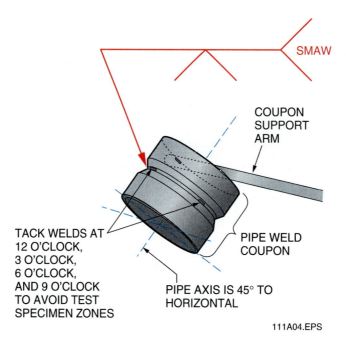

SMAW

COUPON SUPPORT ARM

PIPE WELD COUPON

TACK WELDS AT 12 O'CLOCK, 3 O'CLOCK, 6 O'CLOCK, AND 9 O'CLOCK TO AVOID TEST SPECIMEN ZONES

PIPE AXIS IS 45° TO HORIZONTAL

111A04.EPS

Criteria for Acceptance:

- Uniform rippled appearance on the weld face
- Craters and restarts filled to the full cross-section of the weld
- Uniform weld size
- Acceptable weld profile in accordance with the ASME *Boiler and Pressure Vessel Code, Section IX*
- Smooth transition with complete fusion at the toes of the weld
- Complete uniform root reinforcement at least flush with the inside of the pipe to a maximum buildup of ⅛"
- No porosity
- No excessive undercut
- No inclusions
- No cracks
- Acceptable guided bend test results

Additional Resources

This module is intended to present thorough resources for task training. The following reference works are suggested for further study. These are optional materials for continued education rather than for task training.

Modern Welding, 2000. A. D. Althouse, C. H. Turnquist, W. A. Bowditch, K. E. Bowditch. Tinley Park, IL: The Goodheart-Willcox Company, Inc.

Welding Pressure Pipe Lines and Piping Systems. Cleveland, OH: The Lincoln Electric Company.

Welding Technology, 2001. J. W. Giachino, W. R. Weeks, G. S. Johnson. Homewood, IL: American Technical Publishers, Inc.

Figure Credits

Module 29101-03

Hayes International, Inc. 101F23
Hornell, Inc. 101F09, 101F10, 101F13
Brad Krauel 101F16
Miller Electric Manufacturing Company
 101P0101
Nederman, Inc. 101F08
Chuck Rogers 101F01, 101F02, 101F05, 101F06,
 101F07
Scott Health and Safety 101F11, 101F12, 101F14
Sellstrom Manufacturing Co., Inc. 101F17
Gerald Shannon 101F03, 101F04, 101P0102,
 101P0103, 101F19
J. R. Yochum 101F21

Module 29102-03

American Torch Tip Co. 102P0205, 102P0206
American Welding Society 102P0210, 102F51
Bug-O Systems 102F28
Controls Corporation of America 102F01
ESAB 102F26
H&M Pipe Beveling 102F29
Koike Aronson, Inc. 102F25
Lenco d/b/a NLC, Inc. 102F24
Magnatech Limited Partnership, Inc. 102F30
Chuck Rogers 102F12, 102P0207
Saf-T-Cart 102P0211
Gerald Shannon 102F04, 102F06, 102F07,
 102P0201, 102F08, 102F10, 102P0202, 102P0203,
 102F13, 102F20, 102F21, 102F22, 102F23,
 102P0208, 102P0209, 102F32, 102F33, 102F34
 through 102F44, 102P0212, 102F45 through
 102F47, 102F49
Smith Equipment 102F48
Thermadyne Industries, Inc. 120P0204, 102F15,
 102F27, 102F31, 102F53, 102F58, 102F59

Module 29103-03

Heck Industries, Inc. 103F24
Magnatech Limited Partnership, Inc. 103F26
Chuck Rogers 103F01, 103F03, 103P0301, 103F04,
 103F05, 103F07, 103F08, 103P0302, 103F09,
 103F10, 103F11, 103F12, 103F13, 103P0303,
 103F15, 103F16, 103F23, 103F25
TWI 103F02
J. R. Yochum 103F08, 103F22

Module 29104-03

American Welding Society (AWS) 104UA0401,
 104F01 through 104F04, 104F07, 104F08,
 104F10, 104F19, 104F24, 104E01, 104A01,
 104A02
Curtis Casey 104P0402
G. A. L. Gage Company 104F12, 104F13
Instron Corp. 104P0403 (top)
Chuck Rogers 104F05, 104P0401, 104F15 through
 104F18
Tinius Olsen Testing Machine Co., Inc. 104P0403
 (bottom two)
J. R. Yochum 104F25

Module 29105-03

Miller Electric Manufacturing Co. 105F01,
 105F02, 105F08, 105F09, 105F10, 105F11,
 105F17, 105P0501
Chuck Rogers 105F03, 105F04, 105F14, 105F15
 (top), 105F16, 105F20, 105F21
Gerald Shannon 105F15 (bottom), 105P0502

Module 29106-03

The Lincoln Electric Company 106F01, 106P0601,
 106A01, 106A02
Phoenix International 106F10
Chuck Rogers 106F07
Gerald Shannon 106F02, 106F08

Module 29107-03

American Welding Society 107F11
Rex Ball 107F18
Gullco International, Ltd. 107P0701 (left)
Miller Electric Mfg. Co. 107F03
Chuck Rogers 107F15
Gerald Shannon 107F01, 107P0701 (right), 107F03
 (top), 107P0702, 107P0703, 107P0704, 107F23

Module 29108-03

Rex Ball 108F14, 108F16
Miller Electric Mfg. Company 108F09 (bottom)
John Murray 108F17, 108F19
Gerald Shannon 108F06, 108F09 (top)
J. R. Yochum 108P0801, 108F07, 108F12, 108F21

Module 29109-03

G. A. L. Gage Company 109F05, 109F08
Koike Aronson, Inc. 109F17 (upper left)
Magnatech Limited Partnership, Inc. 109F21
 (middle and bottom)
Mathey Dearman 109F19, 109F22
Chuck Rogers 109F06, 109F10, 109F11, 109F12,
 109F18, 109F20, 109F21 (top), 109F27
Gerald Shannon 109F04 (top)
L. S. Starrett Co., Inc. 109F03, 109F04 (bottom)
Sumner Manufacturing Co., Inc. 109F17 (lower
 left and right)

Module 29110-03

Rex Ball 110F05
Chuck Rogers 110F11, 110F12, 110F13
Gerald Shannon 110F02, 110F04
J. R. Yochum 110F15

Module 29111-03

Koike Aronson, Inc. 111P1101
Miller Electric Mfg. Company 111F04 (bottom)
Gerald Shannon 111F01, 111F02, 111F03, 111F04
 (top), 111F12, 111F13, 111F14
Sumner Manufacturing Co., Inc. 111F10
J. R. Yochum 111F05, 111F08

Index

Accidents, 1.2, 1.15. *See also* Job-site accidents
Acetylene
  cutting torch tips, 2.17-2.19
  and fire and explosion prevention, 2.3, 7.2
  and flame control, 2.37, 2.38
  flow rates, 2.19, 2.32
  and oxyfuel cutting equipment, 2.6-2.8
Acetylene cylinders
  clamshell valve cap, 2.7
  gauges of, 2.12
  markings and sizes, 2.8, 2.9
  and oxyfuel cutting, 2.6-2.8
  ring guard cap, 2.10
  safety fuse plugs, 2.8
  standard safety cap, 2.7, 2.8, 2.10
  valves, 2.8, 2.10
Actual throat, and fillet welds, 7.20
Acute angles, 02.46
  defined, 02.56
Addendum, defined, 9.28
Addendum sheets, and welding codes, 9.2
Adjacent angles, 02.46
  defined, 02.56
Aerial work, and safety, 01.32-01.43
Air-purifying respirators, 1.7, 1.8-1.10
  limitations, 1.12
  National Institute for Occupational Safety and Health (NIOSH) standards, 1.9
  powered air-purifying respirators, 1.9-1.10
Alcohol abuse, and job-site accidents, 01.3, 01.7
Alcohol use, 1.2
Alloys
  defined, 6.17
  and electrodes, 6.1
  and oxyfuel cutting, 2.1
Alphanumeric designations for welding positions, 4.23-4.25
Alternate welds, 9.24
Alternating current (AC)
  defined, 5.25
  and shielded metal arc welding equipment, 5.1, 5.2-5.3

and transformer welding machines, 5.7
Alternator-type welding machine, 5.2
Aluminum, cleaning, 3.3, 3.4
Ambient temperature
  defined, 9.28
  and distortion, 9.18
American Bureau of Shipping (ABS), 6.2
American National Standards Institute (ANSI) requirements, eye and face protection, 01.26
American National Standards Institute (ANSI) standards
  for foot protection, 1.5
  pipe test, 4.27
  for ventilation, 1.1, 1.6
  for welding safety, 1.1
  for weld quality, 4.2-4.3
American Petroleum Institute (API)
  pipe bevel specifications, 11.3-11.4
  welding codes, 4.2, 4.3, 9.1
American Society of Mechanical Engineers (ASME)
  and code changes, 9.2
  Code for Pressure Piping, 4.2, 9.1
  and electrode storage, 6.12
  pipe bevel specifications, 11.3-11.4
  and welder performance qualification tests, 4.29
American Society of Mechanical Engineers (ASME) Boiler and Pressure Vessel Code
  contents of, 4.2, 9.1
  and electrode selection, 6.10
  and electrode traceability requirements, 6.14
  filler metal specifications, 6.2
  and open-root V-groove pipe welds, 11.1, 11.22, 11.24-11.26
  and open-root V-groove welds, 10.1, 10.7, 10.18-10.21
  and welder performance qualification tests, 4.25, 4.27
American Welding Society (AWS)
  and discontinuities, 4.5
  electrode classification system, 4.23, 6.2-6.5, 6.7

electrode selection, 6.10
and electrode storage, 6.12
filler metal specification system, 6.2
plate test, 4.27
and safety in cutting containers, 1.18
and welder performance qualification tests, 4.28-4.29
welding codes, 4.2, 4.23, 4.25, 4.27, 9.2, 9.3
welding consumables specifications, 6.2-6.7, 6.8
Amperage
  defined, 5.25
  and duty cycle, 5.11, 5.12
  and equipment selection, 5.14
  and welding current, 5.5-5.6
Angle grinders, 3.3-3.4, 3.5
Angles, 02.46
  defined, 02.56
Apachi, 2.11
Apparatus, defined, 01.56
Arc blow, 7.9, 7.10
  defined, 7.29
Arc length, and effect on beads, 7.12, 7.14
Arcs. *See also* Arc strikes; Striking an arc
  defined, 01.56, 5.25
  properties of, 5.4
  and shielded metal arc welding equipment, 5.1-5.2
Arc shifting, 7.10
Arc strikes
  defined, 7.29
  and discontinuities, 4.5, 4.11
  and open-root v-groove pipe welds, 11.8
  and scratching method, 7.7
Arc voltage, 5.5, 9.3
Arc wandering, 7.10
Arc welding
  defined, 01.56
  and safety, 01.12
Area
  of a cylinder, 02.50, 02.51
  of circles, 02.50
  defined, 02.56
  of rectangles, 02.49